ADVANCES IN PHOTOCHEMISTRY
Volume 23

ADVANCES IN PHOTOCHEMISTRY
Volume 23

Editors

DOUGLAS C. NECKERS
Center for Photochemical Sciences, Bowling Green State University,
Bowling Green, Ohio

DAVID H. VOLMAN
Department of Chemistry, University of California, Davis, California

GÜNTHER VON BÜNAU
Physikalische Chemie, Universität Siegen, Germany

A WILEY-INTERSCIENCE PUBLICATION
JOHN WILEY & SONS, INC.
New York · Chichester · Weinheim · Brisbane · Singapore · Toronto

This book is printed on acid-free paper. ∞

Copyright ©1997 by John Wiley & Sons, Inc. All rights reserved.

Published simultaneously in Canada.

No part of this publication may be reproduced, stored in a retrieval system or transmitted in any form or by any means, electronic, mechanical, photocopying, recording, scanning or otherwise, except as permitted under Sections 107 or 108 of the 1976 United States Copyright Act, without either the prior written permission of the Publisher, or authorization through payment of the appropriate per-copy fee to the Copyright Clearance Center, 222 Rosewood Drive, Danvers, MA 01923, (508) 750-8400, fax (508) 750-4744. Requests to the Publisher for permission should be addressed to the Permissions Department, John Wiley & Sons, Inc., 605 Third Avenue, New York, NY 10158-0012, (212) 850-6011, fax (212) 850-6008, E-Mail: PERMREQ @ WILEY.COM.

Library of Congress Cataloging in Publication Data:
Library of Congress Catalog Card Number: 63-13592
ISBN 0-471-19289-9

Printed in the United States of America

10 9 8 7 6 5 4 3 2 1

CONTRIBUTORS

Peter F. Bernath
Department of Chemistry
University of Waterloo
Waterloo, Ontario
Canada N2L 3G1

Marcel Drabbels
FOM Institute for Atomic
  and Molecular Physics
Kruislaan 407
1098 SJ Amsterdam
The Netherlands

Martin Goez
Fachbereich Chemie
Martin-Luther-Universität
Halle-Wittenberg
Kurt-Mothes-Straße 2
D-06120 Halle/Saale, FRG

Massimo Guardigli
Dipartimento di Chimica
"G. Ciamician" dell'Università
via Selmi 2
40126 Bologna, Italy

Edward C. Lim
Department of Chemistry
The Univesity of Akron
Akron, OH 44325

Ilse Manet
Dipartimento di Chimica
"G. Ciamician" dell'Università
via Selmi 2
40126 Bologna, Italy

Christopher G. Morgan
Department of Chemistry
University of California
Santa Barbara, CA 93106

Nanda Sabbatini
Dipartimento di Chimica
"G. Ciamician" dell'Università
via Selmi 2
40126 Bologna, Italy

Alec M. Wodtke
Department of Chemistry
University of California
Santa Barbara, CA 93106

PREFACE

Volume 1 of *Advances in Photochemistry* appeared in 1963. The stated purpose of the series was to explore the frontiers of photochemistry through the medium of chapters written by pioneers who are experts. As editors we have solicited articles from scientists who have strong personal points of view, while encouraging critical discussions and evaluations of existing data. In no sense have the articles been simply literature surveys, although in some cases they may have also fulfilled that purpose.

In the introduction to Volume 1 of this series, the editors noted developments in a brief span of prior years that were important for progress in photochemistry: flash photolysis, nuclear magnetic resonance, and electron spin resonance. A quarter of a century later, in Volume 14 (1988), the editors noted that since then two developments had been of prime significance: the emergence of the laser from an esoteric possibility to an important light source; the evolution of computers to microcomputers in common laboratory use of data acquisition. These developments strongly influenced research on the dynamic behavior of excited state and other transients. We can look forward to significant developments to be included by the end of another quarter century.

With an increased sophistication in experiment and interpretation, photochemists have made substantial progress in achieving the fundamental objective of photochemistry: elucidation of the detailed history of a molecule that absorbs radiation. The scope of this objective is so broad and the systems to be studied are so many that there is little danger of exhausting the subject. We hope that this series will reflect the frontiers of photochemistry as they develop in the future.

DOUGLAS C. NECKERS
DAVID H. VOLMAN
GÜNTHER VON BÜNAU

Bowling Green, Ohio
Davis, California
Siegen, Germany

CONTENTS

Spectroscopy and Photochemistry of Polyatomic Alkaline Earth
Containing Molecules   1
    PETER F. BERNATH

Photochemically Induced Dynamic Nuclear Polarization   63
    MARTIN GOEZ

Photophysics of Gaseous Aromatic Molecules: Excess Vibrational
Energy Dependence of Radiationless Processes   165
    EDWARD C. LIM

Lanthanide Complexes of Encapsulating Ligands as Luminescent
Devices   213
    NANDA SABBATINI, MASSIMO GUARDIGLI, AND ILSE MANET

Advances in the Measurement of Correlation in Photoproduct Motion   279
    CHRISTOPHER G. MORGAN, MARCEL DRABBELS,
    AND ALEC M. WODTKE

Index   351

Cumulative Index, Volumes 1–23   357

ADVANCES IN PHOTOCHEMISTRY
Volume 23

SPECTROSCOPY AND PHOTOCHEMISTRY OF POLYATOMIC ALKALINE EARTH CONTAINING MOLECULES

Peter F. Bernath

Department of Chemistry, University of Waterloo, Waterloo, Ontario, Canada N2L 3G1

CONTENTS

I. Introduction, 2
II. Experimental methods, 5
    A. Flames, 5
    B. Broida oven, 6
    C. Molecular beams, 10
III. Chemistry and photochemistry, 14
IV. Electronic structure, 18
V. Survey of molecules, 23
    A. Hydroxides, MOH, 23
    B. Alkoxides, MOR, 32
    C. Monocarboxylates, MO_2CR, 33
    D. Monoformamidates, MO(NH)CH, 33
    E. Monoisocyanates, MNCO, 35
    F. Monoazides, MN_3, 36

Advances in Photochemistry, Volume 23, Edited by Douglas C. Neckers, David H. Volman, and Günther von Bünau
ISBN 0-471-19289-9 © 1997 by John Wiley & Sons, Inc.

G. Monoisocyanides, MNC and monocyanides, MCN, 39
H. Monohydrosulfides, MSH, 43
I. Monothiolates, MSR, 44
J. Monoamides, MNH_2, 46
K. Monoalkylamides, MNHR, 48
L. Monomethyls, MCH_3, 50
M. Monoacetylides, MCCH, 51
N. Monocyclopentadienides, MC_5H_5, 53
O. Monomethylcyclopentadienides, $MC_5H_4CH_3$, 53
P. Monopyrrolates, MC_4H_4N, 54
Q. Monoborohydrides, MBH_4, 55
VI. Conclusions, 56
Acknowledgments, 56
References, 57

I. INTRODUCTION

The interaction of metals with organic molecules is one of the primary themes of modern chemistry [1]. This interest is based both on the possibility of important applications and on the fascinating variety of molecules that can be synthesized. Synthetic chemists working at the interface between organic and inorganic chemistry create new species and characterize them primarily by nuclear magnetic resonance (NMR) and X-ray crystallography. Many of their creations have found great utility in, for example, the Ziegler–Natta polymerization of ethylene to make polyethylene [1]. Metal–ligand interactions are also important in the chemistry of life processes at, for example, the active sites of enzymes. This new field of bioinorganic chemistry continues to expand.

All of this chemistry occurs in either solution or the solid state and is often influenced by the presence of a solvent. Currently, the tools of modern chemical physics are used to try to understand metal–ligand chemistry in the gas phase, free from the effects of solvents. The focus has been on understanding the chemistry, photochemistry, and spectroscopy of relatively small systems. For reasons of sensitivity, the primary tool for these investigations is the mass spectrometer. Sometimes lasers are used to vaporize a metal or to excite and to ionize the species of interest. The experimental techniques range from traditional high-pressure mass spectrometry to Fourier transform ion cyclotron resonance.

The photochemistry and spectroscopy of simple metal–ligand ions has been studied, for example, by the Brucat, Duncan, and Farrar research groups. In the work of Lessen et al. [2], the vanadium ion (V^+) was created by laser vaporization and allowed to react with H_2O. The photodissociation of $V^+(H_2O)$ was driven by a tunable dye laser. The spectra were recorded by monitoring the appearance of V^+ with a time-of-flight (TOF) mass spectrometer as the laser was scanned. Similar work in the Duncan group has provided spectra of $Mg^+(H_2O)$ [3] and $Ca^+(H_2O)$ [4], while the Farrar group has looked at $Sr^+(H_2O)$ and $Sr^+(NH_3)$ [5, 6]. These molecular ions are isoelectronic with the neutral molecules such as $CaNH_2$ and $SrCH_3$ that are discussed in this chapter.

Molecular ions such as $Ca^+(H_2O)$ have relatively weak metal–ligand binding energies of about 25 kcal mol^{-1} [7] compared to the stronger bond present in the isoelectronic $CaNH_2$ system. In $Ca^+(H_2O)$, the H_2O ligand perturbs the Ca^+ energy levels much less than the NH_2^- ligand perturbs Ca^+ in the $CaNH_2$ molecule. The electrostatic interaction of NH_2^- with Ca^+ is much stronger than the ion–dipole interaction of Ca^+ with H_2O, although the molecular symmetry (C_{2v}) is the same for $CaNH_2$ and $Ca^+(OH_2)$.

This chapter will cover the monovalent neutral polyatomic derivatives of the alkaline earth elements: Be, Mg, Ca, Sr, Ba, and Ra. The alkaline earth metals are naturally divalent in the solid state. Well-known examples of alkaline earth compounds are the widely used Grignard reagents, which have an empirical formula R—Mg—X (R is an alkyl group and X is a halogen) [1]. More recently, the Hanusa group at Vanderbilt has explored the organometallic chemistry of divalent Ca, Sr, and Ba derivatives [8]. In contrast to this "normal" chemistry, the monovalent derivatives such as CaOH are unstable in the solid state because they readily disproportionate:

$$2\ CaOH(s) \to Ca(s) + Ca(OH)_2(s) \qquad (1)$$

In fact, "proof" of the nonexistence of MgCl(s) is discussed in undergraduate chemistry textbooks as an exercise in calculating lattice energies using Madelung constants.

The monovalent derivatives of the alkaline earth metals are free radicals that are stable in the thermodynamic sense in the gas phase or when isolated in inert matrices. Molecules such as CaOH have strong bonds (dissociation energy, $D_{Ca-O} = 92$ kcal mol^{-1}) [9, 10] but are very reactive species because of the unpaired electron localized on the Ca atom. In spite of the transient nature of these monovalent derivatives, it has proved possible to develop an extensive gas-phase inorganic chemistry for Mg, Ca, Sr, and Ba. No monovalent polyatomic derivatives are known for the Be or the Ra members of the alkaline earth group, except for BeOH. There are two experimental reports on the BeOH [11, 12] molecule in the literature but one is apparently erroneous [12]. For technical reasons, the widest variety of molecules are known for Ca and Sr.

In addition to the usual divalent molecules and solids, for example, $Ca(OH)_2$, and the monovalent molecules, for example, CaOH, a third type of molecule has been predicted by Kong and Boyd [13]. They calculate that the H—Ca—O isomer of CaOH is a minimum on the potential energy surface. The H—Ca—O molecule is, in fact, a divalent derivative of Ca with an ionic electron distribution, $H^-Ca^{2+}O^-$. Perhaps the ultraviolet (UV) photochemical isomerization of CaOH in a rare gas matrix will yield HCaO.

The metal monohydroxides CaOH and SrOH are the simplest monovalent polyatomic derivatives of the alkaline earths. Both CaOH and SrOH have a surprisingly long history in view of their high chemical reactivity. While CaOH and SrOH can only be stored when isolated in rare gas matrices [14], substantial steady-state concentrations exist in a variety of energetic environments.

In 1823, Herschel [15] in the *Transactions of the Royal Society of Edinburgh* published his observations of the colors of flames produced by the introduction of alkaline earth salts. The green color obtained with barium salts is due to BaOH and the reddish color characteristic of strontium salts is caused by SrOH. The red colors of fireworks can also be attributed to emission from SrOH [16]. It was not until the 1950s that modern flame studies [17, 18] identified the molecules that are responsible for the alkaline earth flame colors. In contrast to the alkaline earths, the flame colors of the alkali elements are produced by atomic emission. The formation of molecules such as CaOH and SrOH, in fact, greatly complicates the use of flame absorption and emission for the determination of the concentrations of alkaline earth elements in analytical chemistry.

The CaOH molecule is also predicted to be of importance in atmospheric chemistry and in astronomy. It has been speculated that the ablation of metals from meteors results in the formation of CaOH [19]. In chemical equilibrium calculations, Tsuji [20] predicted that CaOH should form in the atmospheres of cool oxygen-rich stars. Pesch [21] identified the CaOH molecule through absorption of the $\tilde{B}^2\Sigma^+ - \tilde{X}^2\Sigma^+$ transition near 550 nm. Pettersen and Hawley [22] confirmed this identification using the $\tilde{A}^2\Pi - \tilde{X}^2\Sigma^+$ bands near 620 nm. The recent availability of microwave and millimeter wave spectra of MgOH [23] and CaOH [24] have resulted in unsuccessful searches by radioastronomers. While the high-temperature environment of a stellar atmosphere encourages the formation of CaOH, the molecule does not readily form in cold interstellar clouds.

Interstellar molecular clouds are the sites of star formation and so are of great astronomical interest. In general, metal-containing molecules are not very abundant in dark molecular clouds because they are depleted onto grains. It was, therefore, with some surprise that the MgNC molecule [25, 26] and its isomer MgCN [27] were identified by radioastronomers in the circumstellar envelope of a carbon star. This finding has inspired the Ziurys [27] group to record the laboratory millimeter wave spectra of a large number of simple derivatives of Mg and Ca such as MgCCH and CaCH$_3$.

A review on the calcium- and strontium-containing polyatomic free radicals appeared in 1991 [28], but there has been significant progress since then. In addition to the radio astronomical and laboratory measurements of pure rotation spectra already mentioned, modern molecular beam technology has been used extensively in recent years. Molecular beam spectroscopy has led to an improved understanding of the electronic and vibronic structure of the larger free radicals such as CaC$_5$H$_5$ and CaC$_4$H$_4$N. It is time to cover the field again with a comprehensive review, emphazing the recent advances.

II. EXPERIMENTAL METHODS

A. Flames

The alkaline earth hydroxide molecules were first made in flames and studied by emission spectroscopy [15]. Metal salt solutions are aspirated into atmospheric pressure flames or the salts are placed on a loop of wire in the flame. The MOH (M is an alkaline earth metal) molecules form through a complicated set of flame reactions [18]. More recent flame work has included laser-induced fluorescence studies of CaOH [29] and SrOH [30]. Some of this work is motivated by the observation that soot formation is suppressed in flames when alkaline earths salts are added [31].

Although flames are convenient sources of MOH molecules, they suffer from serious drawbacks for spectroscopic and dynamical studies. The high temperature (~ 2000 K) of flames causes numerous vibrational and rotational levels to be populated resulting in very dense spectra. The high pressure (1 atm) broadens the rotational lines (>0.1 cm^{-1}) and increases the overlap of the lines. In addition, resonant laser-induced fluorescence is difficult to detect because of quenching and the overwhelming presence of nonresonant fluorescence caused by rapid collisional energy transfer. The luminescence of the flame itself also interferes with measurements.

B. Broida Oven

The key discovery that opened up the field was made by Harris and co-workers [32–35] in the early 1980s. They found that alkaline earth monohydroxides and monoamides could be made readily in a flow reactor called a Broida oven [36]. The Broida oven is a relatively cool (~ 500 K), low-pressure (~ 5 torr) source of high-temperature molecules that is suitable for spectroscopic studies. The work of Harris and our own work owes a great debt to the pioneering efforts of Broida in developing the source and demonstrating its potential.

The Broida oven is a remarkable source of molecules. It offers a large concentration of free radicals ($\sim 10^{13}$ molecules cm^{-3}) isolated in a flow of inert carrier gas such as argon. In contrast to a molecular beam, a Broida oven also offers a large flux of molecules (~ 1 g h^{-1}) that could be used for preparative chemistry, although this aspect of the technology has never been exploited. The continuous injection of a room temperature carrier gas is responsible for both the large flux and the relatively low temperature of the source. Molecules produced in a furnace are sometimes called high-temperature molecules and the Broida oven offers the possibility of studying them near room temperature.

The molecular species in a Broida oven can often be detected through their chemiluminescent emission [32]. It is particularly convenient to monitor this emission in the early stages of a low-resolution analysis. The information that can be extracted from a chemiluminescent spectrum recorded with a monochromator is, however, limited. More typically, the molecules are detected by laser-induced fluorescence using either pulsed or continuous wave (CW) dye lasers.

The molecules in a Broida oven are produced by the reaction of a metal vapor with an appropriate oxidant (Fig. 1). The metal (Mg, Ca, Sr, or Ba) is vaporized in a resistively heated crucible and entrained in a flow of Ar gas. The oxidant is added at the top through an oxidant ring. The reaction of the metal with an oxidizer, for example,

$$Ca + H_2O \rightarrow CaOH + H \tag{2}$$

often produces a low-pressure chemiluminescent flame above the oxidant injection ring (Fig. 1). The mechanism for this chemical reaction (2) is discussed below. Typical pressures are 5 torr of Ar carrier gas, 1 mtorr of metal vapor (Ca), and 10 mtorr of oxidant (H$_2$O). The resulting product (CaOH) has a concentration of less than 1 mtorr, perhaps as high as 10^{13} molecules cm^{-3} in the most favorable cases.

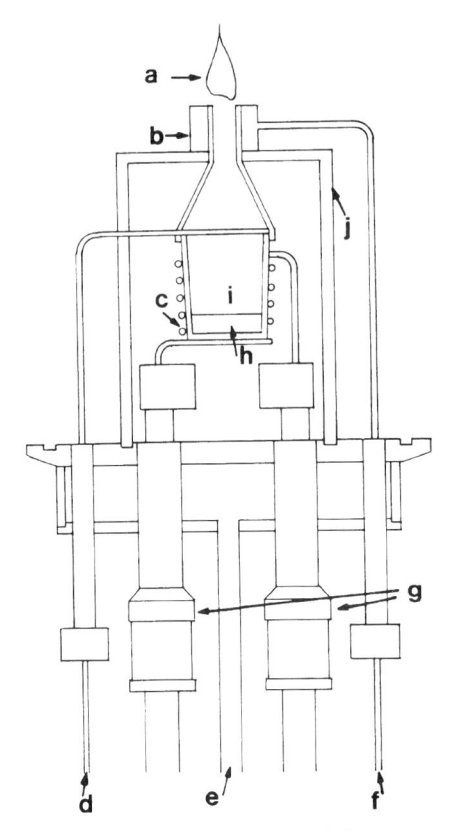

Figure 1. The baseplate for the Broida oven metal flow reactor. This baseplate is attached to the bottom of a vacuum chamber (Fig. 2). (a) chemiluminescent flame; (b) oxidant injection ring; (c) tungsten wire heating basket; (d) Ar carrier gas inlet; (e) cooling water; (f) oxidant gas inlet; (g) electrical feeds for heating; (h) alkaline earth metal; (i) alumina crucible; and (j) alumina heat shield. [Reprinted with permission from ref. 28. Copyright 1991 American Association for the Advancement of Science.]

During the course of our investigations, we found that if the metal atoms are excited by a laser to the metastable $^3P_1$ electronic state, then the concentration of product molecules is dramatically increased,

$$Ca^*(^3P_1) + H_2O \rightarrow CaOH + H, \tag{3}$$

where the asterisk denotes electronic excitation. Reaction 2 is endothermic

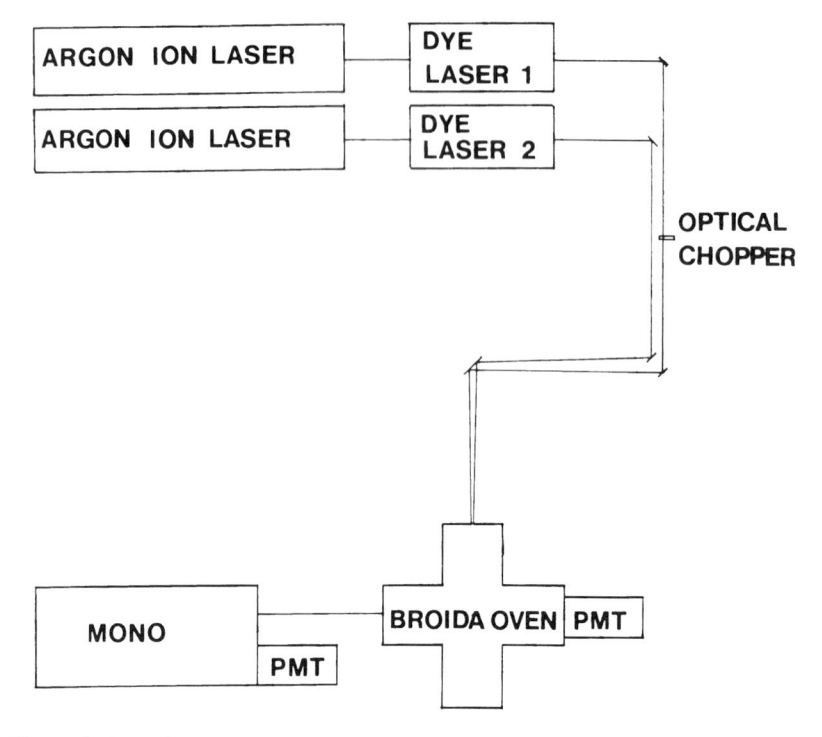

```
┌─────────────────┐   ┌──────────┐
│ ARGON ION LASER │───│   DYE    │────────────┐
└─────────────────┘   │ LASER 1  │            │
                      └──────────┘            │
┌─────────────────┐   ┌──────────┐            │
│ ARGON ION LASER │───│   DYE    │──────────┐ │
└─────────────────┘   │ LASER 2  │          │ │
                      └──────────┘          │ │
                                      ⇐ OPTICAL
                                        CHOPPER
```

Figure 2. Experimental block diagram for Broida oven experiments in the Bernath laboratory. Two Ar ion lasers pump dye lasers that are used to make and to detect the free radicals in the oven. The laser-induced fluorescence is monitored with photomultiplier tubes (PMTs) mounted either directly on the oven or on the exit port of a monochromator (MONO). [Reprinted with permission from ref. 28. Copyright 1991 American Association for the Advancement of Science.]

as written but Reaction 3 can occur directly because the additional energy stored in Ca* makes Reaction 3 exothermic. The use of excited Ca atoms is particularly important for the formation of organometallic molecules such as CaC_5H_5 since ground-state Ca atoms will not react with the C_5H_6 precursor. Almost all of the molecules discussed in this chapter are produced through photochemistry.

Exploiting the enhanced reactivity of Ca* often requires the use of two tunable dye lasers (Fig. 2), one to excite the metal atoms to the metastable $^3P_1$ state and the second to detect the product molecules. (Excited metal atoms can also be produced in an electrical discharge or in a laser-vaporized plume.) The two laser beams are introduced into the Broida oven chamber through Brewster angle windows on the top or the side. The first laser (dye laser 1) is tuned to the $^3P_1-{}^1S_0$ atomic transition at 457.1, 657.3, 689.2 and, 791.1 nm for Mg, Ca, Sr, and Ba, respectively. This first laser is a typically operated broadband (~ 1-cm^{-1} line width) although an étalon is sometimes used to increase the spectral power density and the stability. The second laser (dye laser 2) can be operated broadband for survey work or in single longitudinal mode (~ 1-MHz bandwidth) for high-resolution work. In our laboratory, the lasers we use are the Coherent model 599, 699–29, and 899–29 dye lasers and a titanium sapphire laser, all pumped by Ar ion lasers. The typical laser powers are 0.5 W at the Broida oven. The two dye laser beams are spatially overlapped and focused into the Broida oven.

The first laser beam can be amplitude modulated (~ 1 kHz) with an optical chopper (Fig. 2), which modulates the concentration of the excited metal atoms, M*. Because the excited metal atoms have a much higher reactivity, the concentration of product molecules is also modulated. The modulated fluorescence excited by the second laser is then detected by a PMT and a lock-in amplifier. A monochromator or an optical filter is used to analyze the emission and to control the optical bandwidth detected by the PMT. This photochemical modulation and synchronous detection of the fluorescent signal is a very powerful technique for increasing the signal-to-noise (S/N) ratio.

There are two typical spectroscopic experiments. In the first, both lasers are in resonance with atomic (laser 1) and molecular (laser 2) transitions and the monochromator (Fig. 2) is scanned to record the laser-induced fluorescence. In the second type of experiment, the monochromator is used as a filter and is not scanned while the second laser wavelength is changed. This second type of experiment is called a laser excitation scan since a fluorescent signal is detected by the PMT only when laser 2 is in resonance with a molecular transition. In this case, the scanning laser can be broadband for survey work or single mode for high-resolution experiments.

The high-resolution spectra recorded with a Broida oven are limited by the effects of Doppler and collisional broadening. For example, in the visible region with a pressure of 10 torr, Doppler broadening dominates and a typical line width for a molecule like CaCCH is about 0.03 cm^{-1}. Nonlinear techniques such as intermodulated fluorescence are feasible in a Broida oven in order to remove the effects of Doppler broadening. The technique of laser excitation spectroscopy with selective fluorescence detection is often necessary to simplify the complex rotational structure of a molecule such as CaCH$_3$. Ultimately, the collisional redistribution of the excited state population in a Broida oven limits the size of molecule for which a rotational analysis is possible. For example, the rovibronic lines of CaC$_5$H$_5$ cannot be resolved in a Broida oven and a molecular beam experiment is necessary.

One of the recent developments has been the use of Broida oven technology in recording millimeter wave pure rotational spectra (Fig. 3). The pure rotational transitions are recorded in absorption using a free space cell. There are two main groups working in this area, the Ziurys group at Arizona State University [37] and the Saito group [25] at the Institute for Molecular Science in Japan.

In the Arizona State University design, the radiation is collimated with Teflon lenses and passes twice through the Broida oven. This double passing is achieved (Fig. 3) by sending the radiation through a wire grid polarizer, then reflecting the radiation with a rooftop prism oriented in such a way to rotate the plane of linear polarization by 90°. The return beam through the cell is now totally reflected by the polarizer and focused onto an InSb hot electron bolometer detector.

The source of microwave radiation at Arizona State University is a set of three phase-locked Gunn oscillators (Fig. 3), which operate in the 65–140-GHz region at power levels of about 50 mW. Higher frequencies are obtained by doubling, tripling, or quadrupling the fundamental frequency in a nonlinear Schottky diode multiplier. The millimeter wave radiation is frequency modulated at 25 kHz by adjusting the reference frequency used in the lock circuit.

The main difficulty in adapting the Broida oven to "low"-frequency millimeter wave measurements is the problem of pressure broadening [38]. Since the Doppler broadening of molecular lines is linearly proportional to the transition frequency, the Doppler effect is negligible in the millimeter wave region. The molecular line widths are dominated by pressure broadening. The effect of pressure broadening depends linearly on the pressure with a typical magnitude of 10 MHz torr^{-1} of gas. This result means that the total pressure in the Broida oven chamber (mainly argon carrier gas) has to be reduced to below 100 mtorr in order to obtain narrow lines and strong peak absorption coefficients. The solution to this problem is to increase the pumping speed by replacing the usual mechanical vacuum pump on the Broida oven with a Roots blower.

C. Molecular Beams

The Broida oven is a nearly ideal source for the spectroscopy of diatomic molecules and small polyatomic molecules such as CaOH. For larger species, however, the spectral congestion is too severe and the collisional relaxation rates are too high to record resonant, rotationally resolved spectra. The solution to this problem is to lower the temperature to eliminate the spectral congestion and to lower the pressure to eliminate the

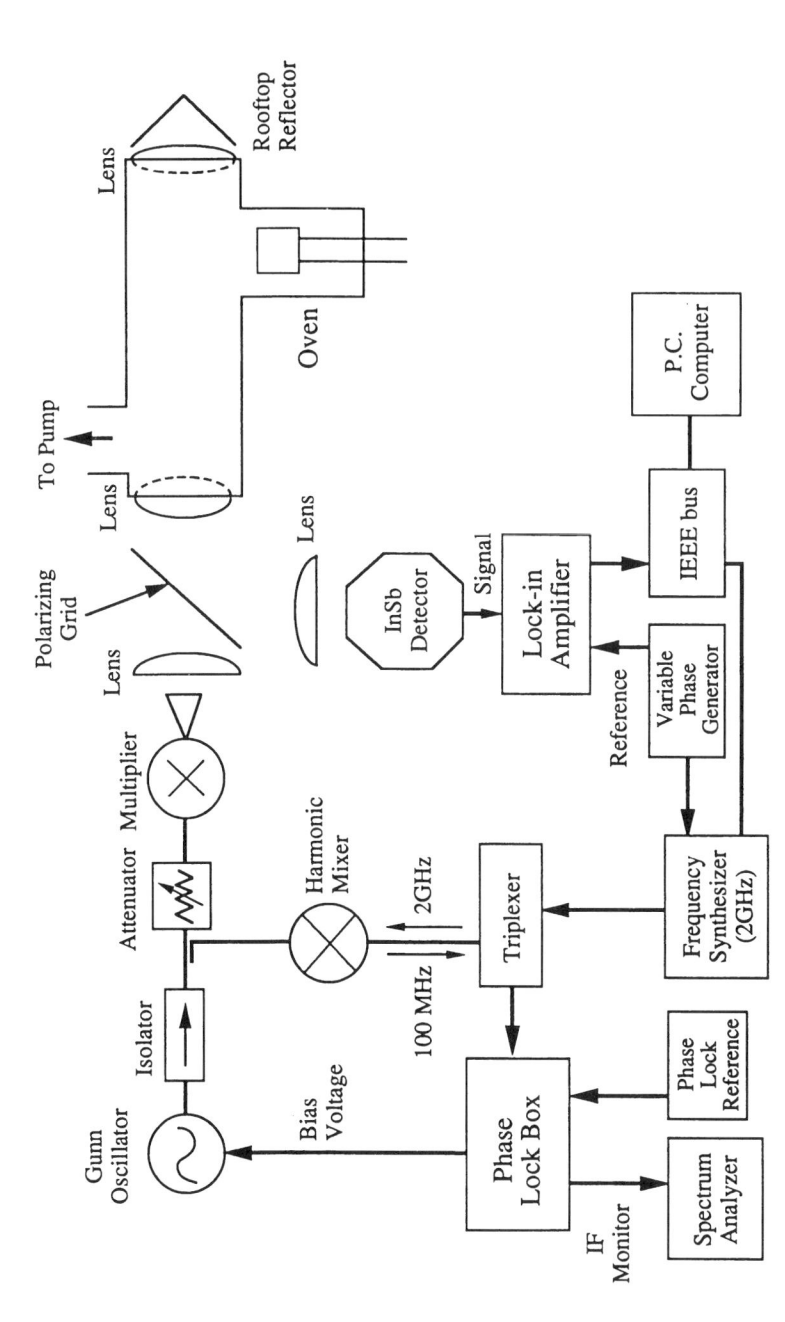

Figure 3. Block diagram of the millimeter wave spectrometer used by the Ziurys group. The instrument uses a phase locked Gunn oscillator as a source of radiation and an InSb detector. [Reprinted with permission from ref. 37. Copyright 1994 American Institute of Physics.]

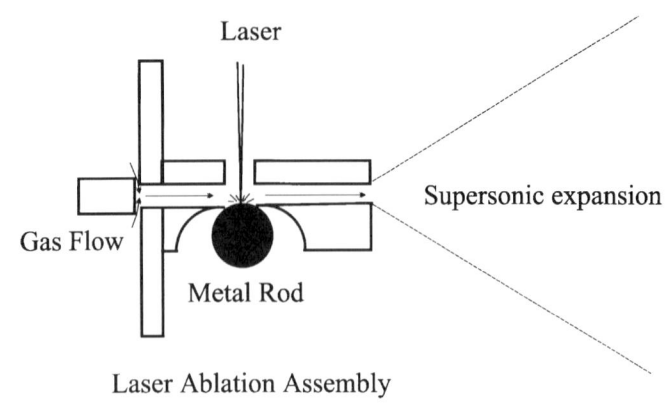

Figure 4. Laser ablation fixture used to make a supersonic molecular jet. The pulsed valve is synchronized with the ablation laser and the gaseous products expand into a vacuum and cool.

collision-induced redistribution of the excited-state population. A molecular beam source has these desirable properties.

The breakthrough experiment was carried out by Whitham et al. [39, 40] in France. They used a "Smalley-type" laser vaporization source (Fig. 4) to provide a molecular beam of Ca atoms entrained in He or Ar gas. The second harmonic (532 nm) from a pulsed Nd:YAG laser was focused (Fig. 4) on a rotating calcium rod. About 500 μs prior to this, a pulsed valve (left side of Fig. 4) is opened and the plume of vaporized metal is entrained in Ar or He gas. The carrier gas is seeded with a few percent of the oxidant such as H_2O. The plume of excited- and ground-state metal atoms are carried down a short channel and react with the oxidant. At the end of the channel, the product molecules such as CaOH expand into the vacuum chamber and cool. After a short expansion, the pressure has dropped so low that the molecules are effectively in a collisionless, ultracold (< 10 K) environment.

The molecular jet of molecules is crossed with a tunable dye laser and the laser-induced fluorescence is collected with a lens and focused on a PMT detector (Fig. 5). In the original experiments, a standard pulsed dye laser was used to match the 10-Hz duty cycle of the pulsed valve and the pulsed Nd:YAG vaporization laser. Although this approach provides a high S/N ratio and wide spectral coverage, the resolution is limited by the laser line width of typically 0.5 cm^{-1} (no étalon) to 0.05 cm^{-1} with an étalon.

The simplest method to obtain high-resolution spectra is to replace the pulsed dye laser with a CW single-mode dye laser. As Steimle and co-workers have demonstrated in a series of beautiful experiments, this

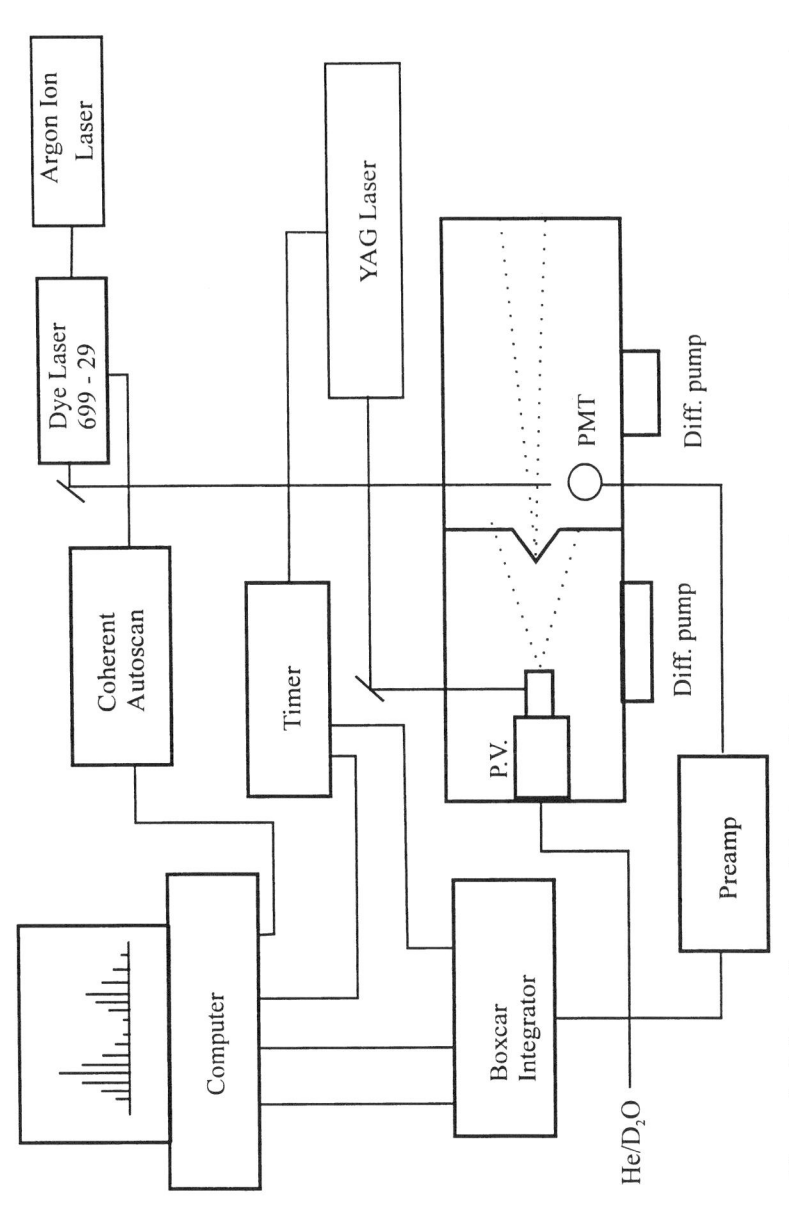

Figure 5. A block diagram of the laser ablation, molecular beam spectrometer at the University of Waterloo. [Reprinted with permission from ref. 74. Copyright 1996 Academic Press.]

technique is very effective. The main drawback is that most of the laser photons are wasted because of the duty cycle mismatch between the pulsed molecular beam source and the CW laser.

An alternate approach used by the Miller group is to pulse amplify a CW dye laser. This approach degrades the laser resolution from about 1 MHz to about 200 MHz but results in a high-power pulsed laser beam with excellent mode quality. This high peak power and high spectral resolution results in excellent spectra with an improved S/N ratio. The only serious drawback of this scheme is the increased complexity and cost of the laser systems.

There is one additional choice that needs to be made in setting up a pulsed molecular beam spectrometer. The laser beam can cross the molecular jet close (~ 1 cm) to the pulsed valve, as implemented by the Miller group. This technique gives a relatively large signal, but there is often a fluorescent background from the laser-vaporized plume of metal atoms (high signal/high noise). In this case, it is useful to use filters or a monochromator to isolate the laser-induced fluorescence signal. The alternative approach is to skim the jet (with either a true skimmer or a large hole) in order to form a molecular beam and to cross the laser and molecular beams at some distance (~ 10 cm) from the pulsed valve (Fig. 5). This arrangement gives a much reduced signal since the molecular concentration is dropping rapidly during the expansion but the fluorescent background is nearly eliminated (low signal/low noise). This latter approach is advocated by the Steimle group.

The application of molecular beam technology has resulted in the detailed rotational analysis of numerous larger alkaline earth derivatives such as CaC_5H_5. From the rotational analysis (and the improved vibrational analysis), one can infer molecular structure. In addition, large electric fields can be easily applied to molecular beams in order to measure dipole moments through the Stark effect. Dipole moments provide information about charge distributions. The alkaline earth derivatives are a unique family of molecules because so much detailed information is available for them. These simple one metal–one ligand molecules are useful as models for more complex species found in inorganic chemistry and surface science.

III. CHEMISTRY AND PHOTOCHEMISTRY

One of the surprising aspects of the chemistry of alkaline earth atoms is that ground-state atoms can react readily in the gas phase with molecules such as H_2O and CH_3OH. In solution, the overall reaction

$$Ca(s) + 2 H_2O(\ell) \rightarrow Ca(OH)_2(s) + H_2(g) \tag{4}$$

is vigorous. In the gas phase, however, the elementary reaction

$$Ca(g) + H_2O(g) \to CaOH(g) + H(g) \qquad (5)$$

does not occur because it is endothermic by 27 kcal mol^{-1}. The CaOH molecule obviously does form in a Broida oven and even chemiluminescence from the $\tilde{B}^2\Sigma^+$ and $\tilde{A}^2\Pi$ states can be seen. What is the source of this extra energy?

The laser excitation of Ca atoms to the $^3P_1$ state can provide an additional 44 kcal mol^{-1} of energy and the direct reaction

$$Ca(^3P_1) + H_2O(g) \to CaOH(g) + H(g) \qquad (6)$$

is now exothermic by 17 kcal mol^{-1}. Thus, the increased production of CaOH when Ca* is available is explained by the opening of the direct reaction channel (6). In a high-pressure CH_4/O_2 flame, there is also considerable energy available and many free radicals such as OH can react with Ca,

$$Ca(g) + OH(g) \to CaOH(g) \qquad (7)$$

in the presence of a third body. In a laser vaporization source, the plume of Ca atoms contains a large fraction of Ca* and Ca$^+$, which readily react to give CaOH in the molecular beam. Thus in many cases the deliberate or inadvertent presence of Ca* or OH accounts for the synthesis of CaOH. In a normal Broida oven, however, the thermal vaporization of Ca at about 1000°C yields a very low concentration of Ca*. It is found that the production of CaOH can be dramatically increased by simply increasing the total pressure in the chamber from 1 to 10 torr by decreasing the pumping speed. This fact points to a mechanism that needs a third body in a rate-controlling step.

Two possible mechanisms [41] are

Mechanism A

$$Ca + H_2O \xrightarrow{Ar} HCaOH \qquad (8)$$

$$HCaOH + Ca \to CaH + CaOH \qquad (9)$$

$$CaH + H_2O \to CaOH + H_2 \qquad (10)$$

$$\overline{2\,Ca + 2\,H_2O \to 2\,CaOH + H_2} \qquad (11)$$

or

Mechanism B

$$Ca + H_2O \xrightarrow{Ar} HCaOH \qquad (8)$$

$$HCaOH + H_2O \to Ca(OH)_2 + H_2 \qquad (12)$$

$$\underline{Ca(OH)_2 + Ca \to 2\,CaOH \qquad (13)}$$

$$2\,Ca + 2\,H_2O \to 2\,CaOH + H_2 \qquad (11)$$

In these reactions, other alkaline earths can be substituted for Ca and other oxidants such as CH_3OH or $HC(O)OH$, which contain the OH group, can be substituted for H_2O.

Both mechanisms have as their first, and rate-controlling step, the insertion of a Ca atom into one of the O—H bonds of water. Additional support for the existence of an HCaOH intermediate comes from the matrix isolation experiments of Margrave and co-workers [14]. When Ca and H_2O were cocondensed in an argon matrix, a $Ca(H_2O)$ complex formed. Upon irradiation of the matrix with light near the metal resonance lines, the Ca atom inserted into H_2O and the infrared (IR) absorption of the HCaOH molecules was detected. Irradiation with UV light then gave the CaOH molecule.

Although we cannot directly detect HCaOH because it probably has a dissociative UV spectrum [14], we can detect another predicted reaction intermediate in some of our experiments. Mechanism A predicts that the CaH molecule will be present in the Broida oven, and with some oxidants we have detected it by laser-induced fluorescence. The CaH molecule is seen when carboxylic acids such as formic acid are used to make the monocarboxylates such as SrO_2CH [42]. Curiously, CaH is not detected [41] when water or alcohols such as CH_3OH are used to make alkoxides such as $CaOCH_3$. More experimental and theoretical work is necessary to establish the chemical mechanisms involved in the reactivity of the alkaline earth atoms.

There have been several studies of the reaction dynamics of the ground and excited states of the alkaline earth atoms with various oxygen-containing molecules under single collision conditions. Although these studies are not directly applicable to the multiple collision regime in the Broida oven, they clarify the dynamics of a single encounter between a metal atom and an oxidant molecule. Oberlander and Parson [43] looked at the reactions of Ca and Sr with water, alcohols, and peroxides. Similar studies

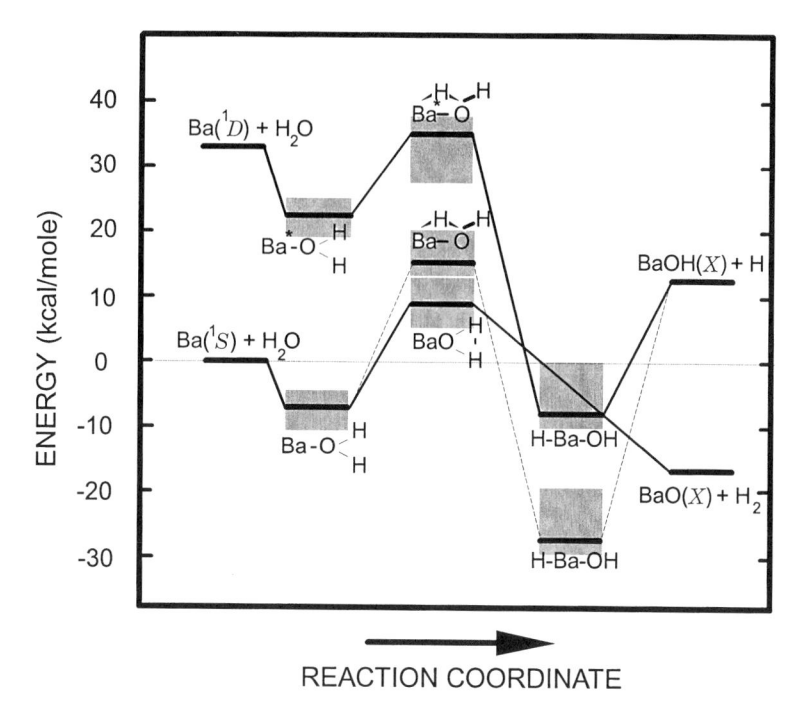

Figure 6. The reaction of Ba and H_2O. The reactions on the ground-state potential surface show the BaO (dashed) and BaOH (solid) product channels. On the excited $Ba(^1D) + H_2O$ surface only the BaOH channel is open. [Reprinted with permission from ref. 44. Copyright 1993 American Institute of Physics.]

of the reactivity of Ba were made by Davis et al. [44] as well as by de Pujo and co-workers [45–47]. The basic conclusions are illustrated in Figure 6.

The ground state of the Ba atom reacts with water by two different mechanisms [44] (Fig. 6). The main channel is the direct exothermic reaction that results in the products BaO and H_2. The minor channel (dotted lines) is the insertion of Ba into the OH bond (H migration) that results in the endothermic BaOH + H products. More than simple energetics is involved, however, because even when the Ba atom is given enough energy to overcome the insertion barrier, the reaction strongly favors the direct BaO product.

The reactivity of excited ^1D atoms with water is very different than the reactivity of the ground state Ba atoms (Fig. 6). The "diradical" character of the excited state strongly favors insertion to give the BaOH product exclusively. Even though the BaO product is favored on energetic grounds, none is produced. The BaOH product angular distributions are consistent with an HBaOH intermediate. The reactions of Ca and Sr with H_2O and H_2O_2 follow a similar pattern, although the dynamics are postulated by Parson and co-workers [48, 49] to be slightly different.

The reactions of Ca, Sr, and Ba with alcohols [43–45] are also different in that only the insertion channel is seen for both ground- and excited-metal atoms. The reaction intermediate is HMOR in all cases and the H atom leaves to give the MOR product. Very recently, the production of a small amount of CaOH was reported in the reaction of Ca with alcohols [50, 51]. The bulky R group suppresses the production of the energetically favored MO + HR products. Under both single collision and multiple collision conditions the most important dynamical event is the insertion of an alkaline earth metal atom in an H—OR bond.

It is presumed that the production of other monovalent derivatives such as $CaNH_2$ or CaC_5H_5 from NH_3 or C_5H_6 may involve similar excited metal atom insertions into N—H or C—H bonds. Unlike the reactions with H_2O and alcohols, however, no studies of the dynamics have been carried out and even the thermochemistry is very uncertain. Clearly, more experimental and theoretical work is necessary before any firm conclusions can be drawn about mechanisms.

IV. ELECTRONIC STRUCTURE

One of the most appealing aspects of the spectroscopy of the monovalent alkaline earth derivatives is that a simple, one-electron, hydrogenic model provides a reasonable picture of the electronic structure. This simple model is not usually applicable to the derivatives of the other elements of the periodic table. In fact, the major impediment to extending the work to the more chemically interesting transition elements is that the spectra will be difficult to interpret because of the presence of many states with large spin and orbital angular momenta.

The electronic structure can best be explained with a correlation diagram for Ca^+, CaF, and $CaNH_2$ as shown in Figure 7. The $CaNH_2$ molecule is ionic like Ca^+F^- so the electronic distribution can be approximated as $Ca^+NH_2^-$. The $CaNH_2$ molecule is planar with C_{2v} symmetry, not pyramidal like NH_3. The Ca^+ ion is isoelectronic with K and has a ^2S ground state with low-lying ^2D and ^2P states. The approach of an F^- ligand changes the atomic ground state into a $4s\sigma\ X^2\Sigma^+$ molecular state in CaF.

ELECTRONIC STRUCTURE 19

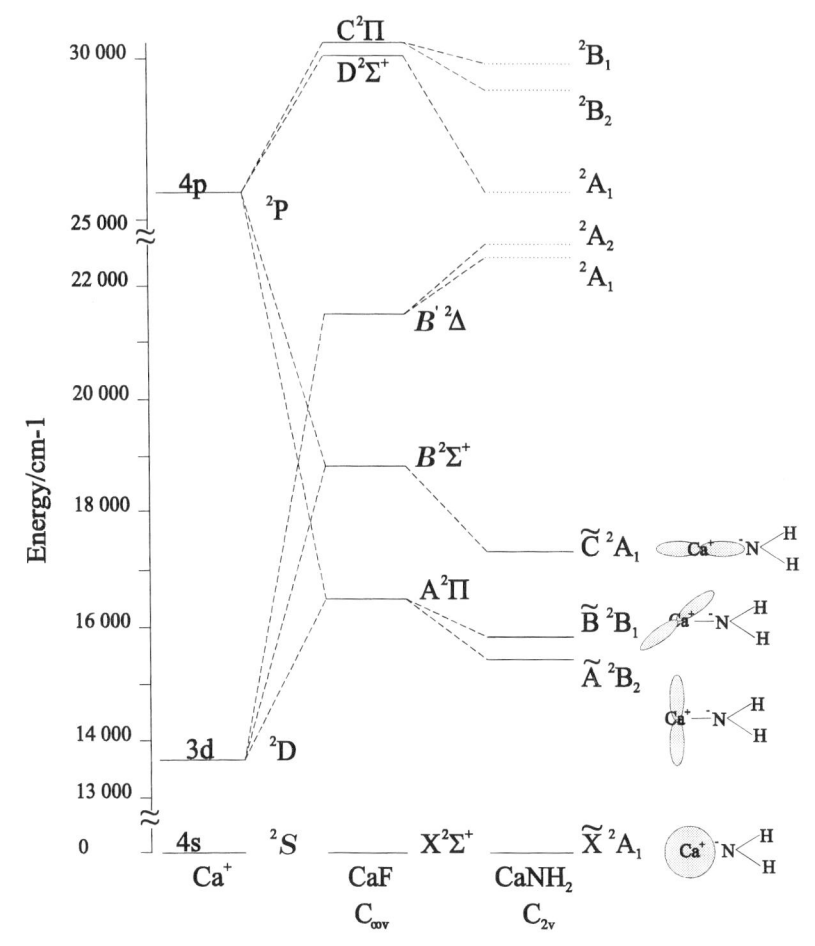

Figure 7. Correlation diagram of the low-lying states of Ca$^+$, CaF, and CaNH$_2$ [53].

The effect of the F$^-$ ligand on the Ca$^+$ excited states is more complicated. The threefold orbital degeneracy of the ^2P state is partly lifted to give $4p\sigma$ and $4p\pi$ orbitals that correlate mainly with the $B^2\Sigma^+$ and $A^2\Pi$ states of CaF, respectively. For a ^2D atomic state, one obtains $3d\sigma$, $3d\pi$, and $3d\delta$ orbitals and $^2\Sigma^+$, $^2\Pi$, and $B'^2\Delta$ molecular states. In addition to lifting the degeneracy, the F$^-$ ligand shifts the location of the states, stabilizing the Ca$^+$ ^2P states more than the ^2D states. Since the atomic l quantum number is no longer good in a molecule, some mixing of atomic orbitals also occurs.

CABH4 GROUND STATE 2A1 FDA

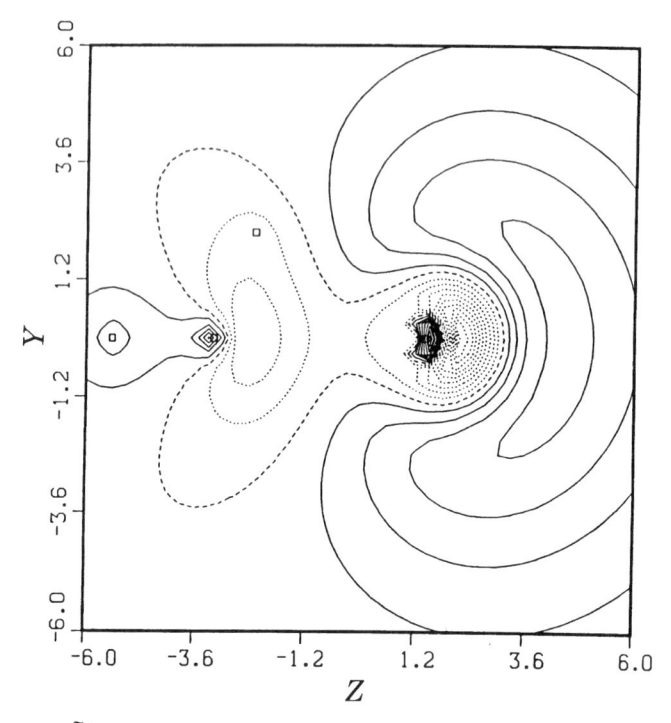

Figure 8. The $\tilde{X}^2A_1$ Feynman–Dyson amplitude of CaBH$_4$ [55]. The Ca atom is to the right and the in-plane BH$_2$ part of the BH$_4$ group is to the left. [Reprinted from ref. 55, with kind permission from Elsevier Science-NL, Sara Burgerhartstraat 25, 1055 KV Amsterdam, The Netherlands.]

In particular, the $A^2\Pi$ state in CaF is about 70% $4p\pi$ and 30% $3d\pi$, while the $B^2\Sigma^+$ state is nearly a 50:50 mixture of $4p\sigma$ and $3d\sigma$ atomic orbitals [52]. This orbital mixing distorts the molecular orbitals in such a way as to keep the unpaired electron as far away as possible from the F$^-$ ligand.

For simplicity, the $A^2\Pi$ state and $B^2\Sigma^+$ state of CaF can be considered to be three pure p orbitals, p_x, p_y for the $A^2\Pi$ state and p_z for the $B^2\Sigma^+$ state. If the F$^-$ ligand is changed to NH$_2^-$, then the degeneracy of the p_x and p_y orbitals is lifted. From a high-resolution rotational analysis of the $\tilde{A}-\tilde{X}$ transition, the first excited state is the in-plane p_y orbital of b_2 symmetry. In Figure 7, the CaNH$_2$ is considered to lie in the plane of the paper so the $\tilde{B}^2B_1$ state is the p_x orbital out of the plane of the molecule

CABH4 EXCITED STATE 2E FDA

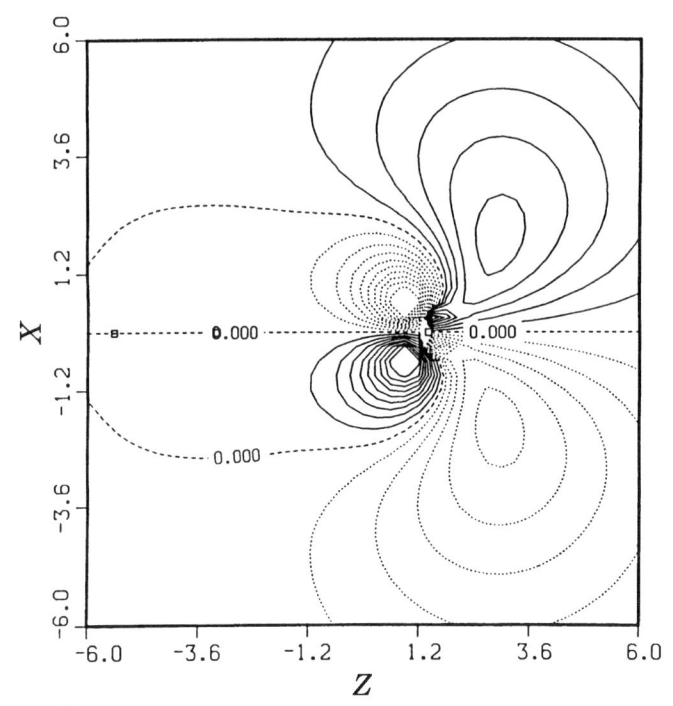

Figure 9. The $\tilde{A}^2E$ Feynman–Dyson amplitude of $CaBH_4$ [55]. [Reprinted from ref. 55, with kind permission from Elsevier Science-NL, Sara Burgerhartstraat 25, 1055 KV Amsterdam, The Netherlands.]

(and the paper). The $\tilde{C}^2A_1$ state is now approximated by the p_z orbital. The $\tilde{A}-\tilde{X}$, $\tilde{B}-\tilde{X}$, and $\tilde{C}-\tilde{X}$ transitions of $CaNH_2$ can be approximated as the $4p$ $(p_x, p_y, p_z) \leftarrow 4s$ resonance transition of Ca^+ perturbed by the presence of an NH_2^- ligand.

These qualitative predictions are clearly too simple but they provide a useful "zeroth-order" model of the electronic structure. For example, the model predicts values for the spin-rotation parameters in the $\tilde{A}$, $\tilde{B}$, and $\tilde{C}$ states of $CaNH_2$ that are in reasonable agreement with experiment [53]. The qualitative predictions of molecular properties can also be compared with the results of ab initio calculations. The first calculations on $CaNH_2$ [54] and related molecules such as $CaBH_4$ were made by Ortiz

22 SPECTROSCOPY AND PHOTOCHEMISTRY OF POLYATOMIC ALKALINE EARTH

CABH4 EXCITED STATE 2A1 FDA

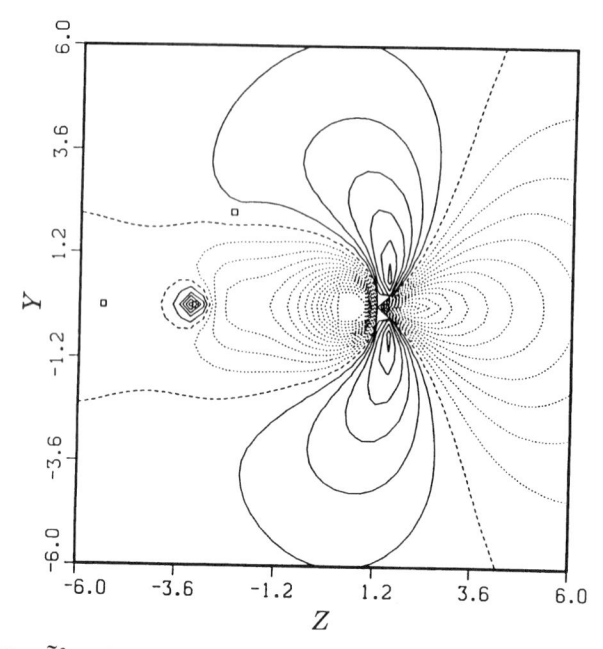

Figure 10. The $\tilde{B}^2A_1$ Feynman–Dyson amplitude of CaBH$_4$. [Reprinted from ref. 55, with kind permission from Elsevier Science-NL, Sara Burgerhartstraat 25, 1055 KV Amsterdam, The Netherlands.]

[55–57] and his results are in general agreement with the simple one-electron picture.

The ab initio electron propagator calculations of Ortiz [54–57] give "Feynman–Dyson amplitudes," which are analogous to orbitals in more conventional calculations. For the CaBH$_4$ molecule [55], the ground $\tilde{X}^2\Sigma^+$ state (Fig. 8) has the general appearance of a distorted $4s$ Ca$^+$ orbital. The Ca$^+$BH$_4^-$ molecule has C_{3v} symmetry with the Ca$^+$ ion (to the right in Fig. 8) bonding to the face of the BH$_4^-$ tetrahedron. In these plots, the solid lines are the positive wave function contours while the dotted lines are negative. Nodal surfaces are indicated by dashed lines. In Figure 9, a similar plot for the first excited ^2E state is presented. This ^2E state correlates to the $A^2\Pi$ state of CaF and the $\tilde{A}^2B_2$ and $\tilde{B}^2B_1$ states of CaNH$_2$ (Fig. 7). Notice that the orbital containing the unpaired electron is a distorted $4p$ orbital. Finally, in Figure 10, the second excited $^2A_1$ state, which correlates to the $B^2\Sigma^+$ state of CaF and the $\tilde{C}^2A_1$ state of CaNH$_2$, is a heavily mixed $4p\sigma$, $3d\sigma$ orbital.

V. SURVEY OF MOLECULES

A. Monohydroxides, MOH

The alkaline earth monohydroxides are the most studied of all of the molecules discussed in this chapter because they are easy to make and are relatively simple to study. Since the early literature has already been reviewed [28], mainly the more recent (since ~ 1990) work will be discussed.

The Ziurys laboratory has recorded the millimeter wave pure rotational spectra for MgOH [23, 58–60], MgOD [60, 61], CaOH [58, 59, 62], CaOD [61], SrOH [58, 59, 63], SrOD [63], BaOH [58, 59, 64], and BaOD [64]. The spectra are simple and show the spin-rotation doublets in the $\tilde{X}^2\Sigma^+$ ground state (Fig. 11) as well as numerous vibrational satellites. A typical spin doublet is shown for the MgOH $N = 13 \leftarrow 12$ transition [23]. The alkaline hydroxides have low-lying vibrational states (Fig. 12) that are populated in the Broida oven and give rise to numerous vibrational satellites as shown in a stick diagram in Figure 13.

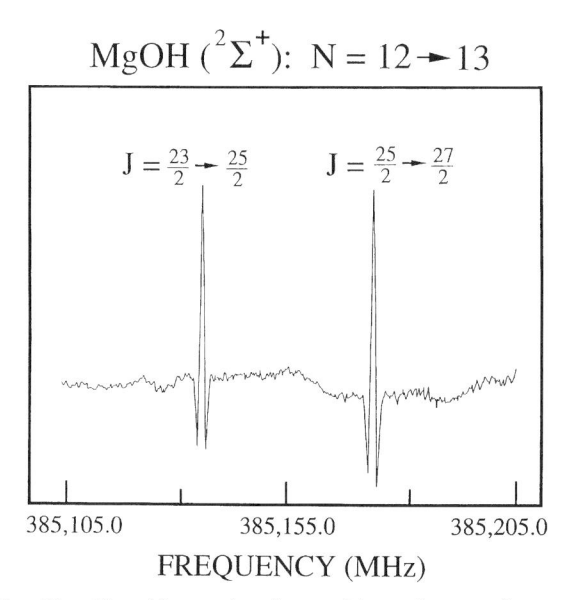

Figure 11. The $N = 13 \leftarrow 12$ rotational transition of ground-state MgOH. The splitting is due to the spin–rotation interaction in the $\tilde{X}^2\Sigma^+$ state. [Reprinted from ref. 23, with kind permission from Elsevier Science-NL, Sara Burgerhartstraat 25, 1055 KV Amsterdam, The Netherlands.]

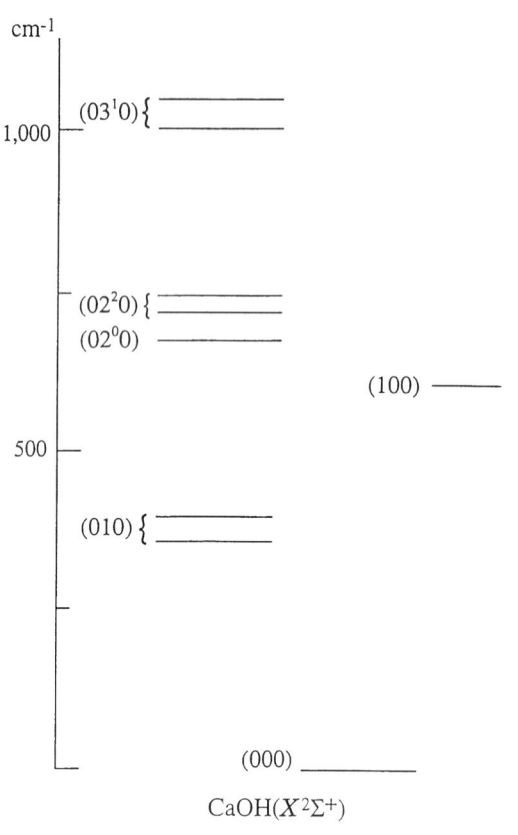

Figure 12. The low-lying vibrational levels of the CaOH $\tilde{X}^2\Sigma^+$ electronic state. [Reprinted with permission from ref. 58. Copyright University of Chicago Press.]

The high-quality pure rotational data, combined with the earlier measurements of the vibrational intervals by laser spectroscopy, allows an improved molecular geometry to be determined. For MgOH, CaOH, SrOH, and BaOH, the geometries are linear in the ground state although the bending potentials [60] are very flat (Fig. 14). Although quasilinear behavior has been ruled out for the heavier members of the family, the situation is more ambiguous for BeOH. The latest ab initio calculations [65] do not predict a barrier to linearity for BeOH in the ground state, in agreement with a linear structure proposed on the basis of electron spin resonance (ESR) experiments by Brom and Weltner [11]. Table 1 gives the r_0 bond lengths for the alkaline earth monohydroxides. The values for BeOH are the r_e predictions of an ab initio calculation by Fernandez [65]. Note that the

SrOH

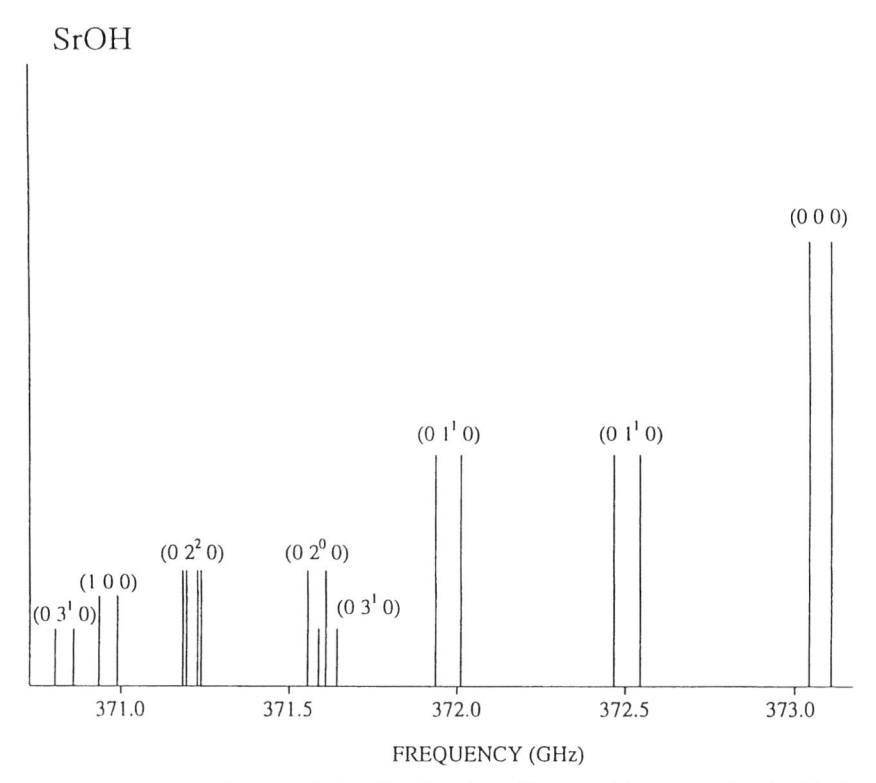

Figure 13. A stick diagram of the vibrational satellite transitions associated with the $N = 25 \leftarrow 24$ rotational transition of SrOH. [Reprinted with permission from ref. 58. Copyright University of Chicago Press.]

first excited electronic state of MgOH [66] and the $\tilde{F}$ state of CaOH [67] have bent geometries.

Millimeter wave spectroscopy with a free space cell such as a Broida oven is more sensitive than lower frequency microwave spectroscopy. However, the higher J transitions monitored by millimeter wave spectroscopy often do not show the effects of hyperfine structure. In the case of CaOH and SrOH, the proton hyperfine structure was measured in beautiful pump–probe microwave optical double resonance experiments in the Steimle group [24, 68]. They adapted the classic atomic beam magnetic resonance experiments to work with a pulsed laser vaporization source and replaced the microwave fields in the A and C regions by optical fields (Fig. 15). These sensitive, high-precision measurements yielded a very small value for the proton Fermi contact parameter (b_F), consistent with ionic bonding and a

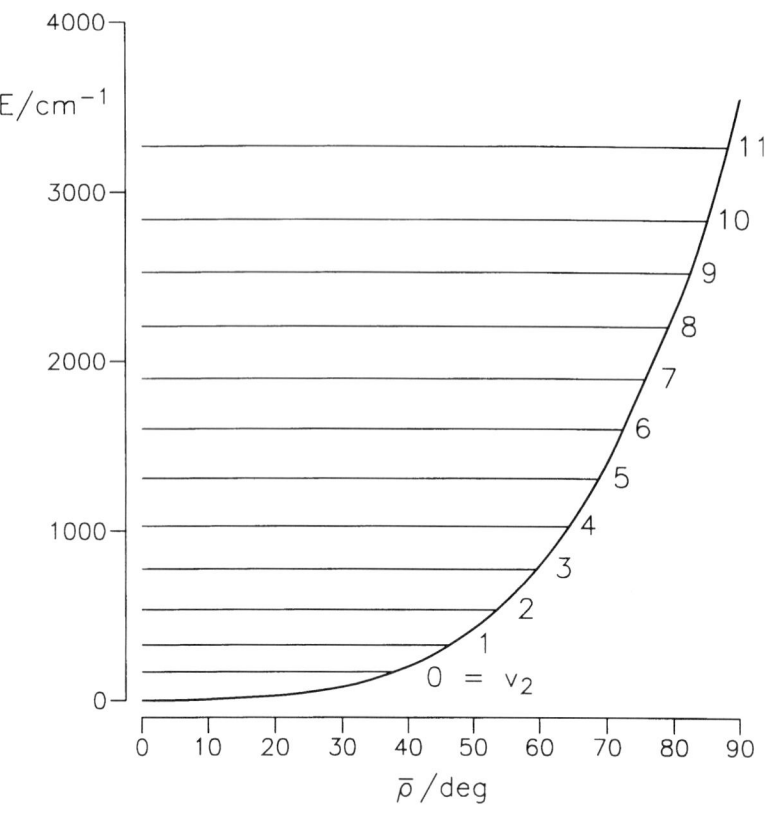

26 SPECTROSCOPY AND PHOTOCHEMISTRY OF POLYATOMIC ALKALINE EARTH

Figure 14. The bending potential energy function of MgOH. [Reprinted from ref. 60, with kind permission from Elsevier Science-NL, Sara Burgerhartstraat 25, 1055 KV Amsterdam, The Netherlands.]

TABLE 1 Bond Lengths (r_0) for MOH Molecules (in Å)

Molecule	r_{MO}	r_{OH}
BeOH[a]	1.378	0.944
MgOH[b]	1.780	0.871
CaOH[b]	1.985	0.922
SrOH[b]	2.111	0.922
BaOH[b]	2.200	0.927

[a] Ab initio r_e values in [65].
[b] Reference 61.

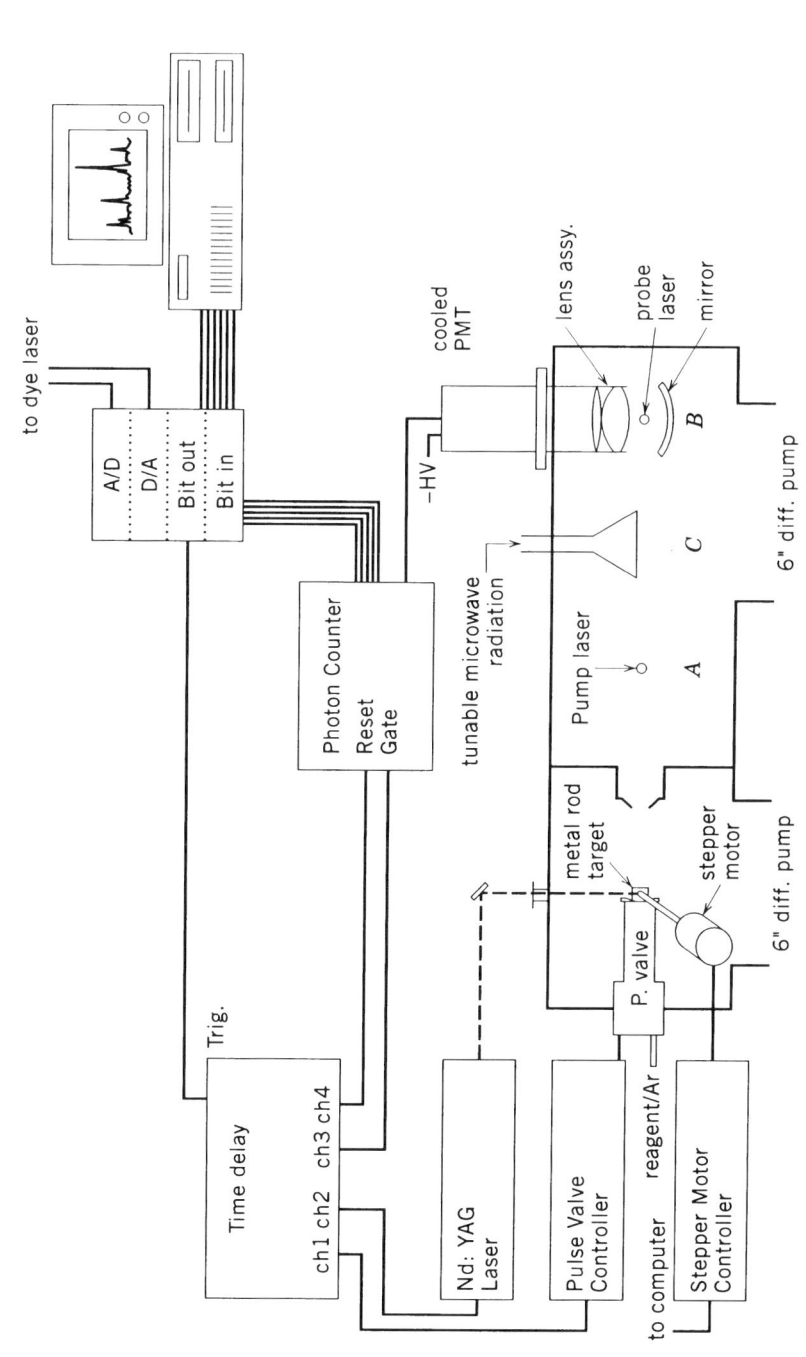

Figure 15. The pulsed molecular beam spectrometer used to record the microwave spectrum of CaNC [108] and CaOH [24]. [Reprinted from ref. 108, Copyright 1994 American Institute of Physics.]

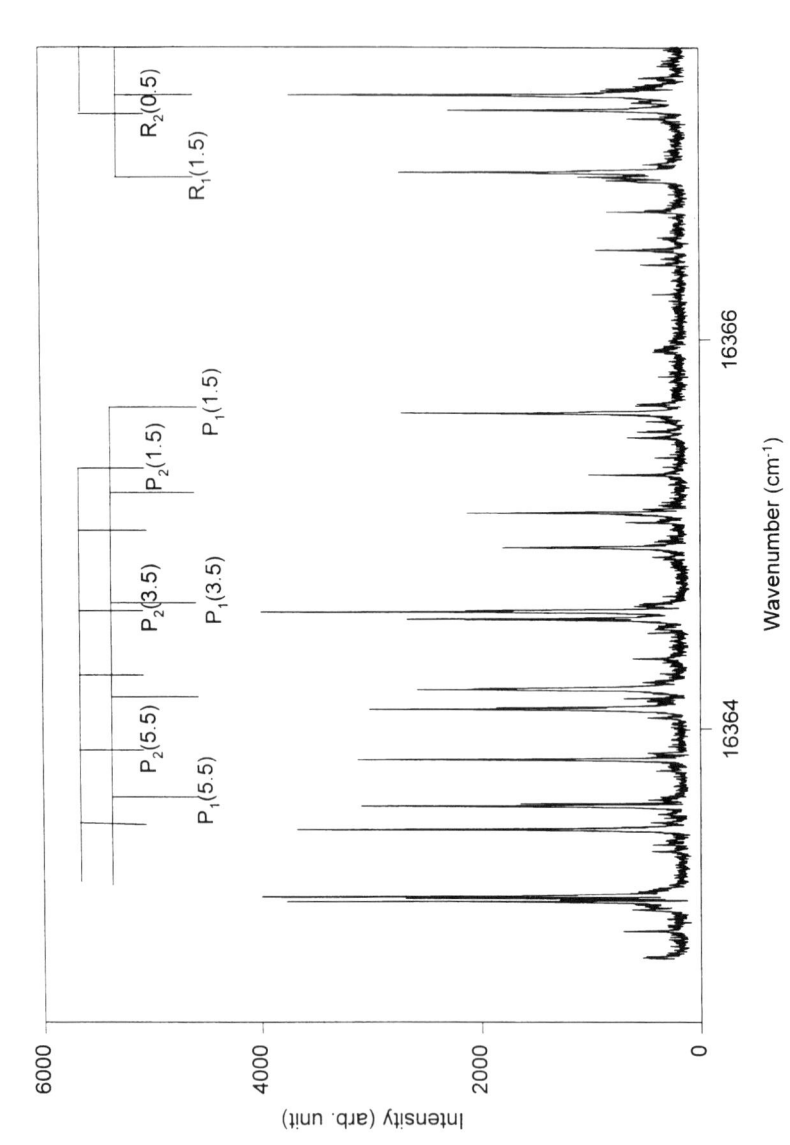

Figure 16. Laser excitation spectrum of the 000–000 band of the $\tilde{B}^2\Sigma^+ - \tilde{X}^2\Sigma^+$ transition of SrOD. [Reprinted with permisssion from ref. 74. Copyright 1996 Academic Press.]

closed-shell OH⁻ ligand. In general, hyperfine structure is very valuable because the hyperfine parameters are determined by the electron distribution.

The Steimle group also measured the dipole moments of CaOH and SrOH in a molecular beam Stark experiment [69]. The surprisingly low values of 1.465 and 1.900 D in the ground $\tilde{X}^2\Sigma^+$ state for CaOH and SrOH, respectively, were found, in moderate agreement with the ab initio calculation of 0.98 D for CaOH [70]. The semiempirical predictions of Mestdagh and Visticot [71] are in good agreement with experiment and explain the small values. The semiempirical scheme is a modified electrostatic Rittner model that allows for the effect of the unpaired electron on the opposite side of the metal away from the OH⁻ ligand (cf. Fig. 8). In other words, the large $Ca^{2+}OH^-$ dipole is nearly canceled by the dipole created by the unpaired electron on the opposite side of the Ca^{2+} atom. A ligand field treatment of the bonding is also available [72].

In recent years, the optical spectroscopy of the metal hydroxides has also bloomed with new analyses of the 0–0 bands of the $\tilde{B}^2\Sigma^+ - \tilde{X}^2\Sigma^+$ transitions of CaOD [73], SrOD [74], and BaOH [47]. The use of a cold molecular beam source results in a simple spectrum and in narrower lines as illustrated in Figure 16 for SrOD [74].

TABLE 2 Vibrational Frequencies for MOH and MOD Molecules (in cm⁻¹)

Molecules	v_1(O—H stretch)	v_2(M—O—H bend)	v_3(M—O stretch)
MgOH		160[a]	750[a]
MgOD		~118[a]	
CaOH	3778[b]	353[c]	609[d]
CaOD	2790[b]	267[e]	605[d]
SrOH		364[f]	527[g]
SrOD		~282[h]	517[i]
BaOH		342[j]	492[j]
BaOD		258[i]	482[j]

[a]Reference 60.
[b]Reference 67.
[c]Reference 85.
[d]Reference 81.
[e]Reference 83.
[f]Reference 77.
[g]Reference 75.
[h]Reference 33.
[i]Reference 74.
[j]Reference 86.

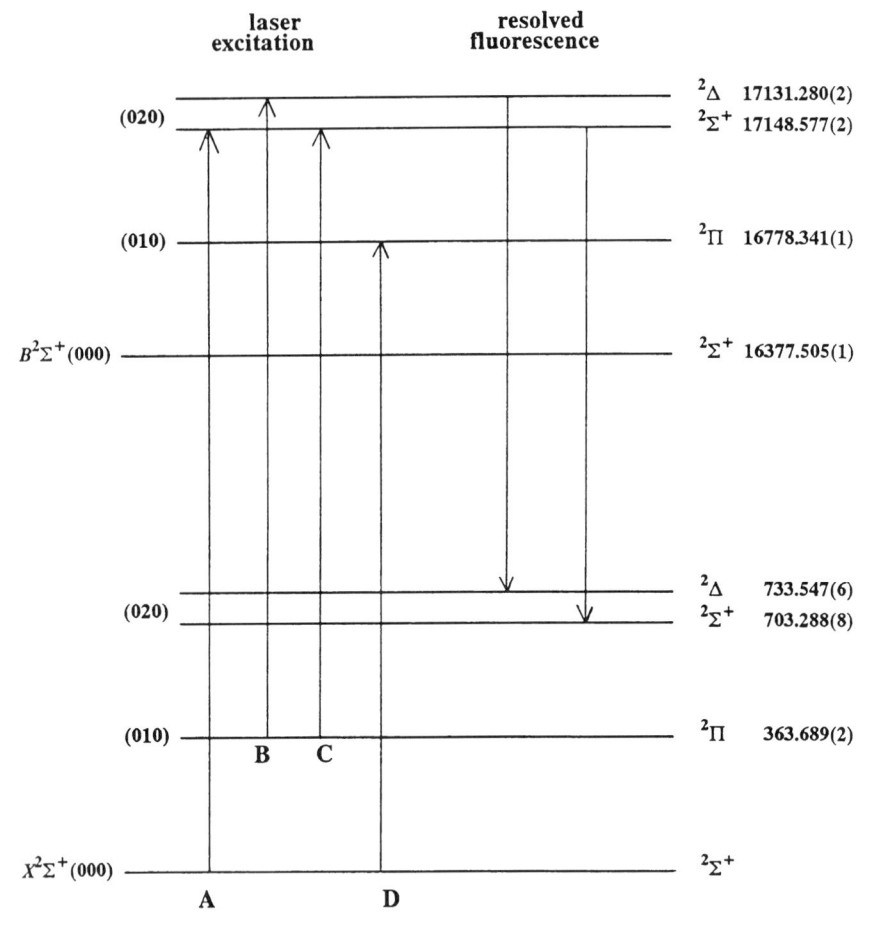

Figure 17. Laser excitation spectra (up arrows) and dispersed laser-induced fluorescence spectra (down arrows) of the $\tilde{B}^2\Sigma^+ - \tilde{X}^2\Sigma^+$ transition of SrOH. [Reprinted with permission from ref. 77. Copyright 1993 NRC Research Press.]

The majority of the work, however, has been carried out in a Broida oven by Coxon and co-workers [75–84]. They concentrated on the study of the vibrational intervals in the ground $\tilde{X}^2\Sigma^+$ states (Table 2) and in understanding the vibronic interactions in the $\tilde{A}^2\Pi$ states. As illustrated in Figure 17, the combination of laser excitation spectroscopy and dispersed laser-induced fluorescence allows the vibronic levels to be mapped out. The $\tilde{A}^2\Pi$ states suffer from the combined influence of the Renner–Teller effect, Fermi resonance and spin–orbit coupling. The Coxon group recorded a large

CaOH $\tilde{C}^2\Delta - \tilde{X}^2\Sigma^+$

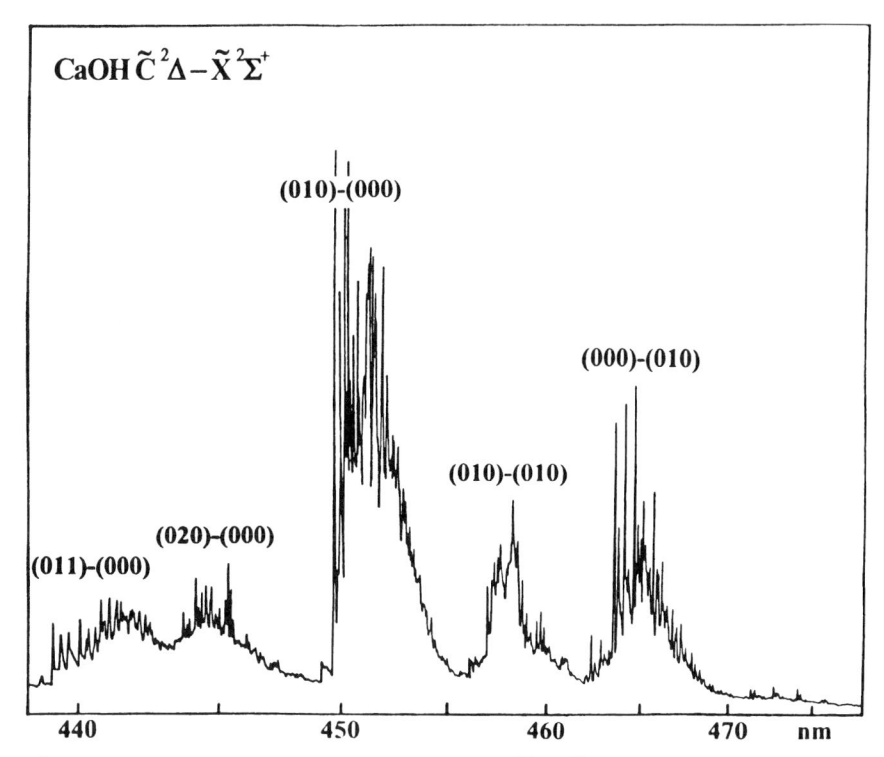

Figure 18. Laser excitation spectrum of the $\tilde{C}^2\Delta - \tilde{X}^2\Sigma^+$ transition of CaOH. [Reprinted with permission from ref. 85. Copyright 1992 American Institute of Physics.]

number of bands in the $\tilde{B}-\tilde{X}$ and $\tilde{A}-\tilde{X}$ transitions of SrOH [75–77], CaOH, and CaOD [78–84].

The Herzberg–Teller effect is another name for the vibronic coupling of electronic and vibrational motion. This effect was used by Jarman and Bernath [85] to locate the missing $\tilde{C}^2\Delta$ state in CaOH via the forbidden $\tilde{C}^2\Delta - \tilde{X}^2\Sigma^+$ transition. In this case, the 0–0 band remains forbidden but the interaction between the bending motion and the electronic motion in the $\tilde{C}^2\Delta$ state make bands such as 010–000 weakly allowed (Fig. 18). The $\tilde{C}^2\Delta$ state also displays the Renner–Teller effect, which is the interaction of vibrational angular momentum with orbital angular momentum in a linear molecule. There is some evidence for a low-lying $\tilde{A}^2\Delta$ state in the spectrum of BaOH, again through the effects of vibronic coupling [86].

Spectroscopic efforts have concentrated on the $\tilde{A}^2\Pi - \tilde{X}^2\Sigma^+$ and $\tilde{B}^2\Sigma^+ - \tilde{X}^2\Sigma^+$ transitions of the MOH molecules although these transitions for

BeOH and MgOH (along with $\tilde{A}-\tilde{X}$ transition of BaOH) are in need of rotational analysis. Very little work is available for other more highly excited states, except in the case of CaOH. Jarman and Bernath [85] located the $\tilde{C}^2\Delta$ state of CaOH and very recently Pereira and Levy [67] found the $\tilde{D}^2\Sigma^+$, $\tilde{E}^2\Sigma^+$, and $\tilde{F}$ states at 28,153, 29,879, and 30,215 cm^{-1}, respectively. In these experiments, pulsed lasers were used with a pulsed laser vaporization source. The $\tilde{F}$ state is the first bent covalent state to be identified for CaOH and it correlates with the $C^2\Pi$ state of CaF (Fig. 7). The first unambiguous measurement of the OH and OD stretching frequency (Table 2) was also possible because of the favorable Franck–Condon factors in the $\tilde{F}$ (bent) → $\tilde{X}$ (linear) transition.

There are also some unpublished optical–optical double resonance data on the $\tilde{G}^2\Pi$ state of CaOH at 32,630 cm^{-1} [87]. The $\tilde{G}^2\Pi$ state correlates with the $E'^2\Pi$ state of CaF. In this work on CaOH (and CaOD), a number of vibronic states were observed for which vibrational assignments are difficult. Similar problems were found by Pereira and Levy [67] for some of their bands.

B. Monoalkoxides, MOR

The first alkaline monoalkoxides were prepared by Wormsbecher and Suenram [88] in 1982. They reacted Ca and Sr vapors with methylnitrite (CH$_3$ONO) in a Broida oven and detected the $\tilde{A}^2E-\tilde{X}^2A_1$ and $\tilde{B}^2A_1-\tilde{X}^2A_1$ transitions of CaOCH$_3$ and SrOCH$_3$. Close similarities with the CaOH and SrOH spectra were noted. Studies of the $\tilde{A}-\tilde{X}$ and $\tilde{B}-\tilde{X}$ transitions of a number of additional alkoxides such as MOCH$_2$CH$_3$, MOC(CH$_3$)$_3$, and so on, were published by Brazier et al. [41] for M = Ca, Sr, and Ba. The rotational analysis of the $\tilde{A}^2E_{3/2}-\tilde{X}^2A_1$ transition [89] proved that SrOCH$_3$ has C_{3v} symmetry with an Sr—O bond length of 2.12 Å in the ground state (cf. Table 1). Additional rotational analyses of CaOCH$_3$ are in progress by Brown [90] and co-workers at Oxford and in the Bernath group [91] at Waterloo.

The alkoxides are the only family of molecules in addition to the monohydroxides for which studies of the molecular dynamics of the formation reactions are available. Davis et al. [44] compared the reactivity of Ba (^1S and ^1D) atoms with water and methanol using crossed molecular beams. Similar experiments with Ba (^1P) atoms with many additional alcohols were published by de Pujo et al. [45]. Alcohols always yielded BaOR products, not the energetically favorable BaOH or BaO molecules. Oberlander and Parsons [43] extended these studies to Ca and Sr metals using a beam-gas configuration. Surprisingly Esteban et al. [50] were able

C. Monocarboxylates, MO₂CR

The low-resolution spectra of Ca and Sr monoformate and monoacetate were first reported in 1985 [42]. A more extensive study of the formates, acetates, propionates, and butanoates [92] appeared in 1990 (Fig. 19). Calcium monoformate is believed to have C_{2v} symmetry with bidentate bonding. The spectra show three electronic transitions similar to CaNH₂

$$Ca^+ \cdots \underset{O}{\overset{O}{\diagup\!\!\!\diagdown}} C-H$$

(Fig. 7) but it is speculated that the first excited $^2A_1$ state lies below the $^2B_1$, $^2B_2$ pair of states, which correlate to the $A\,^2\Pi$ state of CaF. Theoretical work and high-resolution analyses need to be carried out on the monocarboxylates to verify the ordering of the electronic states.

D. Monoformamidates, MO(NH)CH

Isoelectronic analogies are very helpful in understanding the geometry and electronic structure of the monovalent alkaline earth derivatives. For example, formamide, HC(O)NH₂, differs from formic acid, HC(O)OH, by the substitution of an NH₂ group for OH. The use of formamide as an oxidant in a Broida oven gives rise to a spectrum [93] with three electronic transitions $\tilde{A}-\tilde{X}$, $\tilde{B}-\tilde{X}$, and $\tilde{C}-\tilde{X}$ (Fig. 20) very similar to the corresponding monoformate derivatives.

It seems likely that the calcium and strontium monoformamidates have bidendate bonding of the ligand similar to the monoformate derivatives.

$$Ca^+ \cdots \underset{\underset{H}{N\diagdown}}{\overset{O}{\diagup\!\!\!\diagdown}} C-H$$

The formamidate anion differs from the formate anion by the isoelectronic substitution of an NH group for O. The metal–ligand

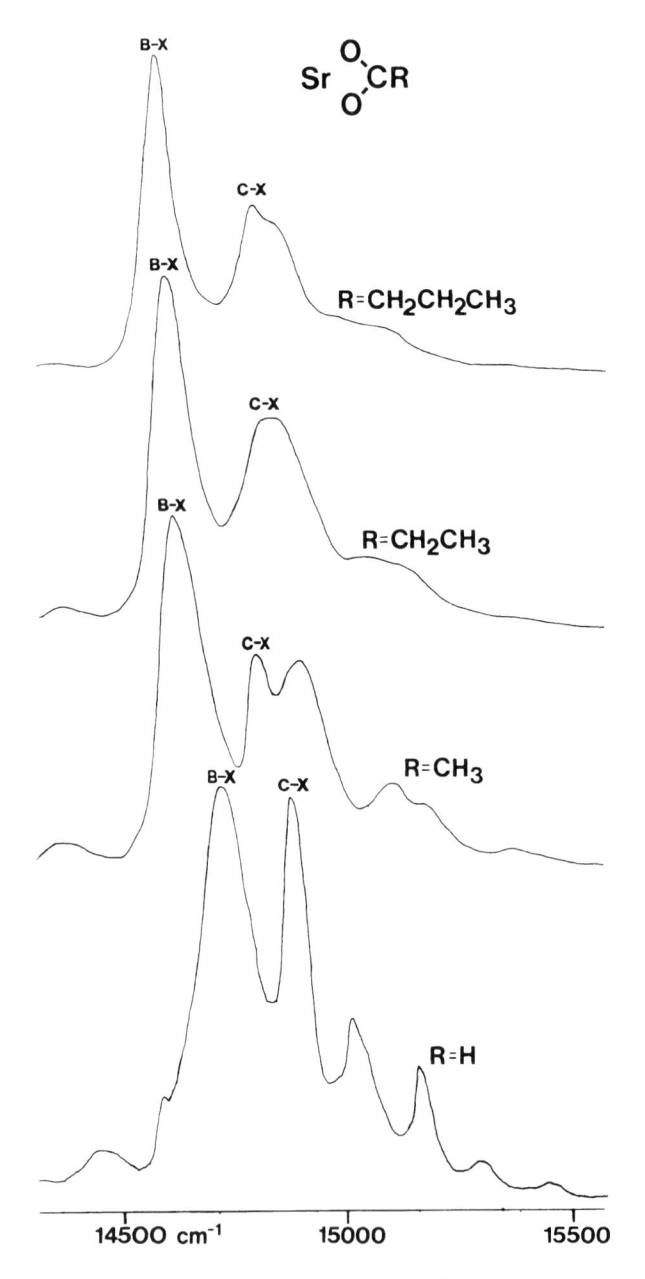

Figure 19. The laser excitation spectra of the $\tilde{B}-\tilde{X}$ and $\tilde{C}-\tilde{X}$ transitions of the strontium monocarboxylate molecules. [Reprinted with permission from ref. 92. Copyright 1990 American Chemical Society.]

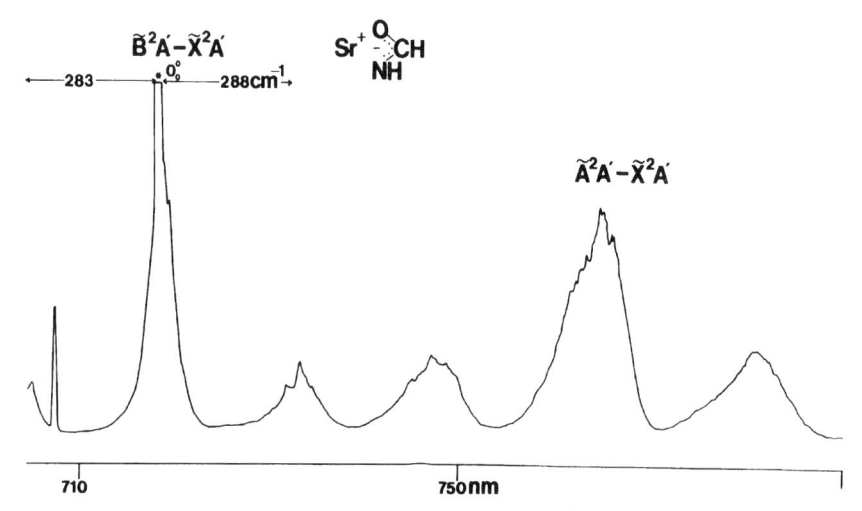

Figure 20. The dispersed fluorescence spectrum of the $\tilde{A}-\tilde{X}$ and $\tilde{B}-\tilde{X}$ transitions of SrOC(NH)H. The laser is exciting the 0–0 band of the $\tilde{B}-\tilde{X}$ transition. [Reprinted with permission from ref. 93. Copyright 1990 American Chemical Society.]

vibrational frequencies of 351 and 288 cm^{-1} in the ground state for the calcium and strontium monoformamidates [93] are similar to the values (349 and 275 cm^{-1}, respectively) for the corresponding monoformates [92]. Again high-resolution spectra and theoretical calculations would be very helpful in confirming the assignments of the low-resolution spectra. Interestingly, while formamidate salts are unknown in normal inorganic chemistry (formates are common), they form readily in the gas phase.

E. Monoisocyanates, MNCO

The spectra of CaNCO and SrNCO were first reported in 1986 based on a low-resolution analysis of the products of the reaction of Ca and Sr with HNCO [94]. The $\tilde{A}^2\Pi-\tilde{X}^2\Sigma^+$ and $\tilde{B}^2\Sigma^+-\tilde{X}^2\Sigma^+$ transitions were clearly due to a linear molecule (cf. CaF in Fig. 7). For NCO derivatives, there is the possibility of linkage isomers. In the initial paper [94], we guessed that the oxygen-bonding structures, MOCN, were lowest in energy (cyanates), but this proved to be erroneous. The NCO ligand binds to Ca and Sr through the N atom to give metal isocyanates.

To determine the structure of CaNCO and SrNCO, a rotational analysis of the $\tilde{A}^2\Pi-\tilde{X}^2\Sigma^+$ transition of the SrNCO molecule was carried out [95].

TABLE 3 Calculated Bond Lengths for CaNCO, CaNNN, and CaNCS (in Å)[a]

Bond Distance	CaNCO	CaNNN	CaNCS
r(CaN)	2.194	2.177	2.247
r(NC or NN)	1.218	1.211	1.215
r(CO, NN, or CS)	1.199	1.188	1.596

[a]Reference 97.

By fixing the N—C bond length to 1.19 Å and the CO bond length to 1.23 Å, the Sr—N bond length of 2.26 Å was determined from the rotational constant (0.04258 cm^{-1}) in the $\tilde{X}^2\Sigma^+$ state. A rotational analysis [96] of the corresponding $\tilde{A}^2\Pi - \tilde{X}^2\Sigma^+$ transition of SrN$_3$ gave an identical Sr—N bond length when the N—N bond lengths in the N$_3^-$ ligand were fixed to 1.18 Å. For SrNCO and SrN$_3$ to have identical metal–ligand bond lengths is consistent only with an isocyanate structure for SrNCO.

The corresponding rotational analyses have not yet been carried out for CaNCO and CaN$_3$ but some ab initio predictions [97] have been made by Chan and Hamilton (Tables 3 and 4). These calculations used the same 6–31 + G* basis set used by Ortiz [54] and are at the $MP2$ level. There is good agreement between the calculated vibrational frequencies and the experimental values (Table 4). The vibrational frequencies of NCO$^-$ in the solid state are 2182, 1211, and 631 cm^{-1} to be compared with the calculated values of 2191, 1327, and 826 cm^{-1} for v_1, v_2, and v_4, respectively (Table 4). No laboratory spectra are available for CaNCS.

F. Monoazides, MN$_3$

The isoelectronic NCO$^-$ and N$_3^-$ anions are often called pseudohalides because they have a chemistry similar to that of F$^-$, Cl$^-$, Br$^-$, and I$^-$. In a similar fashion, we find that CaN$_3$, SrN$_3$, CaNCO, and SrNCO all have similar spectra and that the electronic structure resembles that of CaF (Fig. 7). As in "regular" chemistry, chemical and isoelectronic analogies are a powerful tool for interpreting our gas-phase inorganic chemistry.

The CaN$_3$ and SrN$_3$ molecules were made in a Broida oven by the reaction of Ca and Sr vapors with hydrazoic acid, HN$_3$ [96]. The reaction is relatively vigorous and, in contrast to the other Ca and Sr reactions, exciting the atomic $^3P_1$ state of the metal increases the reactivity only slightly. The $\tilde{B}^2\Sigma^+ - \tilde{X}^2\Sigma^+$ and $\tilde{A}^2\Pi - \tilde{X}^2\Sigma^+$ transitions could be identified

TABLE 4 Calculated and Observed Vibration Frequencies of CaNCO, CaNNN, and CaNCS (in cm^{-1})

Mode	CaNCO Calculated[a]	CaNCO Experimental[b]	CaNNN Calculated[a]	CaNNN Experimental	CaNCS Calculated[a]
v_1 (σ) sym NXX str	2214	2200	2191	2114	2019
v_2 (σ) antisym NXX str	1344		1327	1364	927
v_3 (σ) Ca—N str	368	390	383	396	317
v_4 (π) NXX bend	624		826		464
v_5 (π) Ca—N—X bend	117		172	43	79

[a]Reference 97.
[b]Reference 94.
[c]Reference 96.

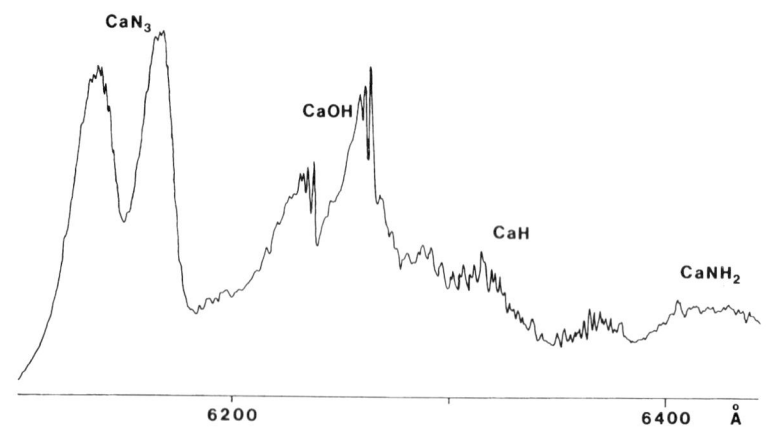

Figure 21. The chemiluminescence spectrum of the Ca + HN$_3$ reaction. [Reprinted with permission from ref. 96. Copyright 1988 American Institute of Physics.]

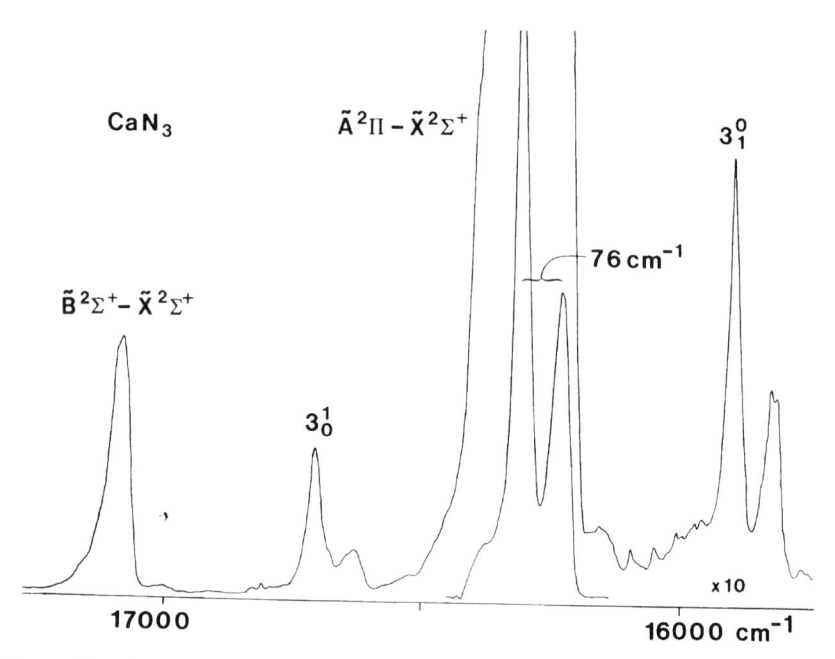

Figure 22. The dispersed fluorescence spectrum of CaN$_3$ [96]. The laser is exciting the 0–0 band of the $\tilde{A}^2\Pi_{3/2}$–$\tilde{X}^2\Sigma^+$ transition of CaN$_3$. The $\tilde{A}^2\Pi$ state has a spin–orbit splitting of 76 cm^{-1}. [Reprinted with permission from ref. 96. Copyright 1988 American Institute of Physics.]

SrN$_3$ $\tilde{A}^2\Pi - \tilde{X}^2\Sigma^+$

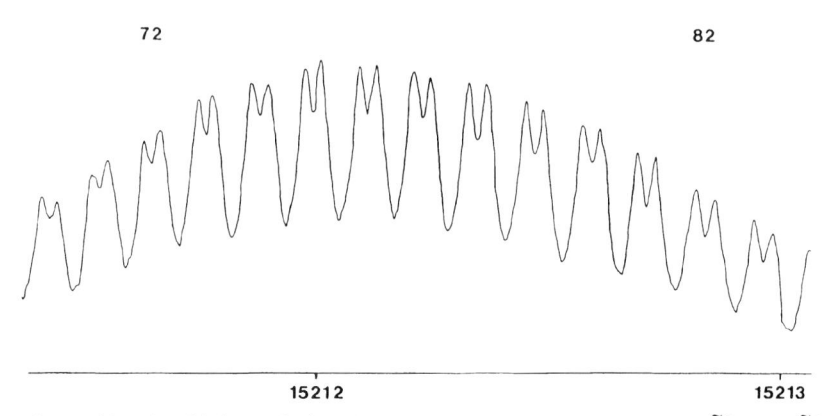

Figure 23. The high-resolution laser excitation spectrum of the $\tilde{A}^2\Pi_{3/2}-\tilde{X}^2\Sigma^+$ transition of SrN$_3$ recorded using a monochromator as a narrow band filter. [Reprinted with permission from ref. 96. Copyright 1988 American Institute of Physics.]

by chemiluminescence (Fig. 21) and by low-resolution laser-induced fluorescence (Fig. 22). A rotational analysis of the $\tilde{A}^2\Pi-\tilde{X}^2\Sigma^+$ transition (Fig. 23) of SrN$_3$ was carried out and confirmed the linear geometry [96].

The vibrational frequencies of CaN$_3$ were measured by fixing the laser on an electronic transition and dispersing the fluorescence with a small monochromator [96]. In this way, it was possible to measure four out of the five expected vibrational frequencies (Table 4). In general, the agreement with the ab initio predictions of Chan and Hamilton [97] is excellent except for the low-frequency metal–ligand bending mode. In this case, the interval of 86 cm^{-1} was measured by laser-induced fluorescence and interpreted as $2\nu_5$ since the selection rules forbid $\Delta v = \pm 1$ transitions for nontotally symmetric vibrational modes like ν_5. It would be expected that the ab initio calculations would have difficulty in predicting such a low frequency in an ionic molecule.

G. Monoisocyanides, MNC and Monocyanides, MCN

In many ways, the alkaline earth monoisocyanides have generated the most widespread interest and still remain the most mysterious of the molecules discussed in this chapter. The CaNC, SrNC, and BaNC molecules were discovered by Pasternack and Dagdigian in 1976 [98] through chemiluminescent emission and low-resolution laser excitation spectra of the products of the reactions of Ca, Sr, and Ba vapors with BrCN. Additional chemiluminescent spectra of the Ca(^1D) reaction with BrCN were recorded by Furio and Dagdigian [99].

The problems (and the most interesting features) found in the spectroscopy of ionic metal cyanides and isocyanides occur because the CN$^-$ ligand is nearly spherical. The CN$^-$ ligand thus has a low barrier–internal rotation in M$^+$CN$^-$ molecules. The equilibrium structure in the ground state can be linear cyanide, MCN, linear isocyanide, MNC, or even a T-shaped side-on complex (e.g., Na$^+$N$^-$). The molecular geometry is evidently determined by
$$\underset{C}{\overset{|||}{}}$$
a sensitive balance between competing factors that are not well understood. Clementi et al. [100] discuss this problem in 1973 and coined the term "polytopic bonding."

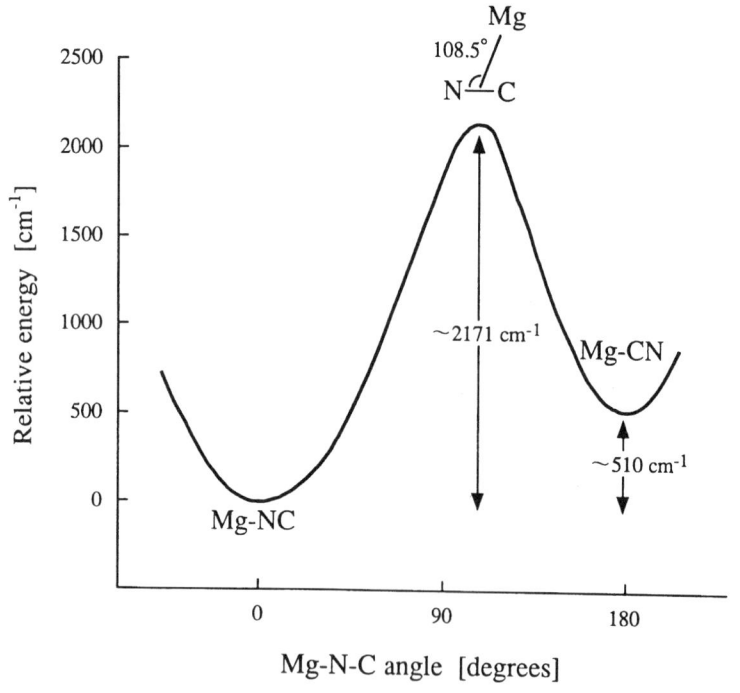

Figure 24. The bending potential surface of MgNC [107] as calculated by Ishii et al. [26]. [Reprinted with permission from ref. 107. Copyright 1996 American Institute of Physics.]

The problem of the molecular geometry of the alkaline earth cyanides was "solved" in 1985 when Bauschlicher et al. [101] calculated that the isocyanide structures, MNC, are the most stable (Fig. 24). They found that the BeNC and MgNC molecules have barriers between the isocyanide and cyanide linkage isomers (Fig. 24) but that the heavier metals, Ca, Sr, and Ba do not. More recent calculations on MgNC [26] support this general picture for the ground $\tilde{X}^2\Sigma^+$ states.

The excited states of CaNC and SrNC are not well understood although high-resolution spectra have been recorded in the ultracold environment of a molecular beam. The low-lying excited states of CaNC and SrNC should be $\tilde{A}^2\Pi$ and $\tilde{B}^2\Sigma^+$ analogous to CaF (Fig. 7). However, additional unassigned broad red emission is seen in both the Broida oven [102] and molecular beam experiments [39].

The Steimle group has extensively analyzed the $\tilde{A}^2\Pi - \tilde{X}^2\Sigma^+$ transition (Fig. 25) of normal CaNC [103] as well as the Ca^{15}NC and CaN^{13}C [104] isotopomers. They obtained a substitution structure with a CaN bond length of 2.207 Å and a CN bond length of 1.119 Å [104] in moderate agreement with the theoretical values [101] of 2.27 and 1.16 Å. A large dipole moment of 6.90 Å was found for the ground $\tilde{X}^2\Sigma^+$ state consistent with an ionic Ca$^+$NC$^-$ structure [103]. The most interesting result, however, was the observation of large positive Λ-doubling constants in the $\tilde{A}^2\Pi$ states consistent with the $\tilde{B}^2\Sigma^+$ state lying about 1000 cm^{-1} below the $\tilde{A}^2\Pi$ state (cf. Fig. 7). The spectra show no clear trace of the $\tilde{B}^2\Sigma^+$ state and the $\tilde{A}^2\Pi$ state is perturbed by some unknown lower state. An interpretation for all of these facts is that the $\tilde{B}^2\Sigma^+$ state is, in fact, a low-lying $\tilde{A}'^2A_1$ state with a T-shaped (or even a linear cyanide, CaCN) geometry. In this case, the $\tilde{A}'^2A_1 - \tilde{X}^2\Sigma^+$ bands would have poor Franck–Condon factors but could give rise to red emission. A careful exploration of the ab initio potential energy surfaces of CaNC should be able to solve this problem [143].

There was excitement in the radioastronomical community with the recent discovery of MgNC in the outer atmosphere of a carbon star, IRC + 10216 [25]. Very few metal-containing molecules are known in cool sources because metals like Mg tend to be condensed in the form of grains. Kawaguchi et al. [25] were able to record a laboratory microwave spectrum of MgNC and comparison with six unidentified radio lines confirmed the presence of MgNC in the dusty envelope of IRC + 10216. Theoretical calculations of Ishii et al. [26] lent support to this claim with predicted equilibrium MgN and NC bond lengths of 1.945 and 1.170 Å, respectively. The vibrational frequencies were predicted as $v_1(\sigma)$ 2188, $v_2(\pi)$ 83, and $v_3(\sigma)$ 539 cm^{-1}. Later millimeter wave work by Anderson and Ziurys [105] on ^{25}MgNC and ^{26}MgNC gave an r_0 structure with bond lengths of 1.924 Å (MgN) and 1.170 Å (NC).

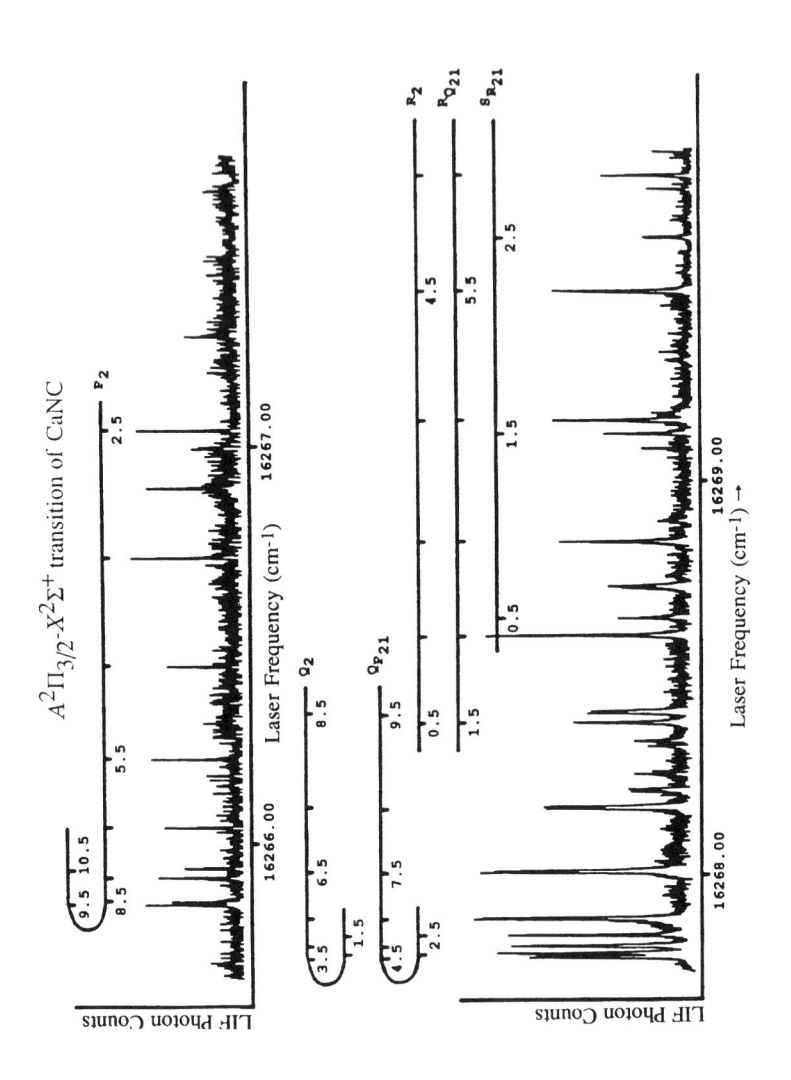

Figure 25. The high-resolution laser excitation spectrum of the $\tilde{A}^2\Pi_{3/2}-\tilde{X}^2\Sigma^+$ transition of jet-cooled CaNC. [Reprinted with permission from ref. 103. Copyright 1992 American Institute of Physics.]

The discovery of MgNC in the source IRC + 10216 sparked a great flurry of work on the rotational spectra of metal cyanides. Millimeter wave measurements were made for CaNC [106] and vibrationally excited states of MgNC [107]. In this later work, the ℓ-type doubling constant was used to estimate the bending frequency (v_2) of 83 cm^{-1} for MgNC. The low-frequency microwave transitions of CaNC were measured in two molecular beam spectrometers in order to determine the hyperfine structure of the $\tilde{X}^2\Sigma^+$ state [108]. In all cases, the data could be interpreted in terms of a linear, ionic isocyanide structure. However, the low barrier–isomerization (Fig. 24) for MgNC or the lack of a barrier in the case of CaNC was evident in the anomalously large centrifugal distortion constants.

A barrier–isomerization of about 2000 cm^{-1} (Fig. 24) allows the metastable cyanide isomer, MgCN, to exist [26]. Anderson et al. [109] recorded the millimeter wave spectrum of the $\tilde{X}^2\Sigma^+$ state of MgCN along with very strong MgNC lines. These data allowed the identification of the MgCN isomer in the carbon star IRC + 10216 by radio astronomers [110].

H. Monohydrosulfides, MSH

The electronic spectra of CaSH and SrSH were first published in 1991 by Fernando et al. [111]. The molecules were made by the reaction of the electronically excited Ca* or Sr* atoms with H_2S. From the three electronic transitions $\tilde{A}^2A'-\tilde{X}^2A'$, $\tilde{B}^2A''-\tilde{X}^2A'$ and $\tilde{C}^2A'-\tilde{X}^2A'$ (Fig. 26) in the spectra, it was evident that CaSH and SrSH were bent molecules. Unlike the very ionic Ca$^+$OH$^-$ molecule that prefers to keep the H(δ^+) atom as far as possible from Ca$^+$, the bonding in CaSH and SrSH is more covalent. The electronic spectra and orbitals in CaSH are very similar to the states of CaNH$_2$ but the lower symmetry (C_s) of CaSH changes the labels of the states ($\tilde{A}^2B_2$ to $\tilde{A}^2A'$, $\tilde{B}^2B_1$ to $\tilde{B}^2A''$ and $\tilde{C}^2A_1$ to $\tilde{C}^2A'$). The ab initio calculations of Ortiz [57] on CaSH confirm the bent geometry and this picture of the electronic structure.

Jarman and Bernath [112] carried out a rotational analysis of the $\tilde{A}^2A'-\tilde{X}^2A'$ transition (Fig. 26). The analysis yielded some fascinating results on the electronic structure in the $\tilde{A}^2A'$ state. Although CaSH is bent and the $A^2\Pi$ state of CaF has now been split into the $\tilde{A}^2A'$ and $\tilde{B}^2A''$ states (Fig. 7), the $\tilde{A}^2A'$ state is strongly affected by orbital angular momentum. Only molecules with a high symmetry such as linear ($C_{\infty v}$) can have orbital angular momentum and spin–orbit coupling. In particular, the large spin–orbit coupling constant ($A \sim 65$ cm^{-1}) of the hypothetical linear CaSH molecule has been transformed to a large ε_{aa} spin–rotation constant (3.446 cm^{-1}) in the $\tilde{A}^2A'$ state. This large ε_{aa} constant in the $\tilde{A}^2A'$ state splits

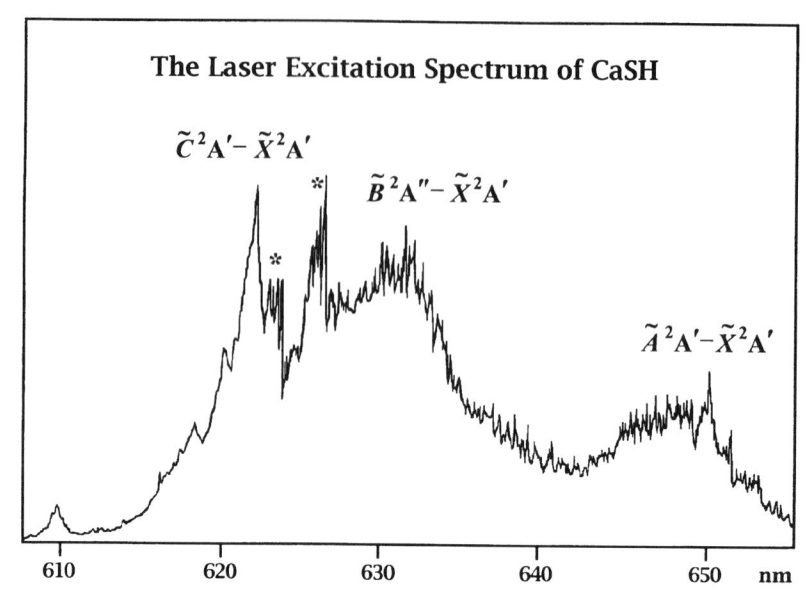

Figure 26. The laser excitation spectrum of the $\tilde{A}-\tilde{X}$, $\tilde{B}-\tilde{X}$, and $\tilde{C}-\tilde{X}$ transitions of CaSH [112]. The asterisks in this case mark CaOH impurity bands. [Reprinted with permission from ref. 112. Copyright 1993 American Institute of Physics.]

the two spin components linearly with increasing K_a (Fig. 27) so that only $K = 0$ obeys Hund's case (b) coupling and the higher K_a values obey Hund's case (a) coupling. K_a is the projection of the total angular momentum $\hat{N}$ (exclusive of electron spin) onto the a inertial axis.

The $\tilde{B}^2A''-\tilde{X}^2A'$ electronic transition of CaSH was rotationally analyzed in a pulsed molecular beam source by Scurlock et al. [113]. The a axis of this asymmetric rotor lies very close to the Ca—S bond and there is a large dipole moment of 5.36 D projected along it in the $\tilde{X}^2A'$ state.

The millimeter wave spectra of CaSH, CaSD, and MgSH were recorded in Ottawa by Taleb-Bendiab et al. [114]. The pure rotational spectra of CaSH and CaSD allow an improved molecular geometry to be determined for the $\tilde{X}^2A'$ state. They found a r_0 Ca—S bond length of 2.564 Å, a S—H bond length of 1.357 Å, and a bond angle of 91°.

I. Monothiolates, MSR

The sulfur analogues of the alkaline earth monoalkoxides have also been made in a Broida oven by the reactions of Ca* and Sr* with thiols [111].

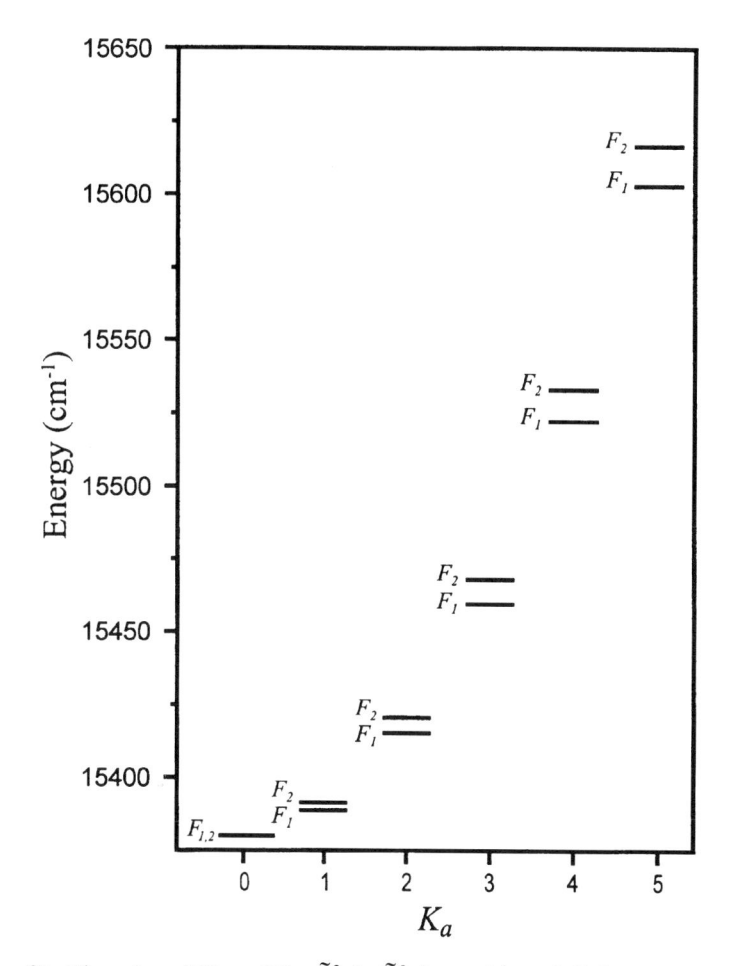

Figure 27. The spin splitting of the $\tilde{A}^2A'-\tilde{X}^2A'$ transition of CaSH as a function of K_a. The large value of ε_{aa} in the $\tilde{A}$ state causes the F_1 and F_2 components to split linearly with K_a. [Reprinted with permission from ref. 112. Copyright 1993 American Institute of Physics.]

The low-resolution laser excitation and fluorescence spectra of the $\tilde{A}-\tilde{X}$, $\tilde{B}-\tilde{X}$ and $\tilde{C}-X$ transitions of CaSCH$_3$, SrSCH$_3$, CaSCH$_2$CH$_3$, and so on, were recorded. The strong resemblance between the CaSH and CaSR spectra suggest that the monothiolates and monohydrosulfides have a similar electronic structure and that the Ca—S—C molecular backbone is bent.

J. Monoamides, MNH$_2$

The CaNH$_2$, SrNH$_2$, and BaNH$_2$ molecules were discovered by Wormsbecher et al. [32, 34] in Santa Barbara using a Broida oven. This work in the Harris group on amides and hydroxides was the crucial first step in the production of the now numerous monovalent derivatives of the alkaline earth metals. The electronic structure of CaNH$_2$ has already been discussed (Fig. 7) and Wormsbecher et al. [34] attempted a rotational analysis of the $\tilde{C}^2A_1 - \tilde{X}^2A_1$ transition.

The CaNH$_2$ and SrNH$_2$ molecules, like CaSH, display three visible electronic transitions: $\tilde{C}^2A_1 - \tilde{X}^2A_1$, $\tilde{B}^2B_1 - \tilde{X}^2A_1$, and $\tilde{A}^2B_2 - \tilde{X}^2A_1$ (Fig. 28). The $\tilde{A} - \tilde{X}$ and $\tilde{B} - \tilde{X}$ transitions of SrNH$_2$ are perpendicular transitions with K_a subband structure that spreads out while the $\tilde{C} - \tilde{X}$ transition has a transition dipole moment that lies along the Sr—N axis and K_a subbands that overlap (Fig. 28). A rotational analysis of the $\tilde{A} - \tilde{X}$ and $\tilde{B} - \tilde{X}$ transitions of SrNH$_2$ by Brazier and Bernath [115] confirmed the symmetries of the $\tilde{A}$ and $\tilde{B}$ states and they found an Sr—N bond length of 2.25 Å and an NH$_2$ bond angle of 104° in the $\tilde{X}$ state.

The molecular beam measurements of the $\tilde{A}^2B_2 - \tilde{X}^2A_1$ and $\tilde{B}^2B_1 - \tilde{X}^2A_1$ transitions of CaNH$_2$ by Whitham et al. [40] were pioneering experiments. Marr et al. [116] used the improved resolution of a CW dye laser to make a complete analysis of the $\tilde{A} - \tilde{X}$ transition. Like CaOH, CaNH$_2$ has a small dipole moment of 1.74 D in the ground $\tilde{X}^2A_1$ state. Marr et al. [116] also measured a large ε_{aa} spin-rotation constant in the $\tilde{A}^2B_2$ state.

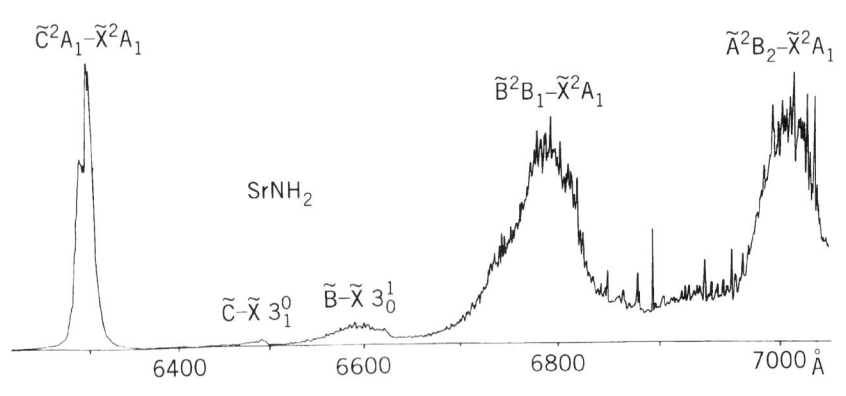

Figure 28. Laser excitation spectra of the $\tilde{A} - \tilde{X}$, $\tilde{B} - \tilde{X}$, and $\tilde{C} - \tilde{X}$ transitions of SrNH$_2$ [115].

TABLE 5 Experimental and ab initio Parameters for the $\tilde{X}^2A_1$ State of CaNH$_2$

	DFT[a]	Experimental[b]	
θ_{HNH}	105.6°	100.5°	102.9°
r_{NH} (Å)	1.02	1.041[c]	1.025[c]
r_{CaN} (Å)	2.13	2.118	2.140

Vibration		DFT (cm^{-1})[a]	Experimental (± 10 cm^{-1})[d]
v_6 Antisym NH bend	b_2	345	320
v_4 Out-of-plane bend	b_1	455	347
v_3 Ca—N Stretch	a_1	544	520
v_2 Sym NH bend	a_1	1586	
v_1 Sym NH stretch	a_1	3448	
v_5 Antisym NH stretch	b_2	3535	

[a] Density functional calculation (DFT), Reference 97.
[b] Reference 117.
[c] Fixed, see text.
[d] From low-resolution Broida oven experiments.

The $\tilde{B}^2B_1 - \tilde{X}^2A_1$ and $\tilde{C}^2A_1 - \tilde{X}^2A_1$ transitions have also been analyzed at Waterloo with a molecular beam spectrometer [117]. All three known electronic transitions have been fitted together to improve the molecular constants. The molecular geometry (Table 5) in the $\tilde{X}^2A_1$ state was found by fixing the N—H bond length to 1.041 Å (from NH$_2^-$) or 1.025 Å (from NH$_2$). Vibrational frequencies were measured by laser-induced fluorescence. For comparison (Table 5), Chan and Hamilton [97] calculated the molecular geometry and vibrational frequencies using density functional theory.

The most interesting aspect of the rotational analyses of the $\tilde{A}-\tilde{X}$, $\tilde{B}-\tilde{X}$ and $\tilde{C}-\tilde{X}$ transitions are the values for the complete set of nine spin–rotation parameters of the $\tilde{A}$, $\tilde{B}$, and $\tilde{C}$ states (ε_{aa}, ε_{bb}, and ε_{cc} for each state). If the orbitals containing the unpaired electron in these three states (Fig. 7) are approximate by a set of p orbitals (p_x, p_y, p_z) then all nine spin–rotation constants can be easily estimated by pure precession relationships.

Van Vleck introduced the pure precession approximation for diatomic molecules. The basic idea is that if the one-electron atomic orbital angular momentum is preserved in a diatomic molecule then the Λ-doubling and spin–rotation constants can be predicted, for example, in a $4p\sigma(^2\Sigma^+)$ and a $4p\pi(^2\Pi)$ state. This pure precession picture can easily be generalized to

TABLE 6 Spin-Rotation Parameters for the $\tilde{A}^2B_2$, $\tilde{B}^2B_1$, and $\tilde{C}^2A_1$ States of CaNH$_2^a$

Parameter		$\tilde{A}^2B_2$	$\tilde{B}^2B_1$	$\tilde{C}^2A_1$
ε_{aa}	Observed	8.236	−7.547	0.998
	Pure precession	8.193	−8.193	0
	Formula[b]	$\dfrac{4AA^{SO}}{\Delta E_{B-A}}$	$\dfrac{4AA^{SO}}{\Delta E_{B-A}}$	0
ε_{bb}	Observed	0.00530	0.00351	−0.0398
	Pure precession	0	0.0539	−0.0547
	Formula	0	$\dfrac{4BA^{SO}}{\Delta E_{C-B}}$	$\dfrac{-4BA^{SO}}{\Delta E_{C-B}}$
ε_{cc}	Observed	0.0558	0.0381	−0.0385
	Pure precession	0.0418	0	−0.0417
	Formula	$\dfrac{4CA^{SO}}{\Delta E_{C-A}}$	0	$\dfrac{-4CA^{SO}}{\Delta E_{C-A}}$

[a] Reference 47.
[b] $A^{SO} = 66.80\,\text{cm}^{-1}$ spin–orbit constant of CaOH.

polyatomic molecules and used to predict the spin–rotation parameters ε_{aa}, ε_{bb}, and ε_{cc} of the interacting electronic states [117]. The results of this model are presented in Table 6. On the whole, the agreement is between experiment and prediction is reasonable with the greatest discrepancies occurring because of vibronic interactions in the $\tilde{B}$ state.

K. Monoalkylamides, MNHR

The preparation and low-resolution spectra of a few calcium and strontium alkylamides were reported in 1987 [118]. By replacing NH$_3$ with monoalkylamines, NH$_2$R [R = —CH$_3$, —C$_2$H$_5$, —CH(CH$_3$)$_2$, and —C(CH$_3$)$_3$] in a Broida oven, the laser-induced fluorescence spectra of the $\tilde{A}$–$\tilde{X}$, $\tilde{B}$–$\tilde{X}$, and $\tilde{C}$–$\tilde{X}$ transitions (see for example, Fig. 29) were recorded. There was a strong similarity with the spectra of CaNH$_2$ (Fig. 7) although the symmetry is lower than C_{2v} for the monoalkylamide derivatives.

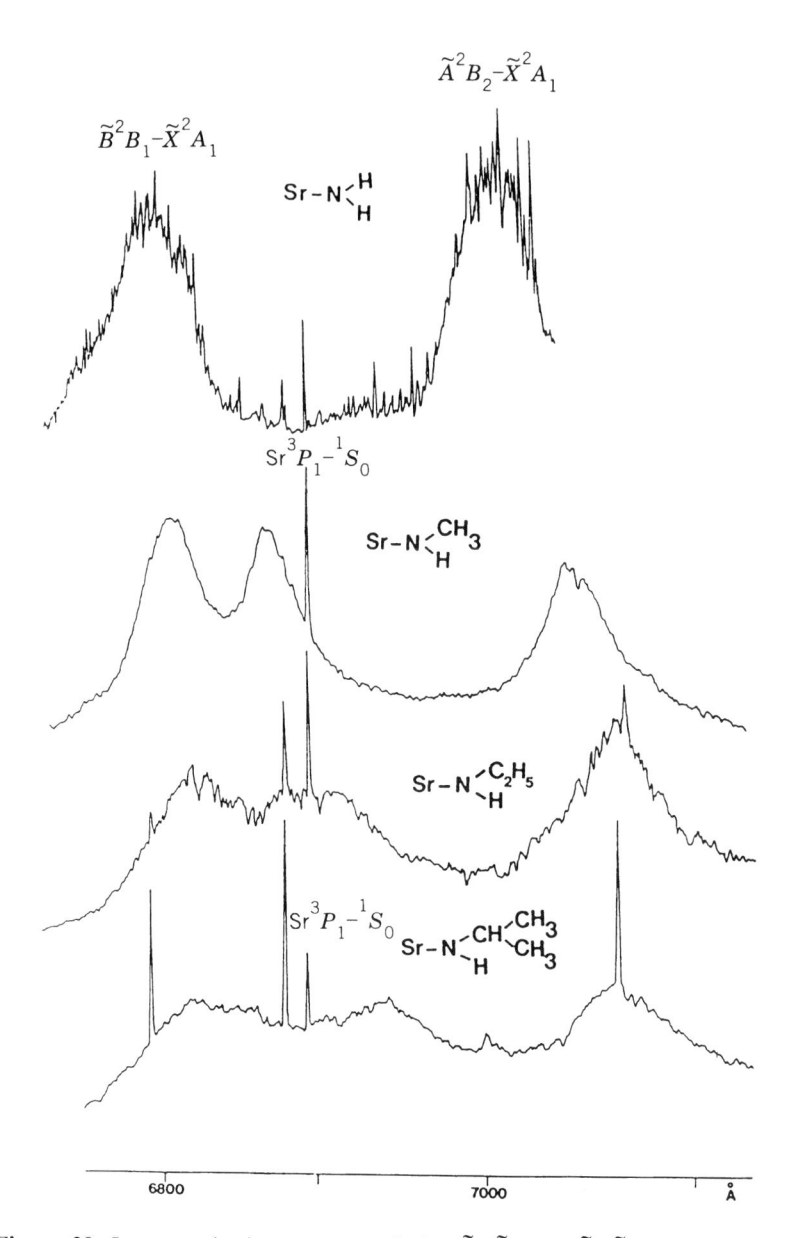

Figure 29. Laser excitation spectra of the $\tilde{A}-\tilde{X}$ and $\tilde{B}-\tilde{X}$ transitions of the strontium monoalkylamides. [Reprinted with permission from ref. 118. Copyright 1987 American Chemical Society.]

L. Monomethyls, MCH$_3$

The monomethyl derivatives are prototypical metal alkyl molecules, which can be made by the reactions of various methylating agents such as CH$_3$—N=N—CH$_3$, Hg(CH$_3$)$_2$, Sn(CH$_3$)$_4$, and CH$_3$CN with excited Mg, Ca, or Sr atoms. The low-resolution spectra of the $\tilde{A}^2E$–$\tilde{X}^2A_1$ and $\tilde{B}^2A_1$–$\tilde{X}^2A_1$ transitions of CaCH$_3$ and SrCH$_3$ were published by Brazier and Bernath [119] in 1987. These molecules have C_{3v} symmetry so that the electronic structure is similar to CaF (Fig. 7) but because of the reduction in symmetry the states are renamed ($^2\Pi \rightarrow {}^2E$, $^2\Sigma^+ \rightarrow {}^2A_1$). In this early work, upper limits for the metal–ligand dissociation energies of 46 and 43 kcal mol^{-1} for CaCH$_3$ and SrCH$_3$, respectively, were estimated from the onset of predissociation [119].

In view of the relatively weak metal–methyl bond, a measurement of the bond length would be useful. The $\tilde{A}^2E$–$\tilde{X}^2A_1$ transition of CaCH$_3$ was rotationally analyzed and the ground-state r_0 Ca—C bond length of 2.349 Å and H—C—H angle of 105.6° (with r_{C-H} fixed to 1.100 Å) were extracted [120]. The calculations of Ortiz [54] give r_{Ca-C} = 2.388 Å, θ_{HCH} = 105.9°, and r_{C-H} = 1.101 Å. In Table 7, the experimental and predicted ground-state vibrational frequencies of CaCH$_3$ are compared. Tyerman et al. [112] calculated the ground-state properties of MCH, MCH$_2$, and MCH$_3$ for the lighter alkali and alkaline earth metals. The Arizona State University group recorded the millimeter wave spectra of MgCH$_3$ [122], CaCH$_3$ [123], SrCH$_3$ [124], and BaCH$_3$.

The MgCH$_3$ $\tilde{A}^2E$–$\tilde{X}^2A_1$ spectrum was analyzed in a laser ablation molecular beam spectrometer [125]. It was noted in this article that the ε_{aa}

TABLE 7 Experimental and Calculated Frequencies of CaCH$_3$ (in cm^{-1})

Mode	Observed Value[a]	Calculated Value[b]	Calculated Value[c]
v_1 (a_1) sym C—H str		3023	3020
v_2 (a_1) sym C—H bend	1085	1203	1104
v_3 (a_1) Ca—C str	419	425	411
v_4 (e) antisym C—H str		3100	3109
v_5 (e) antisym H—C—H bend		1483	1443
v_6 (e) Ca—C—H bend	318	385	356

[a]Reference 119.
[b]Reference 97.
[c]Reference 121.

spin–rotation parameter in the $\tilde{A}$ state for $CaCH_3$ [120] seemed to have an anomalous sign when compared to the values obtained for $MgCH_3$, $ZnCH_3$, and $CdCH_3$. In fact, the original assignment of the excited state K values by Brazier and Bernath [120] was not correct because they did not see the first rotational lines of the branches. A reassignment of the $\tilde{A}^2E-\tilde{X}^2A_1$ transition of $CaCH_3$ by Marr et al. [126] using data from a molecular beam spectrometer corrected this problem and yielded a dipole moment of 2.62 D in the ground state. The power of the molecular beam method is illustrated clearly by this example.

M. Monoacetylides, MCCH

The monoacetylide derivatives are also important examples of simple organometallic molecules. The reactions of excited Ca and Sr with acetylene in a Broida oven resulted in the detection of the CaCCH and SrCCH molecules [127]. Table 8 compares the observed and calculated $\tilde{X}^2\Sigma^+$ vibrational frequencies. This work was rapidly followed by a rotational analysis of a vibrational band in the $\tilde{A}^2\Pi-\tilde{X}^2\Sigma^+$ transition of CaCCH [128]. It was assumed that we had found the 0–0 band, but the low-resolution molecular beam experiment of Whitham et al. [39] showed that we had analyzed a hot band.

Recently, several bands of the $\tilde{A}^2\Pi-\tilde{X}^2\Sigma^+$ transition of CaCCH were analyzed, first in a molecular beam source [129] and then in a Broida oven [130]. The pure rotational spectra of MgCCH [131], CaCCH [132], and SrCCH [133] are also available. If the C—H and C≡C bond lengths are fixed to 1.506 and 1.204 Å, then the Ca—C bond length is 2.349 Å, identical to the $CaCH_3$ value [132]. An ab initio calculation by Chan and Hamilton

TABLE 8   Ground-State Vibrational Frequencies of the CaCCH Molecule (in cm^{-1})

Mode	Experimental[a]	Calculated[b]
v_1 (σ) C—H str		3507
v_2 (σ) C≡C str		2466
v_3 (σ) Ca—C str	399	383
v_4 (π) C—C—H bend		701
v_5 (π) Ca—C—C bend	90	197

[a]Reference 127.
[b]Reference 97.

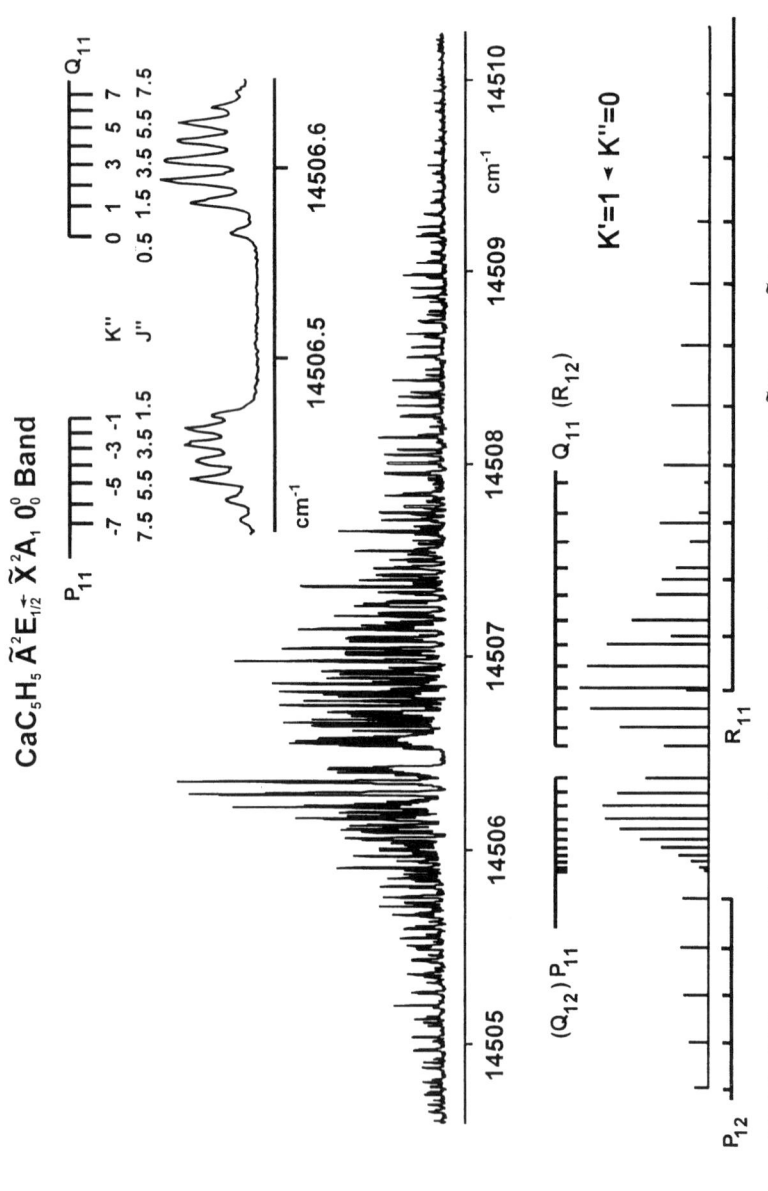

Figure 30. The high-resolution laser excitation spectrum of the 0_0^0 band of the $\tilde{A}^2E_{1(1/2)}$–$\tilde{X}^2A_1$ transition of CaC_5H_5. A simulation of the $K = 1$–0 subband is provided in the lower panel. [Reprinted with permission from ref. 139. Copyright 1995 American Institute of Physics.]

[97] gives $r(CH) = 1.071$ Å, $r(C\equiv C) = 1.240$ Å and $r(Ca—C) = 2.368$ Å. Ab initio calculations are also available for MgCCH, MgCCH$^+$, MgCH$_3$ and MgC$_2$ [134]. The dipole moment of CaCCH (3.01 D) is also close to the value for CaCH$_3$ (2.62 D) in agreement with simple electrostatic models [129]. The laser excitation spectrum of the $\tilde{A}^2\Pi - \tilde{X}^2\Sigma^+$ transition of MgCCH has also been recorded [135]. Curiously, the spectra of CaCCH and SrCCH were originally recorded in an attempt to make the carbides CaC$_2$ and SrC$_2$ [127], and these species have still eluded detection.

N. Monocyclopentadienides, MC$_5$H$_5$

The sandwich complexes are among the most celebrated molecules in organometallic chemistry. The discovery of the structure of the ferrocene molecule, Fe(C$_5$H$_5$)$_2$, by Wilkinson was rewarded with a Nobel prize. The possibility of making half-sandwich complexes in the gas phase was very enticing and, inspired by a report of a metal vapor synthesis of magnesocene, Mg(C$_5$H$_5$)$_2$, we looked for CaC$_5$H$_5$ and SrC$_5$H$_5$ in a Broida oven [136]. The reaction of laser excited metal atoms with the cyclopentadiene precursor C$_5$H$_6$ resulted in the low-resolution spectra of the $\tilde{A}^2E_1 - X^2A_1$, and $\tilde{B}^2A_1 - \tilde{X}^2A_1$ transitions of CaC$_5$H$_5$ and SrC$_5$H$_5$.

The half-sandwich complexes have C_{5v} symmetry and the states are labeled in a similar way as for the metal monomethyls of C_{3v} symmetry. The spectra recorded by the Miller group [137, 138] for jet-cooled CaC$_5$H$_5$ allow a much improved vibrational analysis. Ultimately, in a spectroscopic tour-de-force, the $\tilde{A}^2E_1 - \tilde{X}^2A_1$ transition of CaC$_5$H$_5$ (Fig. 30) was rotationally analyzed and the Ca–ring distance of 2.333 Å was determined [139]. This metal–ring distance is very close to the Ca—C bond distance in the CaCH$_3$ and CaCCH molecules.

O. Monomethylcyclopentadienides, MC$_5$H$_4$CH$_3$

The first detection of a derivative of the monomethylsubstituted ligand C$_5$H$_4$CH$_3$ was carried out by Robles et al. [138] at Ohio State University. The substitution of a methyl group for an H atom on the ring lowers the symmetry to C_s from C_{5v} and lifts the double degeneracy in the $\tilde{A}^2E$ state (cf. Fig. 7). The A' and A'' components that correlate to the ^2E state of CaC$_5$H$_5$ are split by 100 cm^{-1}. The torsional splittings of the CH$_3$ group were also analyzed in the molecular beam experiment.

P. Monopyrrolates, MC$_4$H$_4$N

The isoelectronic analogy again proves to be very useful when the CH group of the C$_5$H$_5^-$ ligand is replaced by N to give the C$_4$H$_4$N$^-$ ligand. Laser excited Ca and Sr atoms reacted with the pyrrole C$_4$H$_5$N precursor in a Broida oven to give CaC$_4$H$_4$N and SrC$_4$H$_4$N [140]. The strong similarity between the monopyrrolates and the monocyclopentadienides led us to conclude that CaC$_4$H$_4$N and SrC$_4$H$_4$N were both half-sandwich complexes.

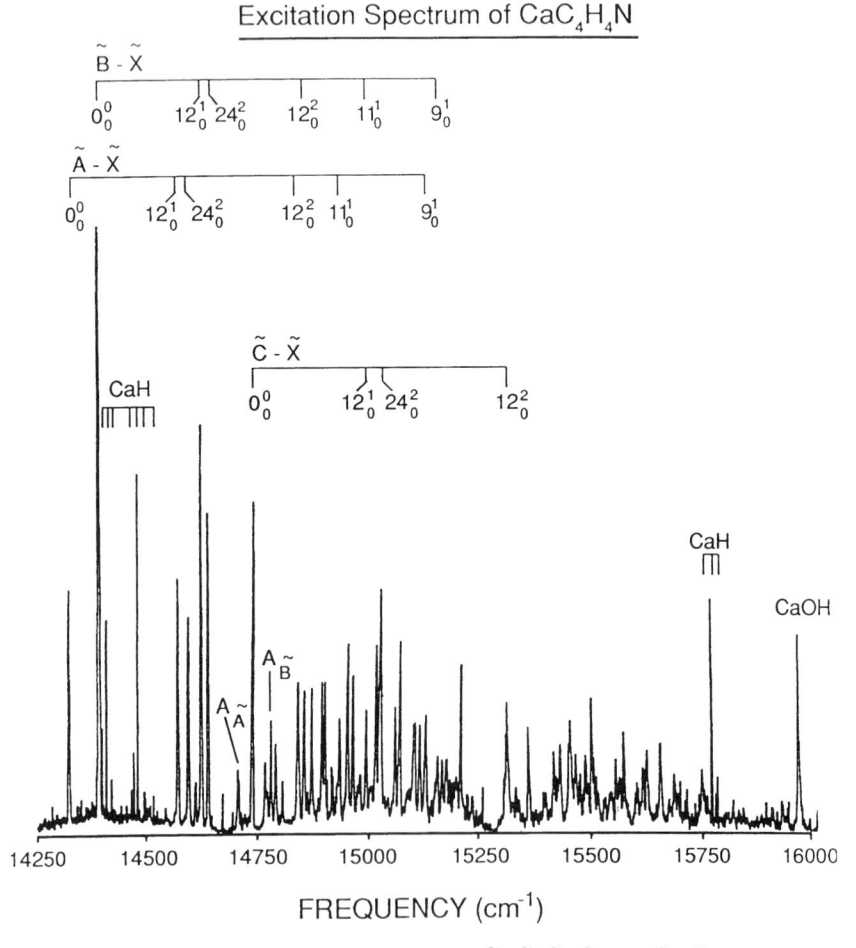

Figure 31. The laser excitation spectrum of the $\tilde{A}-\tilde{X}$, $\tilde{B}-\tilde{X}$, and $\tilde{C}-\tilde{X}$ transitions of jet-cooled CaC$_4$H$_4$N. [Reprinted with permission from ref. 138. Copyright 1992 American Chemical Society.]

Although this geometric structure turned out to be correct, there were problems with interpretation of the electronic structure in this early work.

Once again, it was jet cooling of CaC_4H_4N that clarified the problem [138]. The higher resolution spectra recorded in the Miller group did show that the ligand had η^5 coordination but that the lower C_s symmetry of CaC_4H_4N split the degenerate $\tilde{A}^2E$ state of CaC_5H_5 into two states. The 68-cm^{-1} splitting (Fig. 31) between these two electronic states of CaC_4H_4N was accidentally close to the expected spin–orbit splitting in a ^2E state and caused the confusion in the early work. A thorough vibrational analysis, made possible by the sharp bands in the cold spectrum, proved that the observed 68 cm^{-1} splitting was not a spin–orbit interval in the CaC_4H_4N molecule.

The MgC_5H_5, $MgC_5H_4CH_3$, and MgC_4H_4N molecules have also been synthesized in a pulsed jet source and their spectra analyzed [141]. The $\tilde{A}^2E_1$ state in MgC_5H_5 is very peculiar since it has no observable spin–orbit splitting. Unlike the metal-centered $\tilde{A}^2E$ states of the other molecules (e.g., CaC_5H_5 [56]) in MgC_5H_5, the unpaired electron is located on the C_5H_5 ring. In this case, the $\tilde{A}^2E_1 - \tilde{X}^2A_1$ is a transition metal-to-ligand charge-transfer transition.

Q. Monoborohydrides, MBH_4

The $CaBH_4$ molecule completes the isoelectronic CaF, CaOH, $CaNH_2$, and $CaCH_3$ series of molecules. The Ca and Sr vapors react spontaneously with

$CaBH_4$ Configurations

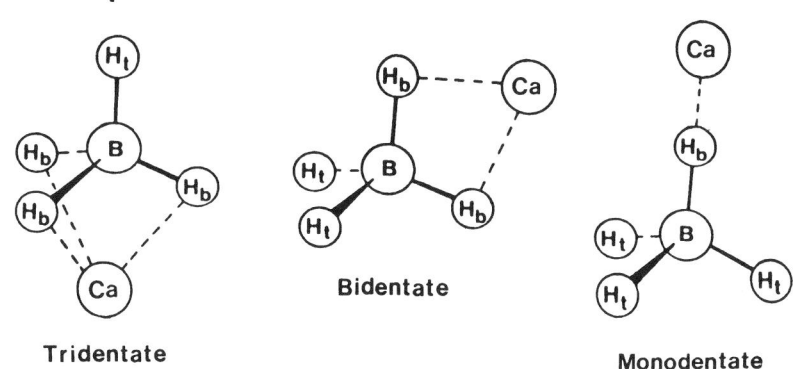

Figure 32. Three possible structures for the $CaBH_4$ molecule. The tridentate structure has the lowest energy. [Reprinted with permission from ref. 142. Copyright 1990 American Chemical Society.]

diborane, B_2H_6, to give $CaBH_4$ and $SrBH_4$. So far, there has been only one experimental paper written about these molecules [142]. The low-resolution laser-induced fluorescence spectra suggest that $CaBH_4$ and $SrBH_4$ have C_{3v} symmetry with the metal atom bonding to the face of the BH_4^- tetrahedron (Fig. 32). In addition, the ^2E and $^2A_1$ states seem to be switched compared to CaF (Fig. 7) so that one detects $\tilde{A}^2A_1-{}^2A_1$ and $\tilde{B}^2E-\tilde{X}^2A_1$ electronic transitions. The ab initio calculations of Ortiz [55] are in general agreement with our observations (although not in the ordering of the states) but more experimental work is necessary.

VI. CONCLUSIONS

The alkaline earth metals form a host of unique monovalent free radicals. Most of these molecules can be formed by the laser-driven chemical reactions of metal vapors with a wide variety of organic and inorganic molecules. This photochemical production of new molecules has led to an extensive gas-phase inorganic chemistry and spectroscopy of alkaline earth derivatives. In recent years, the Broida oven source has been displaced by the pulsed molecular beam spectrometer. The chemical dynamics and photochemistry of these new molecules are still at a very early stage of investigation.

ACKNOWLEDGMENTS

I thank my students and collaborators for their contributions to the work summarized here. At Arizona and Waterloo, this work was supported by NSF, Petroleum Research Fund, Office of Naval Research, Research Corporation, Natural Sciences and Engineering Research Council of Canada, and the Phillips Lab., Edwards AFB. I thank Z. Morbi, C. Zhao, F. Davis, T. Miller, T. Steimle and L. Zuirys for assistance with figures and for reprints and preprints. I also thank W. Chan and I. Hamilton for providing the results of their calculations in advance of publication.

REFERENCES

1. F. A. Cotton and G. Wilkinson, *Advanced Inorganic Chemistry*, 5th ed., Wiley-Interscience, New York, 1988.
2. D. E. Lessen, R. L. Asher, and P. J. Brucat, *J. Chem. Phys.* **93**, 6102 (1990).

3. K. F. Willey, C. S. Yeh, D. L. Robins, J. S. Pilgrim, and M. A. Duncan, *J. Chem. Phys.*, **97**, 8886 (1992).
4. C. T. Scurlock, S. H. Pullins, J. E. Reddic, and M. A. Duncan, *J. Chem. Phys.*, **104**, 4591 (1996).
5. M. H. Shen and J. M. Farrar, *J. Chem. Phys.*, **94**, 3322 (1991).
6. M. Sodupe and C. W. Bauschlicher, Jr., *Chem. Phys. Lett.*, **212**, 624 (1993).
7. C. W. Bauschlicher, Jr., M. Sodupe, and H. Partridge, *J. Chem. Phys.*, **96**, 4453 (1992).
8. T. P. Hanusa, *Chem. Rev.*, **93**, 1023 (1993).
9. E. Murad, *J. Chem. Phys.*, **75**, 4080 (1981).
10. C. W. Bauschlicher, Jr., S. R. Langhoff, and H. Partridge, *J. Chem. Phys.*, **84**, 901 (1986).
11. J. M. Brom, Jr., and W. Weltner, Jr., *J. Chem. Phys.*, **64**, 3894 (1976).
12. A. Antic-Jovanovic, V. Bojovic, and D. Pesic, *Spectrosc. Lett.*, **21**, 757 (1988); private communication with R. Colin.
13. J. Kong and R. J. Boyd, *J. Chem. Phys.*, **103**, 10070 (1995); **104**, 4055 (1996).
14. J. W. Kauffman, R. H. Hauge, and J. L. Margrave, *High Temp. Sci.*, **18**, 97 (1984).
15. J. F. W. Herschel, *Trans. R. Soc. Edinburgh*, **9**, 445 (1823).
16. R. F. Barrow and E. F. Caldin, *Proc. Phys. Soc. London Sect. B*, **62**, 32 (1949).
17. C. G. James and T. M. Sugden, *Nature (London)*, **175**, 333 (1955).
18. C. Th. J. Alkemade, Tj. Hollander, W. Snelleman, and P. J. Th. Zeegers, *Metal Vapours in Flames*, Pergamon, Oxford, UK, 1982.
19. E. Murad, W. Snider, and S. W. Benson, *Nature (London)*, **289**, 273 (1981).
20. T. Tsuji, *Astron. Astrophys.*, **23**, 411 (1973).
21. P. Pesch, *Astrophys. J.*, **174**, L155 (1972).
22. B. R. Pettersen and S. L. Hawley, *Astron. Astrophys.*, **217**, 187 (1989).
23. W. L. Barclay, Jr., M. A. Anderson, and L. M. Ziurys, *Chem. Phys. Lett.*, **196**, 225 (1992).
24. C. T. Scurlock, D. A. Fletcher, and T. C. Steimle, *J. Mol. Spectrosc.*, **159**, 350 (1993).
25. K. Kawaguchi, E. Kagi, T. Hirano, S. Takano, and S. Saito, *Astrophys. J.*, **406**, L39 (1993).
26. K. Ishii, T. Hirano, U. Nagashima, B. Weis, and K. Yamashita, *Astrophys. J.*, **410**, L43 (1993).
27. L. M. Ziurys, A. J. Apponi, M. Guélin, and J. Cernicharo, *Astrophys. J.*, **445**, L47 (1995).
28. P. F. Bernath, *Science*, **254**, 665 (1991).
29. S. J. Weeks, H. Haraguchi, and J. D. Winefordner, *J. Quant, Spectrosc. Radiat. Transfer*, **19**, 633 (1978).
30. P. A. Bonczyk, *Applied Opt.*, **28**, 1529 (1989).
31. P. A. Bonczyk, *Combust. Sci. Tech.*, **59**, 143 (1988).

32. R. F. Wormsbecher, M. Trkula, C. Martner, R. E. Penn, and D. O. Harris, *J. Mol. Spectrosc.*, **97**, 29 (1983).
33. J. Nakagawa, R. F. Wormsbecher, and D. O. Harris, *J. Mol. Spectrosc.*, **97**, 37 (1983).
34. R. F. Wormsbecher, R. E. Penn, and D. O. Harris, *J. Mol. Spectrosc.*, **97**, 65 (1983).
35. R. C. Hilborn, Q. Zhu, and D. O. Harris, *J. Mol. Spectrosc.*, **97**, 73 (1983).
36. J. B. West, R. S. Bradford, J. D. Eversole, and C. R. Jones, *Rev. Sci. Instrum.*, **46**, 164 (1975).
37. L. M. Ziurys, W. L. Barclay, Jr., M. A. Anderson, D. A. Fletcher, and J. W. Lamb, *Rev. Sci. Instrum.*, **65**, 1517 (1994).
38. P. F. Bernath, *Spectra of Atoms and Molecules*, Oxford, New York, 1995.
39. C. J. Whitham, B. Soep, J.-P. Visticot, and A. Keller, *J. Chem. Phys.*, **93**, 991 (1990).
40. C. J. Whitham and Ch. Jungen, *J. Chem. Phys.*, **93**, 1001 (1990).
41. C. R. Brazier, L. C. Ellingboe, S. Kinsey-Nielsen, and P. F. Bernath, *J. Am. Chem. Soc.*, **108**, 2126 (1986).
42. C. R. Brazier, P. F. Bernath, S. Kinsey-Nielsen, and L. C. Ellingboe, *J. Chem. Phys.*, **82**, 1043 (1985).
43. M. D. Oberlander and J. M. Parson, *J. Chem. Phys.*, **105**, 5806 (1996).
44. H. F. Davis, A. G. Suits, Y. T. Lee, C. Alcaraz, and J.-M. Mestdagh, *J. Chem. Phys.*, **98**, 9595 (1993).
45. P. de Pujo, O. Sublemontier, J.-P. Visticot, J. Berlande, J. Cuvellier, C. Alcaraz, T. Gustavsson, J.-M. Mesdagh, and P. Meynadier, *J. Chem. Phys.*, **99**, 2533 (1993).
46. C. Alcaraz, J.-M. Mestdagh, P. Meynadier, P. de Pujo, J.-P. Visticot, A. Binet, and J. Cuvellier, *Chem. Phys. Lett.*, **156**, 191 (1989).
47. T. Gustavsson, C. Alcaraz, J. Berlande, J. Cuvellier, J.-M Mestdagh, P. Meynadier, P. de Pujo, O. Sublemontier, and J.-P. Visticot, *J. Mol. Spectrosc.*, **145**, 210 (1991).
48. M. Oberlander, R. P. Kampf, and J. M. Parson, *Chem. Phys. Lett.*, **176**, 385 (1991).
49. B. S. Cheong and J. M. Parson, *J. Chem. Phys.*, **100**, 2637 (1994).
50. M. Esteban, M. Garay, J. M. Garcia-Tijero, E. Verdasco, and A. Gonzalez Urena, *Chem. Phys. Lett*, **230**, 525 (1994).
51. J. M. Mestdagh, J.-P. Visticot, and P. F. Bernath, *Chem. Phys. Lett.*, **237**, 568 (1995).
52. S. F. Rice, H. Martin, and R. W. Field, *J. Chem. Phys.*, **82**, 5023 (1985).
53. Z. Morbi, Ph.D. Thesis, University of Waterloo, 1997.
54. J. V. Ortiz, *J. Chem. Phys.*, **92**, 6728 (1990).

REFERENCES

55. J. V. Ortiz, *J. Am. Chem. Soc.*, **113**, 1102 (1991).
56. J. V. Ortiz, *J. Am. Chem. Soc.*, **113**, 3593 (1991).
57. J. V. Ortiz, *Chem. Phys. Lett.*, **169**, 116 (1990).
58. L. M. Ziurys, D. A. Fletcher, M. A. Anderson, and W. L. Barclay, Jr., *Astrophys. J. Suppl.*, **102**, 425 (1996).
59. D. A. Fletcher, M. A. Anderson, W. L. Barclay, Jr., and L. M. Ziurys, *J. Chem. Phys.*, **102**, 4334 (1995).
60. P. R. Bunker, M. Kolbuszewski, P. Jensen, M. Brumm, M. A. Anderson, W. L. Barclay, Jr., L. M. Ziurys, Y. Ni, and D. O. Harris, *Chem. Phys. Lett.*, **239**, 217 (1995).
61. B. P. Nuccio, A. J. Apponi, and L. M. Ziurys, *J. Chem. Phys.*, **103**, 9193 (1995).
62. L. M. Ziurys, W. L. Barclay, Jr., and M. A. Anderson, *Astrophys. J.*, **384**, L63 (1992).
63. M. A. Anderson, W. L. Barclay, Jr., and L. M. Ziurys, *Chem. Phys. Lett.,* **196**, 166 (1992).
64. M. A. Anderson, M. D. Allen, W. L. Barclay, Jr., and L. M. Ziurys, *Chem. Phys. Lett.,* **205**, 415 (1993).
65. B. Fernandez, *Chem. Phys. Lett.,* **259**, 635 (1996).
66. Y. Ni, D. O. Harris, and T. Steimle, private communication.
67. R. Pereira and D. H. Levy, *J. Chem. Phys.*, **105**, 9733 (1996).
68. D. A. Fletcher, K. Y. Jung, C. T. Scurlock, and T. C. Steimle, *J. Chem. Phys.*, **98**, 1837 (1993).
69. T. C. Steimle, D. A. Fletcher, K. Y. Jung, and C. T. Scurlock, *J. Chem. Phys.*, **96**, 2556 (1992).
70. C. W. Bauschlicher, Jr., S. R. Langhoff, T. C. Steimle, and J. E. Shirley, *J. Chem. Phys.*, **93**, 4179 (1990).
71. J. M. Mestdagh and J. P. Visticot, *Chem. Phys.*, **155**, 79 (1991).
72. A. R. Allouche and M. Aubert-Frécon, *J. Mol. Spectrosc.*, **163**, 599 (1994).
73. R. A. Hailey, C. N. Jarman, W. T. M. L. Fernando, and P. F. Bernath, *J. Mol. Spectrosc.*, **147**, 40 (1991).
74. C. Zhao, P. G. Hajigeorgiou, P. F. Bernath, and J. W. Hepburn, *J. Mol. Spectrosc.*, **176**, 268 (1996).
75. P. I. Presunka and J. A. Coxon, *Chem. Phys.*, **190**, 97 (1995).
76. P. I. Presunka and J. A. Coxon, *J. Chem. Phys.*, **101**, 201 (1994).
77. P. I. Presunka and J. A. Coxon, *Can. J. Chem.*, **71**, 1689 (1993).
78. J. A. Coxon, M. Li, and P. I. Presunka, *J. Mol. Spectrosc.*, **150**, 33 (1991).
79. M. Li and J. A. Coxon, *Can. J. Phys.*, **72**, 1200 (1994).
80. J. A. Coxon, M. Li, and J. A. Coxon, *J. Mol. Spectrosc.*, **164**, 118 (1994).
81. J. A. Coxon, M. Li, and P. I. Presunka, *Mol. Phys.*, **76**, 1463 (1992).
82. M. Li and J. A. Coxon, *J. Chem. Phys.*, **97**, 8961 (1992).

83. M. Li and J. A. Coxon, *J. Chem. Phys.*, **102**, 2663 (1995).
84. M. Li and J. A. Coxon, *J. Chem. Phys.*, **104**, 4961 (1996).
85. C. N. Jarman and P. F. Bernath, *J. Chem. Phys.*, **97**, 1711 (1992); see also Z. J. Jakubek and R. W. Field, *J. Chem. Phys.*, **98**, 6574 (1993).
86. W. T. M. L. Fernando, M. Douay, and P. F. Bernath, *J. Mol. Spectrosc.*, **144**, 344 (1990).
87. R. Hailey, C. Jarman, and P. F. Bernath, *J. Chem. Phys.* (in press). R. Hailey, M.Sc. Thesis, University of Arizona, Tucson, 1991.
88. R. F. Wormsbecher and R. D. Suenram, *J. Mol. Spectrosc.*, **95**, 391 (1982).
89. L. C. O'Brien, C. R. Brazier, and P. F. Bernath, *J. Mol. Spectrosc.*, **130**, 33 (1988).
90. J. Brown, and T. Steimle private communication.
91. C. Zhao and P. F. Bernath, work in progress.
92. L. C. O'Brien, C. R. Brazier, S. Kinsey-Nielsen, and P. F. Bernath, *J. Phys. Chem.*, **94**, 3543 (1990).
93. A. M. R. P. Bopegedera, W. T. M. L. Fernando, and P. F. Bernath, *J. Phys. Chem.*, **94**, 3547 (1990).
94. L. C. Ellingboe, A. M. R. P. Bopegedera, C. R. Brazier, and P. F. Bernath, *Chem. Phys. Lett.*, **126**, 285 (1986).
95. L. C. O'Brien and P. F. Bernath, *J. Chem. Phys.*, **88**, 2117 (1988).
96. C. R. Brazier and P. F. Bernath, *J. Chem. Phys.*, **88**, 2112 (1988).
97. W.-T. Chan and I. P. Hamilton, private communication.
98. L. Pasternack and P. J. Dagdigian, *J. Chem. Phys.*, **65**, 1320 (1976).
99. N. Furio and P. J. Dagdigian, *Chem. Phys. Lett.*, **115**, 358 (1985).
100. E. Clementi, H. Kistenmacher, and H. Popkie, *J. Chem. Phys.*, **58**, 2460 (1973).
101. C. W. Bauschlicher, Jr., S. R. Langhoff, and H. Partridge, *Chem. Phys. Lett.*, **115**, 124 (1985).
102. M. Douay and P. F. Bernath, *Chem. Phys. Lett.*, **174**, 230 (1990).
103. T. C. Steimle, D. A. Fletcher, K. Y. Jung, and C. T. Scurlock, *J. Chem. Phys.*, **97**, 2909 (1992).
104. C. T. Scurlock, D. A. Fletcher, and T. C. Steimle, *J. Chem. Phys.*, **101**, 7255 (1994).
105. M. A. Anderson and L. M. Ziurys, *Chem. Phys. Lett.*, **231**, 164 (1994); also, M. Guélin, M. Forestini, L. M. Ziurys, M. A. Anderson, J. Cernicharo, and C. Kalhane, *Astron. Astrophys.* **297**, 183 (1995).
106. T. C. Steimle, S. Saito, and S. Takano, *Astrophys. J.*, **410**, L49 (1993).
107. E. Kagi, K. Kawaguchi, S. Takano, and T. Hirano, *J. Chem. Phys.*, **104**, 1263 (1996).
108. C. T. Scurlock, T. C. Steimle, R. D. Suenram, and F. J. Lovas, *J. Chem. Phys.*, **100**, 3497 (1994).
109. M. A. Anderson, T. C. Steimle, and L. M. Ziurys, *Astrophys. J.*, **429**, L41 (1994).

REFERENCES

110. L. M. Ziurys, A. J. Apponi, M. Guélin, and J. Cernicharo, *Astrophys. J.*, **445**, L47 (1995).
111. W. T. M. L. Fernando, R. S. Ram, L. C. O'Brien, and P. F. Bernath, *J. Phys. Chem.*, **95**, 2665 (1991).
112. C. N. Jarman and P. F. Bernath, *J. Chem. Phys.*, **98**, 6697 (1993).
113. C. T. Scurlock, T. Henderson, S. Bosely, K. Y. Jung, and T. C. Steimle, *J. Chem. Phys.*, **100**, 5481 (1994).
114. A. Taleb-Bendiab, F. Scappini, T. Amano, and J. K. G. Watson, *J. Chem. Phys.*, **104**, 7431 (1996).
115. C. Brazier and P. F. Bernath, in preparation.
116. A. J. Marr, M. Tanimoto, D. Goodridge, and T. C. Steimle, *J. Chem. Phys.*, **103**, 4466 (1995).
117. Z. Morbi, C. Zhao, and P. F. Bernath, *J. Chem. Phys.*, **106**, 4860 (1997).
118. A. M. R. P. Bopegedera, C. R. Brazier, and P. F. Bernath, *J. Phys. Chem.*, **91**, 2779 (1987).
119. C. R. Brazier and P. F. Bernath, *J. Chem. Phys.*, **86**, 5918 (1987).
120. C. R. Brazier and P. F. Bernath, *J. Chem. Phys.*, **91**, 4548 (1989).
121. S. C. Tyerman, G. K. Corlett, A. M. Ellis, and T. A. Claxton, *Theochem.*, **364**, 107 (1996).
122. M. A. Anderson and L. M. Ziurys, *Astrophys. J.*, **452**, L157 (1995).
123. M. A. Anderson and L. M. Ziurys, *Astrophys. J.*, **460**, L77 (1996).
124. M. A. Anderson, J. S. Robinson, and L. M. Ziurys, *Chem. Phys. Lett.*, **257**, 471 (1996).
125. R. Rubino, J. M. Williamson, and T. A. Miller, *J. Chem. Phys.*, **103**, 5964 (1995).
126. A. J. Marr, F. Grieman, and T. C. Steimle, *J. Chem. Phys.*, **105**, 3930 (1996).
127. A. M. R. P. Bopegedera, C. R. Brazier, and P. F. Bernath, *Chem. Phys. Lett.*, **136**, 97 (1987).
128. A. M. R. P. Bopegedera, C. R. Brazier, and P. F. Bernath, *J. Mol. Spectrosc.*, **129**, 268 (1988).
129. A. J. Marr, J. Perry, and T. C. Steimle, *J. Chem. Phys.*, **103**, 3861 (1995).
130. M. Li and J. A. Coxon, *J. Mol. Spectrosc.*, **176**, 206 (1996); **180**, 287 (1996); **183**, 250 (1987).
131. M. A. Anderson and L. M. Ziurys, *Astrophys. J.*, **439**, L25 (1995).
132. M. A. Anderson and L. M. Ziurys, *Astrophys. J.*, **444**, L57 (1995).
133. B. P. Nuccio, A. J. Apponi, and L. M. Ziurys, *Chem. Phys. Lett.*, **247**, 283 (1995).
134. D. E. Woon, *Astrophys. J.*, **456**, 602 (1996); *J. Chem. Phys.*, **104**, 9495 (1996).
135. G. K. Corlett, A. M. Little, and A. M. Ellis, *Chem. Phys. Lett.*, **249**, 53 (1996).
136. L. C. O'Brien and P. F. Bernath, *J. Am. Chem. Soc.*, **108**, 5017 (1986).
137. A. M. Ellis, E. S. J. Robles, and T. A. Miller, *J. Chem. Phys.*, **94**, 1752 (1991).

138. E. S. J. Robles, A. M. Ellis, and T. A. Miller, *J. Am. Chem. Soc.*, **114**, 7171 (1992).
139. T. M. Cerny, J. M. Williamson, and T. A. Miller, *J. Chem. Phys.*, **102**, 2372 (1995).
140. A. M. R. P. Bopegedera, W. T. M. L. Fernando, and P. F. Bernath, *J. Phys. Chem.*, **94**, 4476 (1990).
141. E. S. J. Robles, A. M. Ellis, and T. A. Miller, *J. Phys. Chem.*, **96**, 8791 (1992).
142. F. S. Pianalto, A. M. R. P. Bopegedera, W. T. M. L. Fernando, R. Hailey, L. C. O'Brien, C. R. Brazier, P. C., Keller, and P. F. Bernath, *J. Am. Chem. Soc.*, **112**, 7900 (1990).
143. S. Nanbu, S. Minamino and M. Aoyagi, *J. Chem. Phys.* **106**, 8073 (1997).

PHOTOCHEMICALLY INDUCED DYNAMIC NUCLEAR POLARIZATION

Martin Goez

Fachbereich Chemie, Martin-Luther-Universität Halle-Wittenberg,
Kurt–Mothes-Straße 2, D-06120 Halle/Saale, FRG

CONTENTS

I. Introduction, 64
II. Theoretical principles, 67
    A. Basic concepts, 67
    B. The radical pair mechanism, 71
        1. What is a radical pair?, 71
        2. Qualitative explanation of the radical pair mechanism, 72
        3. Spin states and relevant interactions of a radical pair, 74
        4. Density matrix treatment of the spin dynamics of radical pairs, 80
        5. Combining spin dynamics and radical pair dynamics, 84
III. CIDNP effects, 91
    A. Explanation by the radical pair mechanism, 91
    B. Relationship between polarizations and line intensities, 93
    C. Estimations of CIDNP intensities and phases, 95
    D. Instrumentation and techniques, 100
    E. Information accessible by CIDNP, 101

Advances in Photochemistry, Volume 23, Edited by Douglas C. Neckers, David H. Volman, and Günther von Bünau
ISBN 0-471-19289-9 © 1997 by John Wiley & Sons, Inc.

IV. Physical studies and methodological developments, 102
   A. Spin dynamics, 102
   B. Radical pair dynamics, 105
   C. Biradicals, 106
   D. Micellar systems, 110
   E. Other polarization mechanisms, 111
   F. New methods and techniques, 112
V. Applications to chemical problems, 114
   A. Electron transfer, 114
      1. Structures of radical ions, 114
      2. Kinetic and mechanistic studies, 118
   B. Hydrogen abstractions, 123
   C. Fragmentations, 127
      1. α-Cleavage of carbonyl compounds, 127
      2. Other cleavage reactions, 129
   D. Secondary reactions of radicals and biradicals, 132
      1. Radical additions, 132
      2. Radical fragmentations, 134
      3. Cycloadditions, cycloreversions, and isomerizations, 135
   E. Inorganic and metal organic substrates, 143
   F. Polymerization initiators, 144
   G. Biologically relevant molecules, 149
      1. Amino acids, peptides, and proteins, 149
      2. Nucleic acids, 152
      3. Photosynthesis, 154
Acknowledgments, 154
References, 154

I. INTRODUCTION

The course and outcome of chemical reactions can be influenced by magnetic interactions. At first glance, this may seem fantastic because the energies of these interactions are negligible (lower by more than three orders of magnitude) compared to the energies contained in chemical bonds. However, there is a very powerful indirect influence that is based on the Pauli principle. As the latter states, the overall wave function of a species must be antisymmetric with respect to electron permutation, so its spin part—which is subject to magnetic interactions—must be antisymmetric if its space part is symmetric and vice versa. Conversely, a change of symmetry of the spin part brought about by these interactions must thus be accompanied by a change of symmetry of the space part, which has all the chemical consequences. It is by virtue of this fact that the tiny magnetic effects can act as switches and can control reactions. As a consequence of the required symmetry change of the spin part of the wave function, magnetic field effects can only occur for reactions involving species of higher electronic spin multiplicity than one.

Although 'spin chemistry,' as the field has been named as a whole, comprises many different experimental effects [1], its essence is therefore a dependence of product yields on internal or external static or dynamic magnetic fields. Basic techniques are

1. *Direct Product Yield Measurements.* In these experiments, detection is usually accomplished by optical spectroscopy, for sensitivity reasons.
   a. Magnetic field effects (MFE), also called magnetic field modulation of reaction yields (MARY) [2]: A static field is applied, and the product yields are determined as functions of its strength.
   b. Magnetic isotope effects (MIE) [3]: The differences in product yields that occur when the same reaction is run with substrates containing magnetic nuclei (e.g., ^{13}C) versus nonmagnetic nuclei (^{12}C) in specific positions of the molecules are analyzed.
   c. Reaction yield detected magnetic resonance (RYDMR) [4]: In addition to a static field, a dynamic (microwave) field is used, and the product yields are analyzed as functions of frequency and intensity of the latter, the strength of the static field being a parameter.
   d. Radiofrequency induced magnetic isotope effects (RIMIE) [5]: Same as 1.b, but in the presence of a microwave field.
2. *Spin Polarization Phenomena.* Strictly speaking, these experiments are also determinations of product yields. However, as magnetic resonance techniques are used for detection, the yields of products *in certain spin states* (more precisely, differences of populations of these spin states) are measured. The observables are anomalous line intensities in the nuclear magnetic resonance (NMR) or electron paramagnetic resonance (EPR) spectra.
   a. Chemically induced dynamic nuclear polarization (CIDNP) [6]: A reaction is carried out in a static magnetic field (usually directly that of the spectrometer), and the NMR spectra of the diamagnetic products are recorded during its course or immediately after its completion.

b. Chemically induced dynamic electron polarization (CIDEP) [7]: The EPR variant of 2.a, the important difference being that observation occurs at an earlier stage of the reaction (intermediate free radicals or radical pairs [8] are monitored).

c. Stimulated nuclear polarization (SNP) [9] and dynamic nuclear polarization (DNP) [9b]: SNP is a CIDNP experiment carried out in the presence of an additional microwave field, which redistributes the electron spin populations in the intermediates. SNP is usually accompanied by DNP, which means a "leaking," due to cross-relaxation, of electron spin polarization to nuclear spin polarization; hence DNP can also occur in systems exhibiting strong CIDEP effects.

As far as applications to chemical problems are concerned, CIDNP is the most frequently employed technique of spin chemistry. Probably, one of the reasons is simply that the detection method, NMR spectroscopy, is familiar ground to a chemist, and that NMR spectrometers are abundant. Certainly, however, two decisive advantages of CIDNP are that it yields very detailed information about chemical reactions, spanning from the precursors of the paramagnetic stages to the final diamagnetic products, while needing almost no modifications of commercially available standard equipment. For these reasons, this chapter is limited to this method.

Although all techniques of spin chemistry can also be applied to thermal reactions (e.g., the first observations of CIDNP were made in such processes), they seem to be tailored for the investigation of photoreactions. For one thing, both singlet and triplet species are accessible in photochemical reactions. Even more advantageous is that the progress of photoreactions can be controlled most conveniently by gating the illumination. As an extension of this, kinetic studies by spin-chemical methods are feasible just as well, when pulsed light sources and time resolved detection are employed. It is thus not surprising that during recent years the interest not only of CIDNP spectroscopy but also of spin chemistry in general has focused almost exclusively on photoreactions.

The theoretical description of spin chemistry is fairly advanced, so that one might say that the field has obviously reached maturity. Consequently, there are already several excellent reviews [10] and monographs [11] dealing in detail with its physical basis. Yet, despite the fact that spin chemistry, and CIDNP in particular, has long arrived at a stage where it can be used as a tool for mechanistic and kinetic research, chemical applications of magnetic field and spin effects have been reviewed much less thoroughly. This chapter intends to fill this gap to some degree. As far as applications to chemistry are concerned, it concentrates on the literature from 1989, when Steiner's in-depth review [1a] of the whole field of spin chemistry appeared, to the beginning of 1996. Rather than provide complete coverage, it focuses on examples where new chemical insight has been gained by CIDNP spectroscopy.

The organization of the material is as follows. In Section II, the underlying theory is treated. Because the present chapter as a whole is aimed at the nonspecialist, this section is kept "user-friendly" in the sense that for quantitative discussion, models that provide physical insight were chosen whenever possible and for these models there is somewhat more calculational detail than usual. Section III explains CIDNP spectroscopy. Section IV reviews studies of a more physically orientated nature, that is dealing with the CIDNP effect as such and not with its applications to chemical research; however, as this chapter aims at the experimentalist rather than at the theoretician, mere model calculations and simulations are not discussed. New experimental techniques and improvements of existing methods are also covered. Section V constitutes the central part of this work. In it, chemical investigations are reviewed for which CIDNP spectroscopy played a key role. Subdivision was done partly according to the type of process studied (e.g., electron transfer), and partly according to the systems investigated (e.g., polymerization initiators).

II. THEORETICAL PRINCIPLES

A. Basic Concepts

The influence of magnetic interactions on chemical reactions depends on several factors.

1. *Spin Conservation.* In elementary chemical acts, such as bond cleavage, bond formation, or electron transfer, magnitude and direction of spin are conserved. This means that both the expectation value of total electron spin (operator S^2) and that of its z projection (operator S_z) remain constant; the same holds for the nuclear spin states. Hence, for example, fragmentation of a molecule M in a singlet state $|S\rangle$, which is characterized by zero total spin and zero z projection of spin, into two particles 1 and 2 yields either two singlet (i.e., diamagnetic) species, or two radicals, the spins of which are paired in a specific manner such that spin conservation is fulfilled (in a way, they are

"oppositely aligned"; cf. Fig. 2, Section II.B.3). Likewise, fragmentation of a triplet molecule, for instance, in the substate $|T_0\rangle$, which has zero z spin projection, can either lead to a singlet plus a triplet species with the same z projection of spin as the starting molecule, or again produce two radicals, this time, however, with spins being paired in a different manner (for a visualization, see also Fig. 2). Actually, in both these examples a description in terms of individual spins of two radicals would be permissible only in the absence of interactions between them. The reason for this rule of spin conservation is simply that elementary chemical processes are much faster than is change of the spin state because the energies involved in bonding are much larger than the magnetic interactions.

2. *Spin Selectivity.* As a consequence of the preceding, reaction rates depend on the electron spin multiplicity. In many cases, the difference is so large that it practically amounts to a yes–no decision. As a typical example, consider an encounter complex, at the reaction distance, of two radicals, which can have either triplet or singlet multiplicity. In most cases, reaction will be impossible for the triplet complex because it would lead to an electronically excited product. For reaction of the singlet complex, there is no such restriction.

3. *Spin Evolution.* Magnetic interactions, which are listed below, can effect radiationless transitions $|X_{\text{initial}}\rangle \to |X_{\text{final}}\rangle$ between states of different multiplicity. The driving term of any such process is given by the coupling matrix element Q, $Q = \langle X_{\text{final}}|\mathbf{H}_{\text{inter}}|X_{\text{initial}}\rangle$, where $\mathbf{H}_{\text{inter}}$ is the operator of the relevant interaction. However, such transitions are only efficient if the magnitude of the mixing matrix element is at least comparable to the energy difference ΔE, $\Delta E = |E_{\text{final}} - E_{\text{initial}}|$, between the two states. For state i, the energy E_i is of course the diagonal matrix element $\langle X_i|\mathbf{H}_{\text{total}}|X_i\rangle$ of the total Hamiltonian $\mathbf{H}_{\text{total}}$. The spin motion under the combined influence of Q and ΔE (spin dynamics) is an oscillation of frequency $\hbar^{-1}\sqrt{(\Delta E)^2 + 4Q^2}$ and amplitude $1/[1 + (\Delta E/2Q)^2]$. Because most magnetic interactions are extremely small, only functions that are degenerate, or nearly degenerate, therefore lead to nonnegligible amplitudes. In an isolated case, the system periodically reverts to its starting state, so the mean expectation value of total spin differs from that before the oscillation began, but there is no intersystem crossing, which would imply a permanent change of multiplicity: if one starts out with one multiplicity, oscillation can never make the mean expectation value of total spin correspond to more than 50% admixture of the other multiplicity. Intersystem crossing occurs only when there are irreversible processes,

for instance, when one state is coupled to a continuum of states, or when a chemical reaction removes only particles in a certain spin state from the system. In any case, however, the strength of the mixing interaction relative to the energy difference of the states is the crucial parameter for the amplitude of the oscillations and through this also for the efficiency of intersystem crossing.

4. *Spin Hamiltonian.* The operators of the interactions pertinent to this work, neglecting anisotropy effects, are,

   a. The electron Zeeman interaction $\mathbf{H}_Z$ of an electron with an external field of strength B_0,

$$\mathbf{H}_Z = -g\beta B_0 \mathbf{S}_z \tag{1}$$

where g is the isotropic g value of the electron, and β is the Bohr magneton. Because the only operator $\mathbf{H}_Z$ contains is the operator for the z component of the electron spin, it is obvious that this component remains unchanged by the Zeeman interaction; however, states of equal z spin projection but different electron-spin multiplicity can be mixed by $\mathbf{H}_Z$.

   b. The hyperfine interaction $\mathbf{H}_{hfc}$ between electron and magnetic nucleus, which is the scalar product of the operators $\mathbf{S}$ and $\mathbf{I}$ for the electron spin and the nuclear spin.

$$\mathbf{H}_{hfc} = a\mathbf{S} \cdot \mathbf{I} = a(\mathbf{S}_x \mathbf{I}_x + \mathbf{S}_y \mathbf{I}_y + \mathbf{S}_z \mathbf{I}_z) \tag{2}$$

with a being the isotropic hyperfine coupling constant.

The nonsecular terms $\mathbf{S}_x \mathbf{I}_x$ and $\mathbf{S}_y \mathbf{I}_y$ in Eq. 2 complicate matters because these products possess off-diagonal elements between states of different z components of *both* electron *and* nuclear spins; they effect opposite spin flips of electron and nucleus, so-called flip–flop transitions. Consequently, large sets of basis functions are necessary. Even a radical with only two protons requires as many as eight spin functions so that the operator matrices, which in this example have dimension 8×8, soon become intractable except on a computer. Fortunately, this difficulty arises mainly in low magnetic fields; in high magnetic fields the difference of the magnetic energies of electrons and nuclei usually causes mixing by the terms $\mathbf{S}_x \mathbf{I}_x$ and $\mathbf{S}_y \mathbf{I}_y$ to be negligible. Hence, under these circumstances only the secular part $\mathbf{S}_z \mathbf{I}_z$ of the hyperfine interaction needs to be considered. Consequently, electron and nuclear spin states can be factorized, and the operator matrix $\mathbf{H}'_{\text{hfc}}$ for the electron spin states can be constructed with the nuclear spin state being treated as a constant parameter for a given nuclear configuration

$$\mathbf{H}'_{\text{hfc}} = \mathbf{S}_z \sum_i a_i \mathbf{I}_{iz} = \mathbf{S}_z \sum_i a_i m_i \qquad (3)$$

where m_i is the spin quantum number of nucleus i. Even a radical with an arbitrary number of magnetic nuclei thus needs two basis functions of its electron spin state only.

As with the Zeeman interaction, the secular term $\mathbf{H}'_{\text{hfc}}$ of the hyperfine interaction cannot alter the z component of the electron spin state (Eq. 3). However, the nonsecular terms can, with concomitant opposite alteration of the z projection of the nuclear spin state. Intersystem crossing, on the other hand, can be effected by *both* secular and nonsecular terms.

c. The exchange interaction $\mathbf{H}_{\text{ex}}$ of two electrons

$$\mathbf{H}_{\text{ex}} = -J(\tfrac{1}{2} + 2\mathbf{S}_1 \cdot \mathbf{S}_2) \qquad (4)$$

$\mathbf{H}_{\text{ex}}$ is the only nonmagnetic interaction that needs to be considered in this chapter. Unless an additional paramagnetic species is present, as is the case in spin-catalysis [12], it induces neither intersystem crossing nor a change of z spin projection but determines the energy separation of levels of different multiplicity, and is therefore a key parameter for every intersystem crossing step. While J can be related to a quantum mechanical exchange integral, it is best treated as an empirical parameter [13]. Because the magnitude of the exchange interaction reflects the strength of the Coulombic repulsion of the two electrons, J strongly decreases with increasing separation of the latter.

d. Spin–orbit coupling $\mathbf{H}_{\text{SO}}$,

$$\mathbf{H}_{\text{SO}} = \zeta \mathbf{L} \cdot \mathbf{S} \qquad (5)$$

where $\mathbf{L}$ is the operator for the orbital momentum, and ζ is the spin–orbit coupling constant. This interaction is responsible for intersystem crossing of excited molecules; it is much stronger than the previously listed magnetic interactions. The operator $\mathbf{H}_{\text{SO}}$ is seen to have the same functional form as $\mathbf{H}_{\text{hfc}}$; hence, the terms $\mathbf{L}_x \mathbf{S}_x$ and $\mathbf{L}_y \mathbf{S}_y$ change the z projection of orbital momentum.

Because orbital momentum, and thus orbital shape and orientation, are involved in intersystem crossing driven by spin–orbit coupling, molecular symmetry and electronic configuration come into play. Consequently, the efficiency of intersystem crossing is usually different for the sublevels of a given multiplicity. This effect is the basis of the so-called triplet mechanism of electron spin polarizations; as far as CIDNP is concerned, $\mathbf{H}_{SO}$ only plays a role for systems with restricted diffusion (biradicals, radical pairs in micelles).

5. *Time Dependence of Perturbations.* Magnetic interactions can cause coherent or incoherent perturbations because they are, in general, tensor interactions that depend on molecular orientation, and are thus stochastically modulated as the molecules tumble. Their average (i.e., isotropic) values act as coherent perturbations (constant mean local fields), and the deviations from the average values as incoherent perturbations (magnetic noise). Intersystem crossing can be effected both by coherent and by incoherent processes but for the systems considered in this work the former usually are much more efficient, which is why the preceding equations for the relevant interactions of the spin Hamiltonian have been formulated for the isotropic case only.

The remainder of Section II is devoted to a detailed treatment of the radical pair mechanism, which is of paramount importance for spin chemistry and practically the only mechanism (cf. Section IV.E) responsible for CIDNP effects.

B. The Radical Pair Mechanism

1. What Is a Radical Pair? Organic chemists are used to looking upon the intermediates of radical reactions as free radicals (i.e., single radicals), and they tend to regard radical pairs as an exotic variety of them. However, radical pairs are in fact the more frequent species because most processes that produce radicals (e.g., homolytical bond cleavage or single electron transfer in homogeneous phase) produce them in pairs. The only exceptions are radical transformations, as, for instance, in radical polymerizations where the chain-propagating species are free radicals. Furthermore, radical pairs must be considered species in their own right because they possess a specific reaction channel: The radicals constituting a pair can react with one another in what is denoted as "the cage"; this process is called geminate reaction. Although combination of two free radicals A˙ and B˙ and geminate reaction of a radical pair $\overline{A^{\cdot}B^{\cdot}}$—the overbar expresses that A$^{\cdot}$ and B$^{\cdot}$ belong to a pair—lead to the same product or products, kinetically there is an important difference. Reactions of two free radicals are second-order processes, whereas geminate reactions are first order.

When are two radicals A$^{\cdot}$ and B$^{\cdot}$ a radical pair and when are they two free radicals? The standard answer, namely, that A$^{\cdot}$ and B$^{\cdot}$ are a pair as long as they are in the cage, reveals a fundamental language barrier between spin chemistry and organic chemistry. For an organic chemist, "in the cage" means that A$^{\cdot}$ and B$^{\cdot}$ are surrounded by solvent molecules that keep them together. A cage is thus a region of space to him, and it would in principle be possible at every moment to decide whether A$^{\cdot}$ and B$^{\cdot}$ are in a cage by simply looking at their spatial positions. In the context of spin chemistry, on the other hand, a cage is a much more abstract concept, a region of time rather: A$^{\cdot}$ and B$^{\cdot}$ still belong to a cage if they separate and meet again at a later time—the Noyes concept [14] of secondary reencounters. Unless the duration of separation is excessive, they retain a memory of one another, the nature of this memory being a correlation of their electron spins (this is what the overbar symbolizes). Consequently, A$^{\cdot}$ and B$^{\cdot}$ can still be a pair even if they are separated by many solvent molecules; on the other hand, the correlation can only be detected once the two radicals meet again by diffusion i.e., the question whether or not A$^{\cdot}$ and B$^{\cdot}$ are a pair $\overline{A^{\cdot}B^{\cdot}}$ can only be decided in retrospect. The spin chemist's definition of a cage is not simply an extension of the organic chemist's definition—a cage in the sense of the latter could not give rise to effects such as CIDNP.

2. Qualitative Explanation of the Radical Pair Mechanism. Because spin is conserved in chemical reactions, radical pairs are born with the electron spin multiplicity of their precursors, singlet or triplet. Geminate reaction, on the other hand, is usually possible in the singlet state only, exceptions being known for back electron transfer of radical ion pairs when the pair energy lies above the triplet energy of one of the reactants [15] (see also the formation of triplet biradicals from spin-correlated radical ion pairs, Section V.D.3). Chance encounters of free radicals provide a second pathway to radical pairs. In this case, any spin state is equally probable; however, as a high percentage of the singlet pairs react immediately, a surplus of triplet pairs survive, and these so-called F pairs therefore behave qualitatively like triplet born pairs. At the moment of generation of a radical pair $\overline{A^{\cdot}B^{\cdot}}$, there exists thus a correlation of the electron spins of the two radicals forming this pair.

Within a time on the order of 10^{-11} s, diffusion will effect a separation of the radicals. While this does not influence the spins directly, the exchange interaction decreases with the interradical distance and vanishes once the radicals are separated by some 10 Å. In this situation, the spins of the two radicals A˙ and B˙ are no longer coupled, that is, singlet and triplet are no longer eigenstates of the magnetic energy. Instead, the system now consists of two independent radicals, two doublets, the spins of which precess independently under the influence of local magnetic interactions. Unless A˙ and B˙ are absolutely identical, with respect to both chemical constitution and nuclear spin state, their precession frequencies and axes will differ, so in time the phase relation between the electron spins changes. Hence, the spins are still correlated but the overall spin function is a superposition of singlet and triplet that varies with time.

If the radicals separate permanently, which for practical purposes means for a time much longer than the electron spin relaxation time, the spin correlation is lost. However, there is a substantial chance that they (i.e., the *same* radicals) will meet again before this happens. As they approach one another, the exchange interaction once more becomes operative and forces the system into the singlet or a triplet state, with a probability that depends on the momentary weight of that state in the superposition.

A visualization of this intersystem crossing mechanism by vector models is given in Sections II.B.3 and II.B.4. In the described process, spins are never flipped, so the z component of spin remains constant. There is also another, less common, variant of the radical pair mechanism that is based on simultaneous electron–nuclear spin flips, and is thus accompanied by a change of the z component of the electron spin. As explained in Section II.B.3, this mechanism can only operate efficiently for a critical interradical separation and in low magnetic fields. While it also relies on diffusion for its generation, reencounters of the radicals are not necessary for its detection because with it the electron Zeeman energy changes. With both mechanisms, the amount of intersystem crossing is typically very small during one diffusive excursion. However, a radical pair can undergo a series of such excursions, the effects of which are cumulative. This important effect was first incorporated into radical pair theory by Adrian [16a,b,c].

The radical pair mechanism thus links "spin dynamics" (the evolution of the spin state of $\overline{A˙B˙}$, hence of spin correlation of the pair, under the influence of magnetic interactions) and "radical pair dynamics" (the diffusion, relative to one another, of the two radicals, and the coupling to chemical processes). Spin dynamics will be treated in Sections II.B.3 and II.B.4, and then combined with radical pair dynamics in Section II.B.5. Although most of the discussion is centered around radical pairs, and the described mechanism of intersystem crossing is generally called "radical pair mechanism," much of the following applies also to biradicals, which can be regarded as a special case of radical pairs, namely, pairs with restricted interdiffusion.

3. Spin States and Relevant Interactions of a Radical Pair. For most of this section, conditions will be considered (i.e., high magnetic fields) that allow factorization of the overall spin function of a radical pair into an electron spin part and a nuclear spin part. Under these circumstances, only the former changes during intersystem crossing, and the latter is a constant parameter.

Two important limiting cases exist.

1. When the two radicals of the pair are separated by more than a few molecular diameters, the exchange interaction vanishes, and the spin Hamiltonian is dominated by the local magnetic interactions. In this case, it is permissible to decribe the spin state by simple product functions

$$|\alpha\alpha\rangle \quad |\alpha\beta\rangle \quad |\beta\alpha\rangle \quad |\beta\beta\rangle \tag{6}$$

   in other words, to distinguish between the two unpaired electrons by assigning a unique spin to each of them. This is borne out by experimental observation: Under these conditions, the EPR spectrum of such a system consists of two subspectra, one for each radical, and by adding a third reaction partner, one of the radicals can be scavenged specifically.

2. When the two radicals are very near one another (at the extreme, in contact), the spin Hamiltonian is dominated by H_{ex}, and the exchange interaction causes the electrons to become indistinguishable. What can still be distinguished is whether the spin part of the wave function is symmetric or antisymmetric with respect to interchange of the electrons, that is, whether it is invariant or changes sign when the electrons are permuted. A group of three symmetric spin functions exists

$$|T_+\rangle = |\alpha\alpha\rangle$$
$$|T_0\rangle = \frac{1}{\sqrt{2}}(|\alpha\beta\rangle + |\beta\alpha\rangle) \tag{7}$$
$$|T_-\rangle = |\beta\beta\rangle$$

   the triplet states, and one antisymmetric spin function,

$$|S\rangle = \frac{1}{\sqrt{2}}(|\alpha\beta\rangle - |\beta\alpha\rangle) \tag{8}$$

the singlet state. This distinction is again reflected by experiment: Singlet molecules do not give rise to EPR signals, whereas molecules in triplet states do, and their EPR spectra are unlike those of separated radicals; the chemical reactivity of singlets and triplets is often totally different.

It is seen that $|\alpha\alpha\rangle$ ($\equiv|T_+\rangle$) and $|\beta\beta\rangle$ ($\equiv|T_-\rangle$) are energy eigenfunctions regardless of whether the two radicals are in contact or far apart, whereas the other two basis functions, which possess zero z component of total spin, change. With decreasing separation of the radicals there must obviously be a gradual transition of these eigenfunctions from the set $|\alpha\beta\rangle$ and $|\beta\alpha\rangle$ to the set $|S\rangle$ and $|T_0\rangle$, such that for a system with some intermediate distance between the radical centers (e.g., a short-chain biradical in a frozen conformation), the eigenfunctions with zero z projection of total spin are mixtures of these two sets. This is the basis of electron spin polarization by the so-called spin-correlated radical pair mechanism [8]. However, this section is concerned with a different problem: Owing to the diffusive motion of the radicals the transitions between the two above-mentioned limiting situations are abrupt. Thus, except for a short time after its creation, the radical pair is no longer found in any of its eigenstates with zero z component of spin but in a superposition state that can be expressed as $c_{\alpha\beta}|\alpha\beta\rangle + c_{\beta\alpha}|\beta\alpha\rangle$ or, alternatively, as $c_S|S\rangle + c_T|T_0\rangle$; the superposition is time dependent.

As basis functions for a radical pair, those of Eqs. 7 and 8 are suited best because the radical pair starts out in one of them and its geminate reaction is (usually) equivalent to filtering away singlet pairs only. Hence, the matrix elements given in the following refer to the functions $|T_+\rangle$, $|S\rangle$, $|T_0\rangle$, and $|T_-\rangle$, in that order. These states are eigenfunctions of the exchange operator $\mathbf{H}_{ex}$, the matrix of which is therefore diagonal.

$$\mathbf{H}_{ex} = J \, \text{diag}(-1, +1, -1, -1) \tag{9}$$

The relation of the exchange integral J appearing in Eq. 9 with quantum mechanical exchange integrals is discussed in detail in [13]. In the context of Eq. 9, J is one-half the singlet–triplet splitting; almost always, J is negative, which means that the singlet state lies below the triplet states.

The distance dependence of J plays a crucial role for intersystem crossing of radical pairs. Usually, an exponential relation between J and the distance r of the radicals is assumed [17, 18]

$$J(r) = -J_0 \exp(-\alpha r) \tag{10}$$

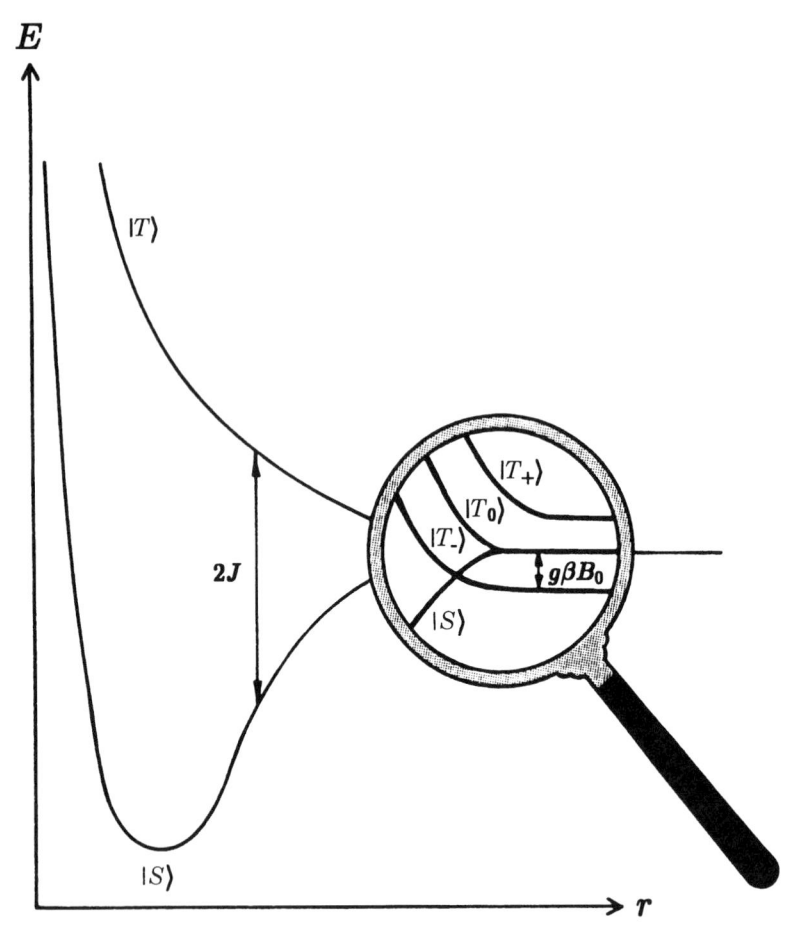

Figure 1. Typical plot of the energies E of singlet and triplet states of a radical pair as functions of the interradical distance r. The inset shows the additional splitting of the triplet levels caused by the Zeeman interaction. [Adapted from ref. [6m] with permission. Copyright © 1995 John Wiley & Sons, Inc.].

Theoretical considerations [17] and experimental results [19–22] suggest that J_0 (in frequency units) is about $10^{16} \cdots 10^{19}$ rad s^{-1}, and $\alpha \approx 2.2 \cdots 2.7$ Å^{-1}. The energies of singlet and triplet states of a radical pair as functions of the interradical separation r are typically displayed as in Figure 1. In such a plot, Coulomb and overlap integrals are set equal to zero, because these quantities do not depend on the multiplicity.

In an external magnetic field, the Zeeman interaction splits the three triplet levels. As the inset of Figure 1 shows, this gives rise to two regions where the singlet and one triplet state are degenerate and where intersystem crossing can thus be most efficient. First, there is a range stretching from a minimum value of r to infinity in which the exchange interaction vanishes and $|S\rangle$ and $|T_0\rangle$ are degenerate. Second, the levels of $|S\rangle$ and $|T_-\rangle$ cross at a specific value of r; in those exceptional cases when J is positive (triplet ground state) this crossing occurs between $|S\rangle$ and $|T_+\rangle$. Consequently, given appropriate mixing matrix elements any radical pair has two pathways of intersystem crossing, one by $S-T_0$ mixing, which can take place at any distance larger than a minimum distance, and one by $S-T_-$ mixing (rarely, $S-T_+$ mixing), which is possible only if the pair spends sufficient time at or around a crucial distance.

The Zeeman interaction $\mathbf{H}_Z$ and the secular part of the hyperfine interaction $\mathbf{H}_{hfc}$ possess nonzero diagonal elements for $|T_+\rangle$ and $|T_-\rangle$, and also nonvanishing off-diagonal elements between $|S\rangle$ and $|T_0\rangle$. It is therefore more convenient to decompose these operators into operators that are completely diagonal and thus describe the level splitting, and operators that are completely off-diagonal, and thus drive intersystem crossing. This decomposition is possible by forming symmetry-adapted linear combinations.

The sum of the operators for the Zeeman interaction of spins 1 and 2,

$$\mathbf{H}_{Z,\text{sym}} = \frac{(\mathbf{H}_{Z,1} + \mathbf{H}_{Z,2})}{2} = \frac{(g_1 + g_2)\beta B_0}{2}(\mathbf{S}_{1z} + \mathbf{S}_{2z}) \tag{11}$$

as well as the sum of the secular terms $\mathbf{H}'_{hfc}$ of the hyperfine interaction

$$\mathbf{H}'_{hfc,\text{sym}} = \frac{(\mathbf{H}'_{hfc,1} + \mathbf{H}'_{hfc,2})}{2} \tag{12}$$

$$= \frac{1}{2}\left(\sum_i a_i m_i + \sum_k a_k m_k\right)(\mathbf{S}_{1z} + \mathbf{S}_{2z})$$

(the index i refers to radical 1, k to radical 2) are completely diagonal in the chosen basis,

$$\mathbf{H}_{Z,\text{sym}} + \mathbf{H}'_{hfc,\text{sym}} = \text{diag}(R, 0, 0, -R) \tag{13}$$

$$R = \frac{1}{2}\left\{(g_1 + g_2)\beta B_0 + \sum_i a_i m_i + \sum_k a_k m_k\right\} \tag{14}$$

Equation 13 expresses the splitting of the triplet levels shown in the inset of Figure 1. In a situation when diffusion has effected a separation of the radical pair A˙B˙ to a distance where the system may be described in terms of individual spins (Eq. 6), R determines the mean precession frequency of A˙ and B˙ in a magnetic field.

The difference of the operators of the Zeeman interaction

$$H_{Z,\text{anti}} = \frac{(H_{Z,1} - H_{Z,2})}{2} = \frac{\Delta g \beta B_0}{2}(S_{1z} - S_{2z}) \qquad (15)$$

where $\Delta g = (g_1 - g_2)$, and of the secular terms of the hyperfine interaction

$$\begin{aligned}H'_{\text{hfc,anti}} &= \frac{(H'_{\text{hfc},1} - H'_{\text{hfc},2})}{2} \\ &= \frac{1}{2}\left(\sum_i a_i m_i - \sum_k a_k m_k\right)(S_{1z} - S_{2z})\end{aligned} \qquad (16)$$

has the matrix

$$H_{Z,\text{anti}} + H'_{\text{hfc,anti}} = \begin{pmatrix} 0 & 0 & 0 & 0 \\ 0 & 0 & Q & 0 \\ 0 & Q & 0 & 0 \\ 0 & 0 & 0 & 0 \end{pmatrix} \qquad (17)$$

with the mixing matrix element Q,

$$Q = \frac{1}{2}\left\{(g_1 - g_2)\beta B_0 + \sum_i a_i m_i - \sum_k a_k m_k\right\} \qquad (18)$$

that only joins $|S\rangle$ and $|T_0\rangle$.

Both the Zeeman interaction (Δg mechanism) and/or the hyperfine interaction (hyperfine mechanism) can thus effect intersystem crossing between $|S\rangle$ and $|T_0\rangle$. As Figure 1 and the pertaining discussion show, this process is only efficient for comparatively large separations of the radicals, where the radical pair A˙B˙ behaves like two individual radicals A˙ and B˙. The origin of $S-T_0$ mixing is seen to be a difference of the precession frequencies of A˙ and B˙. A convenient visualization of this intersystem crossing mechanism is given by vector models [6c, 23], as shown in Figure 2. While these models are not true representations of the actual quantum mechanical situation (their limitations have been discussed in the literature [1a, 2l, 24]), they are highly useful in practice. A slightly less suggestive, but exact, vector model will be derived Section II.B.4.

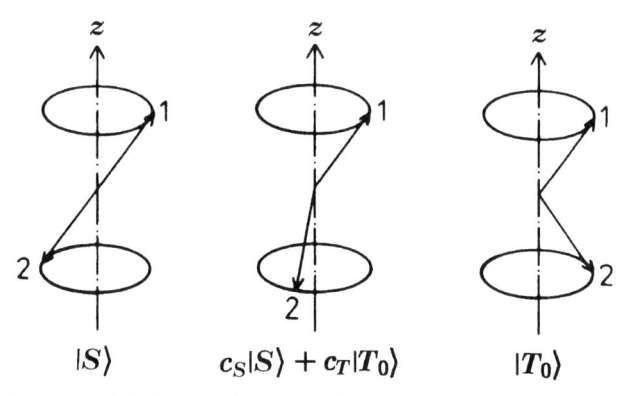

Figure 2. Vector model of the spin states of a radical pair; singlet state $|S\rangle$ (left), state $|T_0\rangle$ (right), and a superposition state (center) resulting from different precession frequencies of the two spins 1 and 2. [Adapted from ref. [6m] with permission. Copyright © 1995 John Wiley & Sons, Inc.]

From the formulas given, it is evident that $S-T_\pm$ mixing can be brought about neither by the Zeeman interaction nor by the secular terms of the hyperfine interaction. However, the nonsecular terms $\mathbf{H}''_{\text{hfc}}$ of the hyperfine interaction,

$$\mathbf{H}''_{\text{hfc}} = \left[\sum_i a_i(\mathbf{S}_{1x}\mathbf{I}_{ix} + \mathbf{S}_{1y}\mathbf{I}_{iy}) + \sum_k a_k(\mathbf{S}_{2x}\mathbf{I}_{kx} + \mathbf{S}_{2y}\mathbf{I}_{ky}) \right] \quad (19)$$

which have vanishing matrix elements between $|S\rangle$ and $|T_0\rangle$ for any nuclear spin configuration as well as vanishing diagonal elements, can cause transitions between $|S\rangle$ and $|T_-\rangle$ or $|T_+\rangle$, with accompanying change of the nuclear spin state. Besides, $\mathbf{H}''_{\text{hfc}}$ also joins $|T_-\rangle$ and $|T_+\rangle$ with $|T_0\rangle$. However, the latter transitions are inefficient except in very low magnetic fields because these three levels are separated by the Zeeman interactions.

The matrix element of $S-T_\pm$ mixing for a pair consisting of one radical without nuclear spins and one radical containing a nucleus of spin I has the value $a\sqrt{[I(I+1) - m(m \pm 1)]/8}$ [25] and is thus only slightly smaller than that of $S-T_0$ mixing by the hyperfine mechanism. Nevertheless, in high magnetic fields, $S-T_\pm$ transitions are only possible for a small range of interradical distances around the level crossing (see Fig. 1), where exchange integral J and Zeeman plus hyperfine terms R (Eq. 14) match. This mechanism is thus found in situations when diffusion is restricted, that is, for biradicals and radical pairs in micelles or very viscous solvents. Even there it is not very efficient in high fields because the angle of intersection of the potential curves is large. The only domain where intersystem crossing between $|S\rangle$ and $|T_-\rangle$ or $|T_+\rangle$ becomes important is in low magnetic fields. Under these conditions, intersystem crossing between $|S\rangle$ and any of the three triplet states is of comparable efficiency. The theoretical treatment of intersystem crossing in these circumstances yields much more complex expressions, so that in most cases solutions can only be obtained numerically; for a detailed discussion see, for example, [2i]. Here, such a discussion does not appear to be warranted because practically all CIDNP experiments reviewed in Sections IV and V of this chapter were carried out in high field.

Concerning the other interactions listed in Section II.B.3, we finally note that spin–orbit coupling plays no role for intersystem crossing of freely diffusing radical pairs, because its efficiency is negligible for most of the life of the pair, that is, unless the two radicals are very near one another [26]. In the case of biradicals, however, this mechanism is important and competes with intersystem crossing by the radical pair mechanism [23, 27].

4. Density Matrix Treatment of the Spin Dynamics of Radical Pairs. We consider intersystem crossing between the singlet state $|S\rangle$ and one triplet state $|T\rangle$. In high magnetic fields and for radical pairs that undergo free diffusion, $|T\rangle$ is equal to $|T_0\rangle$, and the nuclear spin configuration is left unchanged by singlet–triplet mixing. In low fields and for systems with restricted diffusion such as micellized radical pairs, or biradicals, $|T\rangle$ denotes either $|T_-\rangle$ (if $J < 0$) or $|T_+\rangle$ (if $J > 0$, which is exceptional), and intersystem crossing is accompanied by a nuclear spin flip. While the latter is not explicitly mentioned in the following, it has the consequence that only one-half of the radical pairs can react, namely those possessing the right nuclear configuration for this spin flip.

The spin functions $|S\rangle$ and $|T\rangle$ are orthonormal. As only these two states are mixed, they form a complete set, and the wave function $|X\rangle$ of the radical pair, which is time dependent, can be described at any moment by a superposition

$$|X(t)\rangle = c_S(t)|S\rangle + c_T(t)|T\rangle \tag{20}$$

of the stationary basis functions $|S\rangle$ and $|T\rangle$, the coefficients carrying the time dependence. In an actual experiment, many radical pairs are observed at the same time, each of them having a different wave function. However, the (time-dependent) expectation value of any property of this system of many radical pairs can be expressed as the product of two kinds of quantities only: matrix elements of the respective operator in the chosen basis $|S\rangle$ and $|T\rangle$, and averaged products of coefficients, for example, $\overline{c_S^*(t)c_T(t)}$ (the overbar denoting ensemble averaging). These quantities can be arranged in a matrix, the density matrix ϱ, of the same dimension as the operator matrices (2 × 2 in the present case). The convention for the matrix elements ϱ_{ij} is

$$\varrho_{ij} = \overline{c_i(t)c_j^*(t)} \tag{21}$$

Introductions to the density matrix formalism in magnetic resonance can be found in [28a] and [28b]. An expectation value $\langle \mathbf{O} \rangle$ of any operator $\mathbf{O}$ is obtained by multiplying the density matrix with the matrix of the operator and forming the trace,

$$\langle \mathbf{O} \rangle = \mathrm{Tr}(\varrho \mathbf{O}) \tag{22}$$

The density matrix thus describes a system as fully as possible.

The basis for the following calculations is provided by Eq. 23:

$$\frac{\partial}{\partial t} \varrho = -\frac{i}{\hbar}(\mathbf{H}\varrho - \varrho\mathbf{H}) \tag{23}$$

which states that the time dependence of the density matrix is given by its commutator with the Hamiltonian $\mathbf{H}$.

In the present case, ϱ is, in explicit notation,

$$\varrho = \begin{pmatrix} \varrho_{SS} & \varrho_{ST} \\ \varrho_{TS} & \varrho_{TT} \end{pmatrix} \tag{24}$$

and the Hamiltonian $\mathbf{H}$ of the system is

$$\mathbf{H} = \begin{pmatrix} +E & Q \\ Q & -E \end{pmatrix} \tag{25}$$

with Q being the mixing matrix element driving intersystem crossing and $2E$ being the splitting of the two levels. The actual forms of Q and E are different for S–T_0 and S–$T_\pm$ intersystem crossing, as discussed in Section II.B.3.

Carrying out the matrix multiplications of Eq. 23, and rewriting the result as a vector equation yields

$$\frac{\partial}{\partial t} \begin{pmatrix} \varrho_{SS} \\ \varrho_{ST} \\ \varrho_{TS} \\ \varrho_{TT} \end{pmatrix} = -\frac{i}{\hbar} \begin{pmatrix} 0 & -Q & Q & 0 \\ -Q & 2E & 0 & Q \\ Q & 0 & -2E & -Q \\ 0 & Q & -Q & 0 \end{pmatrix} \begin{pmatrix} \varrho_{SS} \\ \varrho_{ST} \\ \varrho_{TS} \\ \varrho_{TT} \end{pmatrix} \tag{26}$$

Physical insight can be gained by a change of variables [29]. First, for a closed system, the trace of the density matrix, that is, the sum of the squared coefficients of all basis functions, must be unity. As long as no chemical reactions are being considered, this condition is fulfilled, so instead of the individual elements ϱ_{SS} and ϱ_{TT} their difference ($\varrho_{SS} - \varrho_{TT}$) can be introduced, thus reducing the number of variables in Eq. 26 by one. Second, any density matrix must be Hermitian, so the off-diagonal elements ϱ_{ST} and ϱ_{TS} are complex conjugates of one another. As ϱ_{ST} and ϱ_{TS} cannot correspond to observables because they are, in general, complex numbers, it is advantageous to use their sum and difference as variables

$$\varrho_{ST} + \varrho_{TS} = 2\text{Re}(\varrho_{ST}) \qquad i(\varrho_{TS} - \varrho_{ST}) = 2\text{Im}(\varrho_{ST}) \qquad (27)$$

which are real quantities. With the new set of variables, Eq. 26 is transformed into

$$\frac{\partial}{\partial t}\begin{pmatrix} \varrho_{SS} - \varrho_{TT} \\ 2\text{Im}(\varrho_{ST}) \\ 2\text{Re}(\varrho_{ST}) \end{pmatrix} = \frac{2}{\hbar}\begin{pmatrix} 0 & -Q & 0 \\ Q & 0 & -E \\ 0 & E & 0 \end{pmatrix}\begin{pmatrix} \varrho_{SS} - \varrho_{TT} \\ 2\text{Im}(\varrho_{ST}) \\ 2\text{Re}(\varrho_{ST}) \end{pmatrix} \qquad (28)$$

The physical significance of the new variables is seen with the aid of Eq. 22, by searching for the operators that filter out the respective terms of ϱ in trace formation. We denote these operators as $\mathbf{O}_a$ for ($\varrho_{SS} - \varrho_{TT}$), $\mathbf{O}_b$ for $2\text{Re}(\varrho_{ST})$, and $\mathbf{O}_c$ for $2\text{Im}(\varrho_{ST})$.

It follows that

$$\mathbf{O}_a = \begin{pmatrix} 1 & 0 \\ 0 & -1 \end{pmatrix} \qquad (29)$$

which in both bases is the matrix of $-(\frac{1}{2} + \mathbf{S}_1 \cdot \mathbf{S}_2)$, that is, that of the exchange operator $\mathbf{H}_{ex}$ divided by one-half the energy of singlet–triplet splitting. Hence, ($\varrho_{SS} - \varrho_{TT}$) is the *population difference between singlet and triplet state*, because $\mathbf{H}_{ex}$ yields the exchange energy, that is, the population of the triplet state weighted with $+J$ plus the population of the singlet state weighted with $-J$. As is further seen from the wave functions of $|T_\pm\rangle$ and $|T_0\rangle$, the former contains a surplus of spins of one sort (e.g., of β spins in the case of $|T_-\rangle$) while the latter does not. Only in intersystem crossing between $|S\rangle$ and $|T_-\rangle$ or $|T_+\rangle$ is the population difference thus accompanied by an *absolute electron spin polarization*. While being intuitively obvious this is also reflected by the fact that the matrix of the operator ($\mathbf{S}_{1z} + \mathbf{S}_{2z}$)

vanishes in the basis $|S\rangle$ and $|T_0\rangle$, whereas it is

$$S_{1z} + S_{2z} = \begin{pmatrix} 0 & 0 \\ 0 & \pm 1 \end{pmatrix}$$

in the bases $|S\rangle$ and $|T_\pm\rangle$, and thus filters out ϱ_{TT} from the density matrix. The meaning of the operators $\mathbf{O}_b$ and $\mathbf{O}_c$,

$$\mathbf{O}_b = \begin{pmatrix} 0 & 1 \\ 1 & 0 \end{pmatrix} \quad \mathbf{O}_c = \begin{pmatrix} 0 & +i \\ -i & 0 \end{pmatrix} \tag{30}$$

depends on the situation considered. With respect to intersystem crossing between $|S\rangle$ and $|T_0\rangle$, $\mathbf{O}_b$ is equal to $(S_{1z} - S_{2z})$, and $\mathbf{O}_c$ is equal to $2(S_{1x}S_{2y} - S_{1y}S_{2x})$. The quantity $2\text{Re}(\varrho_{ST})$ is therefore an *opposite electron spin polarization* of both electrons. On the other hand, $2\text{Im}(\varrho_{ST})$ describes a *phase correlation* that is not directly observable. In contrast, for intersystem crossing between $|S\rangle$ and $|T_\pm\rangle$ $\mathbf{O}_b$ equals $\pm\sqrt{2}(S_{1x} - S_{2x})$, the leading minus sign being valid if the triplet function is $|T_+\rangle$, and $\mathbf{O}_c$ equals $\sqrt{2}(S_{1y} - S_{2y})$, so in this case $2\text{Re}(\varrho_{ST})$ and $2\text{Im}(\varrho_{ST})$ are *opposite transversal magnetizations* of the two electrons, with orthogonal phases.

After clarifying the meaning of the variables in the density matrix of Eq. 28, the solution of its equations of motion shall be given. The result, which can be obtained by standard methods, is

$$\begin{pmatrix} \varrho_{SS} - \varrho_{TT} \\ 2\text{Im}(\varrho_{ST}) \\ 2\text{Re}(\varrho_{ST}) \end{pmatrix} \tag{31}$$

$$= \begin{pmatrix} \dfrac{E^2}{V^2} + \dfrac{Q^2}{V^2}\cos(\Omega t) & -\dfrac{Q}{V}\sin(\Omega t) & \dfrac{QE}{V^2}[1 - \cos(\Omega t)] \\ \dfrac{Q}{V}\sin(\Omega t) & \cos(\Omega t) & -\dfrac{E}{V}\sin(\Omega t) \\ \dfrac{QE}{V^2}[1 - \cos(\Omega t)] & \dfrac{E}{V}\sin(\Omega t) & \dfrac{Q^2}{V^2} + \dfrac{E^2}{V^2}\cos(\Omega t) \end{pmatrix} \cdot \begin{pmatrix} \varrho_{SS} - \varrho_{TT} \\ 2\text{Im}(\varrho_{ST}) \\ 2\text{Re}(\varrho_{ST}) \end{pmatrix}_{t=0}$$

with the abbreviations

$$V = \sqrt{Q^2 + E^2} \quad \Omega = \frac{2V}{\hbar}$$

While for constant Q and E and for given initial conditions, Eq. 31 allows straightforward calculation of the components of ϱ at any time t, a physical picture of the change of ϱ can be obtained by rewriting Eq. 28 as a cross-product of two vectors, as was first pointed out by Adrian [29],

$$\frac{\partial}{\partial t}\begin{pmatrix} \varrho_{SS} - \varrho_{TT} \\ 2\text{Im}(\varrho_{ST}) \\ 2\text{Re}(\varrho_{ST}) \end{pmatrix} = \begin{pmatrix} 2E/\hbar \\ 0 \\ 2Q/\hbar \end{pmatrix} \times \begin{pmatrix} \varrho_{SS} - \varrho_{TT} \\ 2\text{Im}(\varrho_{ST}) \\ 2\text{Re}(\varrho_{ST}) \end{pmatrix} \quad (32)$$

which is seen to be a set of Bloch equations (without relaxation terms). This can be utilized to construct a vector model that allows one to visualize the spin dynamics of an ensemble of radical pairs in a concrete manner while being physically exact: The density matrix of this system is depicted as a vector $\vec{\varrho}$. Its Cartesian components denote

$$\vec{\varrho} = \begin{pmatrix} \text{population difference between singlet and triplet states} \\ \text{phase correlation (not directly observable)} \\ \text{opposite polarization of both electron spins} \end{pmatrix} \quad (33)$$

in the case of intersystem crossing between $|S\rangle$ and $|T_0\rangle$, whereas for intersystem crossing between $|S\rangle$ and $|T_-\rangle$, or $|T_+\rangle$, they mean

$$\vec{\varrho} = \begin{pmatrix} \text{population difference between singlet and triplet states} \\ \text{opposite } x \text{ magnetization of both electron spins} \\ \text{opposite } y \text{ magnetization of both electron spins} \end{pmatrix} \quad (34)$$

Mixing interaction Q and level splitting E form the z and x components of a vector $\mathscr{V}$, the y component being zero. The time evolution of the density matrix is then a precession of $\vec{\varrho}$ around the axis $\mathscr{V}$.

5. Combining Spin Dynamics and Radical Pair Dynamics. In a very general form, spin dynamics and radical pair dynamics can be treated simultaneously with the so-called stochastic Liouville equation [10b],

$$\frac{\partial}{\partial t}\varrho = -\frac{i}{\hbar}[\mathbf{H}, \varrho] + D\Gamma\varrho + \mathbf{K}\varrho + \mathbf{R}\varrho \quad (35)$$

Equation 35 is an extension of Eq. 23. The additional terms describe diffusion ($D\Gamma\varrho$), chemical reactions ($\mathbf{K}\varrho$), and relaxation ($\mathbf{R}\varrho$). While Eq. 35 is the physically most consistent description of the radical pair mechanism, it can only be solved numerically. A review is given in [1a].

In the case of intersystem crossing between $|S\rangle$ and $|T_0\rangle$, much more physical insight can be obtained by an approximate treatment [30]. Use is

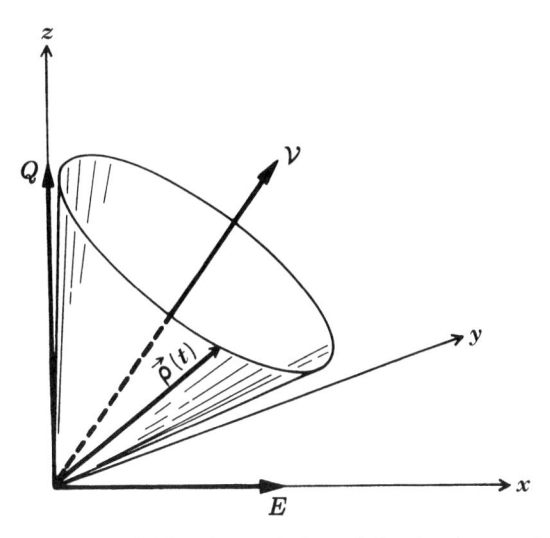

Figure 3. Exact vector model for the evolution of the density matrix $\vec{\varrho}$ of a radical pair. The meaning of the components of $\vec{\varrho}$ is given by Eq. 33 or Eq. 34, depending on the intersystem crossing mechanism. For further explanation, see the text. [Adapted from ref. [10i] with permission. Copyright © 1995 John Wiley & Sons, Inc.]

made of the fact that the exchange integral J is a strong function of the distance between the radical pairs (Eq. 10). Hence, during a diffusive excursion there are almost instantaneous transitions between a region called the "exchange region" for which the level splitting $2J$ is much larger than the mixing matrix element Q (Eq. 18), and a region extending to infinity for which J is negligible compared to Q.

Within the exchange region the axis $\mathscr{V}$ in the vector model of Figure 3 practically coincides with the x axis of the coordinate system. As is obvious from the figure, the precession of $\vec{\varrho}$ around $\mathscr{V}$ therefore leaves unchanged the population difference between $|S\rangle$ and $|T_0\rangle$; only electron spin polarization and phase correlation are mixed in this region. Under these circumstances ($J \gg Q$), Eq. 31 is reduced to

$$\vec{\varrho}(t) = \begin{pmatrix} 1 & 0 & 0 \\ 0 & \cos(2Jt/\hbar) & -\sin(2Jt/\hbar) \\ 0 & \sin(2Jt/\hbar) & \cos(2Jt/\hbar) \end{pmatrix} \cdot \vec{\varrho}(0) \qquad (36)$$

Depending on the strength of the exchange interaction and the average time τ the radicals spend in the exchange region, two limiting cases can be distinguished [29]. For weak exchange, that is, $2J\tau/\hbar \ll 1$, the rotation matrix of Eq. 36 turns into a unit matrix, so $\vec{\varrho}$ does not change at all. For strong exchange ($J\tau/\hbar \gg 1$), on the other hand, electron spin polarization $2\text{Re}(\varrho_{ST})$ and phase correlation $2\text{Im}(\varrho_{ST})$ are destroyed, because during the dwell time in the exchange region the vector $\vec{\varrho}(t)$ is rotated several times, and, as the rotation angles are statistically distributed, is spread out in the yz plane.

Outside the exchange region ($J \approx 0$), the axis $\mathscr{V}$ coincides with the x axis, so only population difference ($\varrho_{SS} - \varrho_{TT}$) and phase correlation $2\text{Im}(\varrho_{ST})$ are mixed. Equation 31 becomes

$$\vec{\varrho}(t) = \begin{pmatrix} \cos(2Qt/\hbar) & -\sin(2Qt/\hbar) & 0 \\ \sin(2Qt/\hbar) & \cos(2Qt/\hbar) & 0 \\ 0 & 0 & 1 \end{pmatrix} \cdot \vec{\varrho}(0) \qquad (37)$$

As is seen from the preceding equations and the pertaining discussion, the amount of intersystem crossing depends primarily on the total time spent outside the exchange region. For an ensemble of radical pairs there is a statistical distribution of these times, which is described by a conditional probability density $f(t, d, r_0)$. For two radicals initially separated by a distance r_0, $f(t, d, r_0)$ gives the probability of a first arrival at a distance d in the period of time between t and $t + dt$. Different boundary conditions (values of r_0 and d) and diffusional models (continuous diffusion vs. diffusion in discrete steps) have been used in the literature [2i, 30–32] (cf. also [1a]). However, for any functional form of $f(t, d, r_0)$ one can calculate an average density matrix $\bar{\bar{\varrho}}$ for an ensemble of radical pairs, in which every pair either has completed one diffusive excursion leading from r_0 to d or has separated forever,

$$\bar{\bar{\varrho}} = \begin{pmatrix} c & -s & 0 \\ s & c & 0 \\ 0 & 0 & p \end{pmatrix} \cdot \vec{\varrho}(0) \qquad (38)$$

where c and s are the cosine and sine transforms of $f(t, d, r_0)$ at the angular frequency of intersystem crossing $2Q/\hbar$, and p is the total probability of reencounter,

$$c = \int_0^\infty \cos(2Qt/\hbar) f(t, d, r_0) dt \qquad (39)$$

$$s = \int_0^\infty \sin(2Qt/\hbar) f(t, d, r_0) dt \qquad (40)$$

$$p = \int_0^\infty f(t, d, r_0) dt \qquad (41)$$

A frequently used form of $f(t, d, r_0)$ has been derived by Noyes [14c],

$$f(t, d, d) = \frac{p(1-p)d}{\sqrt{4\pi D}} t^{-3/2} \exp\left[-\frac{(1-p)^2 d^2}{4Dt}\right] \quad (42)$$

The parameter D is the interdiffusion coefficient. Equation 42 is valid for $r_0 = d$ (hence, if the processes in the exchange region can be neglected) and diffusion occurring by steps of length $\lambda_D (\lambda_D < d)$ and in the absence of attractive or repulsive forces between the radicals. With this model, the probability p is, to a very good approximation, given by [33]

$$p \approx 1 \Big/ \left(1 + \frac{2\lambda_D}{3d}\right) \quad (43)$$

and the Fourier transforms are

$$c = \cos[(1-p)\delta] \exp[-(1-p)\delta] \quad (44)$$
$$s = \text{sgn}(Q) \sin[(1-p)\delta] \exp[-(1-p)\delta] \quad (45)$$

with $\text{sgn}(Q)$ meaning the sign of Q, and

$$\delta = \sqrt{\frac{2|Q|d^2}{D}} \quad (46)$$

For a small mixing matrix element Q and a large diffusion coefficient (i.e., nonviscous solvents), an expansion of Eqs. 44 and 45 to first order is sufficient, which leads to $c \approx 1 - (1-p)\delta$, and $s \approx \text{sgn}(Q) \cdot (1-p)\delta$.

When chemical reactions that occur preferentially from one multiplicity (i.e., geminate reactions) are to be described, it is necessary to separate the variables ϱ_{SS} and ϱ_{TT} again. This would give a vector $\vec{\varrho}$ of four components; however, unless CIDEP phenomena are of interest, the electron spin polarization $2\text{Re}(\varrho_{ST})$ can be neglected, because, as the discussions of Eqs. 36 and 37 have shown, this component can only be generated within the exchange region, and there it does not play any role in the limits of weak or strong exchange. To avoid unwieldy expressions, a vector $\vec{\varrho}'$ of three components,

$$\vec{\varrho}'(t) = \begin{pmatrix} \varrho_{SS} \\ \varrho_{TT} \\ 2\text{Im}(\varrho_{ST}) \end{pmatrix}(t) \quad (47)$$

is, therefore, used for the remainder of this section.

The reencounter formalism [10b, 30], which was introduced into radical pair theory by Adrian [16c], allows the treatment of geminate reactions in a formal, yet fairly transparent, way. For simplicity, it will be assumed in the following that only singlet pairs react; extension to reactions of both singlets and triplets is straightforward. Let λ be the probability of geminate reaction of singlet pairs. Denote the density matrix at the moment of the nth encounter as $\vec{\varrho}'_{n,\text{before}}$. In this encounter, an amount F_n of geminate product is obtained,

$$F_n = (\lambda\ 0\ 0\ 0)\vec{\varrho}'_{n,\text{before}} \tag{48}$$

and the density matrix $\vec{\varrho}'_{n,\text{before}}$ is changed into $\vec{\varrho}'_{n,\text{after}}$

$$\vec{\varrho}'_{n,\text{after}} = \hat{\mathsf{R}}\vec{\varrho}'_{n,\text{before}} \tag{49}$$

with $\hat{\mathsf{R}}$ being a diagonal matrix. In Pedersen's treatment [30], weak exchange was assumed, so $\hat{\mathsf{R}}$ was chosen as diag$(1 - \lambda, 1, 1)$. This form of $\hat{\mathsf{R}}$ has been criticized on the grounds that the removal of singlet radical pairs from the system must also be accompanied by a decrease of the phase correlation $2\text{Im}(\varrho_{ST})$, and it was suggested that $\hat{\mathsf{R}}$ should be chosen as diag$(1 - \lambda, 1, \sqrt{1 - \lambda})$ [21]. In the case of strong exchange, $\hat{\mathsf{R}}$ is diag$(1 - \lambda, 1, 0)$.

After the encounter, those radicals that have not reacted enter on a diffusive excursion, during which intersystem crossing can take place. The evolution of the density matrix up to the moment of the next encounter is described by the action of a mixing matrix $\hat{\mathsf{M}}$, for example, the one given in Eq. 37, on $\vec{\varrho}'_{n,\text{after}}$, which also takes into account that a fraction of the radical pairs separates permanently. In the new basis of Eq. 47, Eq. 37 becomes

$$\hat{\mathsf{M}} = \begin{pmatrix} (1+c)/2 & (1-c)/2 - s/2 & 0 \\ (1-c)/2 & (1+c)/2 + s/2 & 0 \\ s & -s & c \end{pmatrix} \tag{50}$$

At the moment of the $(n+1)$th encounter, one thus has, dropping the subscript "before,"

$$\vec{\varrho}'_{n+1} = \hat{\mathsf{M}}\hat{\mathsf{R}}\vec{\varrho}'_n \tag{51}$$

which yields a quantity F_{n+1} of geminate product (cf. Eq. 48). Summing up over all reencounters, one obtains

$$F_{tot} = \sum_{k=0}^{\infty} F_k = (\lambda \ 0 \ 0 \ 0) \left[\sum_{k=0}^{\infty} (\hat{M}\hat{R})^k \right] \vec{\varrho}_0' \quad (52)$$

where $\vec{\varrho}_0$ is the density matrix at the moment of generation of the pairs. The infinite series of matrices of Eq. 52 converges because the diagonal elements are smaller than unity; its value is formally identical with the sum of a geometric series,

$$\sum_{k=0}^{\infty} (\hat{M}\hat{R})^k = (\hat{E} - \hat{M}\hat{R})^{-1} \quad (53)$$

where the exponent -1 denotes matrix inversion.

With $\hat{R}$ being either $\text{diag}(1 - \lambda, 1, 1)$ or $\text{diag}(1 - \lambda, 1, 0)$, spin evolution and chemical reactivity can be separated [21] by introducing [30] a quantity F^* that gives the amount of geminate product formed for $\lambda = 1$ from radical pairs with a triplet precursor without initial phase correlation, and a quantity Λ, the spin independent reaction probability,

$$\Lambda = \lambda \cdot \sum_{k=0}^{\infty} p^k (1 - \lambda)^k = \frac{\lambda}{1 - p(1 - \lambda)} \quad (54)$$

As long as this separation is possible, arbitrary initial populations of the singlet and triplet state of the pair can be treated by using [30]

$$F(|T_0\rangle) = \frac{\Lambda F^*}{1 + F^*(1 - \Lambda)} \quad (55)$$

$$F(|S\rangle) = \frac{\Lambda}{1 + F^*(1 - \Lambda)} \quad (56)$$

$$F(\text{RI}) = \frac{\Lambda(1 + F^*)/2}{1 + F^*(1 - \Lambda)} \quad (57)$$

where RI denotes F pairs.

Tedious manipulation of Eq. 50 yields, for $\hat{R}$ equal to $\text{diag}(1 - \lambda, 1, 1)$,

$$F^* = \frac{p[1 - c + p(c^2 + s^2 - c)]}{2 - p(1 + 3c) + p^2(c^2 + s^2 + c)} \quad (58)$$

For the fast-exchange case [$\hat{R} = \text{diag}(1 - \lambda, 1, 0)$], a much simpler expression is obtained,

$$F^* = \frac{p(1-c)}{2 - p(1+c)} \quad (59)$$

These relations are independent of the diffusion model used. When c and s are given by Eqs. 44–46, an expansion to first order, which is permissible in solvents of low viscosity, in both cases leads to the same, very simple, expression,

$$F^* = \frac{p\delta}{2} \quad (60)$$

Extensions of this model have been developed, for instance, to take into account the exchange interaction [32]. However, the facts that such refined models require more parameters, that the uncertainty of these parameters is usually rather large, and that the observables, product yields or spin polarizations, are only global effects, tend to reduce their practical value.

Intersystem crossing between $|S\rangle$ and $|T_-\rangle$ or $|T_+\rangle$ is somewhat more involved because, as is evident from Figure 1, the level splitting $2E$ is nonzero both for small separations of the radicals ($2E \approx 2J$) and for large separations ($2E = \pm g\beta B_0$, the negative sign being valid for $S-T_{+1}$ mixing, i.e., positive J). In both limits, the mixing matrix element is much smaller than $2E$, so intersystem crossing is negligible, as the vector model of Figure 3 shows: The axis $\mathscr{V}$ is oriented along the positive or negative x axis, and the population difference between singlet and triplet, that is, the x component of the vector $\vec{\varrho}$, is not changed by rotation of $\vec{\varrho}$ about this axis. Only around a critical distance R_c, for which $J = \pm g\beta B_0$, is the level splitting small enough to allow efficient intersystem crossing ($\mathscr{V}$ aligned approximately parallel or antiparallel with the z axis of Fig. 3); this range is rather short because the exchange interaction varies so strongly with distance. In the limit of dominating exchange interaction as well as in that of dominating Zeeman splitting, the y and z components of $\vec{\varrho}$ are mixed. Hence, randomization of these components occurs in the same way as in the strong-exchange situation of $S-T_0$ intersystem crossing, and no transversal electron spin magnetization is observed. In the case of biradicals or radical pairs in micelles, quantitative estimates of the efficiency of $S-T_\pm$ intersystem crossing efficiency have only been obtained numerically; for freely diffusing radical pairs, a closed-form expression has been derived by an asymptotic solution of the stochastic Liouville equation [34].

III. CIDNP EFFECTS

As stated above, CIDNP denotes the transient occurrence of anomalous line intensities in NMR spectra recorded during chemical reactions or shortly after their completion. The phenomenon was first observed in 1967 by Bargon, Fischer and Johnsen [35a] in thermal decompositions of peroxides and azo compounds, and, independently, by Ward and Lawler [35b] in the reactions of alkyl lithium with alkyl halides. It was immediately realized that the line anomalies are caused by populations of the nuclear spin states in the reaction products that deviate from the Boltzmann populations. After initial attempts of interpreting CIDNP by electron–nuclear cross-relaxation, the radical pair mechanism was developed in 1969 by Kaptein and Oosterhoff [36a], and independently by Closs [36b].

A. Explanation by the Radical Pair Mechanism

The basic reaction scheme for all CIDNP experiments is shown in Chart I. $R_1^{\bullet} R_2^{\bullet}$ may also mean a biradical.

CIDNP arises through the interplay of three processes (see labels ①–③ in Chart I), the third of which is only necessary in the case of $S-T_0$-type CIDNP.

① Generation of a radical pair or biradical in a definite electron spin state. This has already been expounded on in Section II.A, item 1.

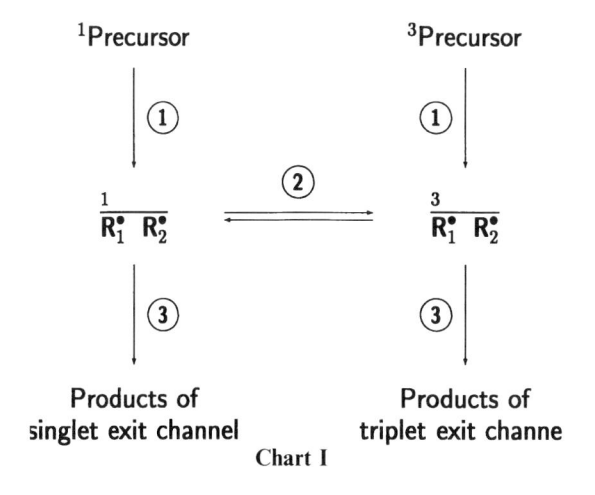

<pre>
       ¹Precursor   ³Precursor
            │   │
           ①│   ①│
            │   ②   │
         ¹R₁• R₂•   ⇌   ³R₁• R₂•
            │   │
           ③│   ③│
            │   │
       Products of   Products of
    singlet exit channel   triplet exit channe
</pre>

Chart I

② Nuclear spin dependence of intersystem crossing (or rather, in the light of Section II.A, item 3, *oscillations* between singlet and triplet with a frequency that is modulated by the nuclear spin state). The mechanisms of this have been discussed in detail in Section II.B.3.

③ Electron spin selective reactions of the radical pairs or biradicals (see Section II.A, item 2).

A prerequisite of $S-T_0$-type CIDNP is that the two electron spin states possess different exit channels leading to different products. A difference in reaction probabilities is sufficient for this, which is usually realized by the competition of a spin-dependent reaction (geminate reaction) and a spin-independent reaction (escape from the cage in the case of radical pairs, scavenging of one radical center in the case of biradicals). Given this condition, those nuclear spin states that decrease the intersystem crossing frequency ω_{isc} lead to preferential removal of pairs of the starting multiplicity from the system, those that increase ω_{isc} to preferential removal of pairs of the other multiplicity. Speaking pictorially, for the former nuclear spin states entry channel to and exit channel from the radical pair or biradical tend to lie on the same side of the diagram in Chart I; for the latter states, there is a slight preference of traversing the diagram diagonally. Consequently, the nuclear spin states are *sorted:* They are distributed among the products in such a way that an overpopulation of a particular nuclear spin state in the products of one exit channel is accompanied by an underpopulation, to precisely the same degree, of that nuclear spin state in the products of the other exit channel. The polarizations from the two exit channels are thus exactly opposite.

The situation is quite different with $S-T_{\pm}$-type CIDNP because nuclear spins are *flipped* in that case. Owing to the coupling of nuclear spin motion and electron spin motion, not only the electron spin state oscillates in such a system but also the nuclear spin state. Since, however, one-half of the pairs or biradicals cannot participate in this because their nuclear spin state does not allow an electron–nuclear flip–flop transition, the oscillation is not symmetrical. Its turning points are zero nuclear spin polarization and 100% nuclear spin polarization of one sign only. In contrast, the distribution of nuclear spin polarizations between singlet and triplet members of the ensemble is symmetrical. As an example, consider an ensemble of biradicals, where each biradical contains a single proton. Let the ensemble be created in the state $|T_-\rangle$, and without initial nuclear spin polarization. Half of the pairs, namely those that have nuclear spin $|\beta\rangle$, cannot undergo flip–flop transitions. The others oscillate between $|T_-\alpha\rangle$ and $|S\beta\rangle$. When all of those happen to be in $|S\beta\rangle$, every nuclear spin of the triplet biradicals and every nuclear spin of the singlet biradicals is $|\beta\rangle$; when all of those are in state $|T_-\alpha\rangle$, there is no nuclear spin polarization, which was the initial condition; half-way in between, 25% of the biradicals are in the singlet state, all of them possessing nuclear spin $|\beta\rangle$, and 75% of the biradicals are in the triplet state, two-thirds of them having nuclear spin state $|\beta\rangle$ and one-third $|\alpha\rangle$, so there is the same overpopulation of nuclear spin state $|\beta\rangle$ in singlet and triplet biradicals; a surplus of nuclei with spin $|\alpha\rangle$ can never arise. The spin flips thus have two consequences. One is that different exit channels are not needed for $S-T_\pm$-type CIDNP to arise. The other is that even in cases when two exit channels are available, the polarizations from them are exactly equal.

B. Relationship between Polarizations and Line Intensities

Sampling the polarizations created by the CIDNP effect with a continuous wave NMR spectrometer is straightforward because with that detection scheme the line intensity of a transition $L^{(rs)}$ between nuclear spin levels r and s is proportional to the population difference $(P_r - P_s)$ of these levels only. However, with pulsed detection, as is now universally employed, $L^{(rs)}$ depends on the population differences across all parallel transitions between levels t and u including the one between r and s, and also in a complicated manner on the tip angle ϑ of the observation pulse. For a weakly coupled system of N spins, one has [37]

$$L^{(rs)} = \sin\vartheta \sum_{(tu)} \left(\cos^2\frac{\vartheta}{2}\right)^{(N-1-\Delta_{rstu})} \left(\sin^2\frac{\vartheta}{2}\right)^{\Delta_{rstu}} \frac{1}{2}(P_t - P_u) \qquad (61)$$

with Δ_{rstu} being the number of spins one must flip to bring the transitions rs and tu to coincide. Equation 61 is only valid when no coherences of the nuclear spins are present before the observation pulse, which is normally the case in CIDNP spectroscopy.

Although Eq. 61 can be used to calculate line intensities from given populations, it is not very transparent. Much more physical insight is obtained by using the well-known product operator formalism of NMR spectroscopy to decompose the polarizations into contributions of different product operators, as was first proposed by Fischer et al. [38]. The advantages are that each of these product operators can immediately be related to an intensity pattern and has a very simple tip angle dependence. An in-depth treatment aimed at the nonspecialist can be found in [39]. Because such a detailed explanation is outside the scope of this chapter, an

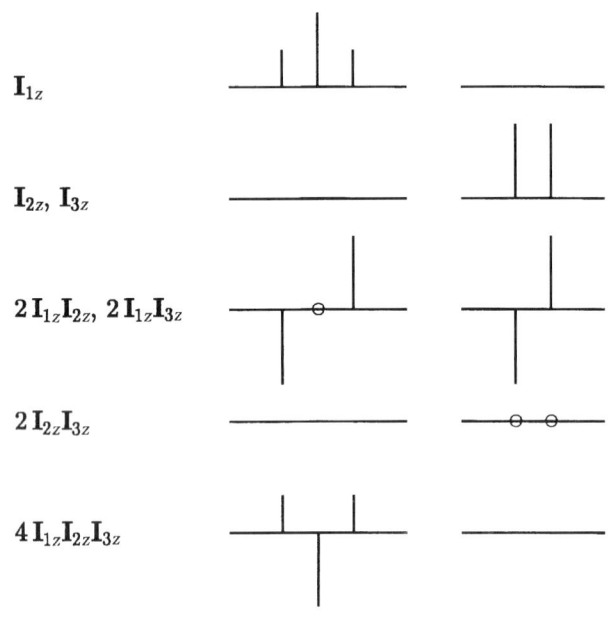

Figure 4. Intensity patterns in an AX$_2$ spin system (left, resonances of A; right, of X$_2$) resulting from different longitudinal spin order. Spin 1 is that of the A nucleus. Circles at line positions mean that the respective transition is unobservable owing to cancellation (e.g., $2I_{2z}I_{3z}$ would be observable in an AMX spin system). For further explanation, see the text.

illustrative example shall suffice. Consider an AX$_2$ system (spins 1, 2, and 3). Any population of its eight nuclear spin levels can be expressed as a linear combination of the following eight orthogonal operators, the nonvanishing elements of which have the values $+1$ or -1 only, and all lie on the diagonal: the unit operator, three operators for longitudinal 1-spin order (I_{1z}, I_{2z}, and I_{3z}), three operators for longitudinal 2-spin order ($2I_{1z}I_{2z}$, $2I_{1z}I_{3z}$, and $2I_{2z}I_{3z}$), and one operator for longitudinal 3-spin order ($4I_{1z}I_{2z}I_{3z}$). Figure 4 displays the characteristic intensity pattern associated with each operator from the latter three categories in this spin system (the unit operator does not give rise to any NMR signal).

When the ratios of populations within the spin system correspond to thermal equilibrium but the total population does not—this situation is called a CIDNP net effect—only 1-spin order is present. Hence, it is meaningful to speak of the polarization of a particular nucleus in this case. The appearance of the signals is the same as in a normal NMR spectrum, the relative intensities within multiplets being unchanged. However, a signal group (e.g., I_{1z} in the figure) as a whole might be scaled by a different factor, positive or negative, than the signal of another nucleus (e.g., I_{2z} and I_{3z}).

2-spin order is called a CIDNP multiplet effect. It is obvious that this type of polarization cannot be ascribed to a particular nucleus. Relative intensities within multiplets are different from those in a normal NMR spectrum (cf. the figure). The two multiplets look alike only if the spin system is of type AX, A_2X_2, A_3X_3, and so on. Also, coupling to other nuclei can cause additional, and possibly different, splittings of the patterns. In any case, however, the integrated intensity of each multiplet is zero, and the symmetry is preserved. Higher spin order can arise in systems of three and more coupled nuclei but the intensity of these higher multiplet effects is generally small.

The tip angle dependence of an n-spin order operator is given by $\sin\vartheta \cos^{n-1}\vartheta$. It is thus immediately seen that the intensity of a CIDNP net effect ($n = 1$) has the same tip angle dependence as the intensities in a normal NMR spectrum, proportionality to $\sin\vartheta$. Furthermore, all multiplet and higher multiplet effects vanish when 90° pulses are used; in contrast, when ϑ is quite small, the cosine terms are nearly unity and almost the same relative line intensities are obtained as in a continuous wave NMR experiment. Finally, odd and even spin orders can be filtered out by adding or subtracting, respectively, two spectra acquired with tip angles of ϑ and $90° - \vartheta$. In practice, this is mostly used to separate net and multiplet effects; for that case, best sensitivity is obtained by setting $\vartheta = 45°$.

C. Estimations of CIDNP Intensities and Phases

Quantitative calculations of the CIDNP effect can be performed as described in Section II.B.5. One has to set up the nuclear spin system of the intermediate radical pair or biradical, choose a diffusional model, compute reaction probabilities for every nuclear spin state by solving the stochastic Liouville equation numerically or approximately, establish a correlation between the nuclear spin states in the paramagnetic intermediates and the nuclear spin states in the products to obtain the populations of the latter, and finally apply Eq. 61 or the formalism of the preceding section to get line intensities. This approach, which for all but the simplest systems is impracticable except on a computer, is often necessary; with the usual uncertainty of the parameters entering the calculations of the radical pair mechanism, a reasonable accuracy can be expected. However, qualitative relationships between signal intensities, especially signal phases, and parameters of the reaction mechanism as well as magnetic properties of the intermediates are even more valuable because they provide insight into the factors that govern the CIDNP effects, and often yield direct access to chemically relevant information in a quick and transparent way.

From the above discussion of Chart I, a sort of topological criterion can be derived for $S-T_0$-type CIDNP. As has emerged, all nuclear spin states that favor intersystem crossing are overpopulated in the products of the exit channel on the opposite side as the entry channel in the scheme, and underpopulated in the products of the exit channel on the same side. This relationship can be described by assigning a parameter μ to the precursor multiplicity, that is, the entry channel ($\mu = +1$, triplet precursor; $\mu = -1$, singlet precursor), and a parameter ε to the exit channel ($\varepsilon = +1$, product formation from singlet pairs or biradicals; $\varepsilon = -1$, from triplet intermediates). Only the product $\mu\varepsilon$ plays a role for the polarizations. Separating the two contributions is valuable nevertheless, because independent information about μ or ε is often available from other experiments or thermodynamic considerations. For instance, in the majority of cases geminate reaction is energetically feasible in the singlet state only. If this condition holds, triplet radical pairs cannot react; they eventually separate and form other products by reactions of the individual radicals, for example, with the solvent. Because this case is so frequent, the products of the singlet exit channel are usually called cage products and the products of the triplet exit channel escape products, and ε is accordingly taken to denote the kind of product formed. However, as exceptions happen (cf. Sections V.A.2 and V.D.3 for formation of cage products from triplet radical pairs) the interpretation of ε as the exit channel is preferable.

The influence of the magnetic parameters on $S-T_0$-type CIDNP can be rationalized by vector models similar to those shown in Figure 2. It is advantageous to draw only projections on the xy plane in coordinate systems rotating with the mean Larmor frequency $R/\hbar$ (Eq 14). After a time t, the two electron spins will have experienced a change of relative phase by an angle $Qt/\hbar$, the mixing matrix element Q being defined in Eq. 18. Their projection on the starting state is given by $\cos^2 Qt/\hbar$, that on the other multiplicity by $\sin^2 Qt/\hbar$. Consider as an example a radical pair of initial multiplicity triplet with one proton contained in radical 1. The hyperfine coupling constant a of this proton shall be positive, and the g value of radical 1 shall be larger than that of radical 2. After t, the electron spins have fallen more out of step (higher amount of singlet character) when the spin state of the proton is $|\alpha\rangle$ than when it is $|\beta\rangle$ (see Fig. 5). Therefore, $|\alpha\rangle$ is overpopulated in a product of the singlet exit channel, and its NMR spectrum displays an absorption line.

Had a been negative, or, alternatively, g_2 been larger than g_1, an emission line would have been obtained; on the other hand, a being negative and, at

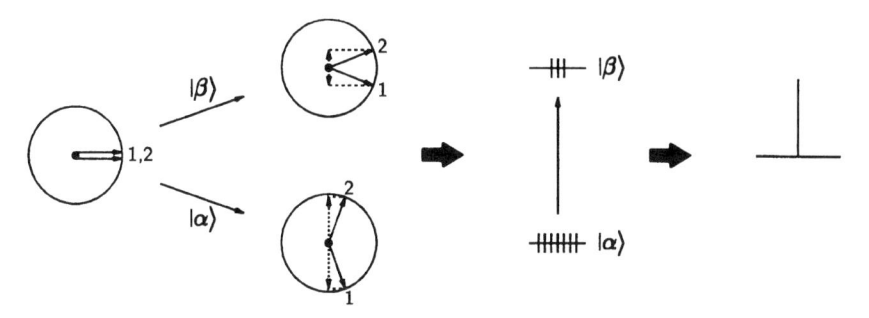

Figure 5. Explanation of an $S-T_0$-type CIDNP net effect with vector models (left), resulting schematic population diagram (center), and NMR spectrum (right). The example describes a radical pair with one proton in radical 1, triplet precursor, product of the singlet exit channel, $g_1 > g_2$, and positive hyperfine coupling constant. For the vector models, a clockwise sense of precession has been chosen, the labels 1 and 2 designate the radical and $|\alpha\rangle$ and $|\beta\rangle$ the nuclear spin state, and the dotted vertical lines in the projections give the amount of singlet character. For further details, see the text.

the same time, g_2 being larger than g_1 would give the same result as in the first example. All this is easily seen by drawing the appropriate vector diagram with the aid of Eq. 18 for the mixing matrix element. These diagrams also show that an $S-T_0$-type CIDNP net effect can only arise if both a and $(g_1 - g_2)$ ar nonzero. Finally, groups of equivalent protons can be treated in the same way by the particle spin approach, which leads to the same qualitative results as above.

The phase Γ_i of an $S-T_0$-type net effect of proton i can thus be predicted with a simple sign rule that was formulated by Kaptein [40],

$$\Gamma_i = \mu \times \varepsilon \times \mathrm{sgn}\Delta g \times \mathrm{sgn} a_i \tag{62}$$

$\Gamma_i = +1$ denotes absorption, $\Gamma_i = -1$ emission; Δg is $g_1 - g_2$. To remove an ambiguity in $\mathrm{sgn}\Delta g$, the radical containing the proton considered must be taken as radical 1.

A more quantitative estimation of the polarization in this model system can be obtained from Eqs. 18, 46, and 60,

$$P_\alpha - P_\beta \propto \sqrt{|Q_\alpha|} - \sqrt{|Q_\beta|} \propto \sqrt{|\Delta g \beta B_0 + a/2|} - \sqrt{|\Delta g \beta B_0 - a/2|} \tag{63}$$

From this, it is obvious that the maximum CIDNP intensity is reached when the nuclear spin independent and the nuclear spin dependent contributions to the intersystem crossing rate are matched, that is, $2|\Delta g \beta B_0| \approx |a|$. In the limiting case of large Δg and high fields, the dependence of the polarization intensities on the magnetic parameters can be easily derived by factoring out the Δg term. Choosing Δg and a as positive to simplify the expressions, one has

$$P_\alpha - P_\beta \propto \sqrt{\Delta g \beta B_0} \times [\sqrt{1 + a/(2\Delta g \beta B_0)} - \sqrt{1 - a/(2\Delta g \beta B_0)}]$$
$$\approx (\Delta g \beta B_0)^{-1/2} \times a/2 \tag{64}$$

where an expansion to first order has been performed to get the final expression. First, Eq. 64 shows that in this limit there is proportionality between hyperfine coupling constant and CIDNP intensity. Second, the CIDNP signal is seen to be inversely proportional to $\sqrt{\Delta g}$. Third, the signal-to-noise ratio S/N in high-field CIDNP experiments is found to be independent of B_0, because the signal is proportional to $B_0^{+1/2}$ (the additional factor B_0^{+1} stemming from the detection principle), as is the noise. This situation is in interesting contrast to normal NMR spectroscopy, where S/N varies as $B_0^{+3/2}$.

If a chemical reaction proceeds via two (or more) successive radical pairs, new aspects arise. The theory of this so-called "pair substitution," also called "memory effect" or "cooperative effect," has been developed by den Hollander [41]. In such a system the polarizations are not simply superpositions of polarizations from the first and second pair. Rather, they can be described as arising in a hypothetical radical pair with the combined magnetic properties of both pairs weighted with their respective lifetimes. Examples have even been reported [41b] where neither pair on its own would give rise to a net effect, one because of vanishing Δg and the other because of negligible a, but in combination they do.

Multiplet effects in $S-T_0$-type CIDNP can also be treated by vector models. As the simplest example, we consider a radical pair possessing two protons i and j in radical 1. Let Δg be zero, to remove the net effects. Furthermore, we take both hyperfine coupling constants a_i and a_j to be positive and assume that $a_i > a_j$. As in the previous case, we choose a triplet precursor and observe the product of the singlet exit channel. The pertaining diagrams and resulting signals are displayed in Figure 6. The assignment of the NMR transitions given there is valid for a positive spin–spin coupling constant J_{ij}. As the figure shows, an E/A multiplet (emission to low field, absorption to high field) is obtained with these parameters.

By drawing up the respective diagrams, it can be seen that inverting the sign of one of the hyperfine coupling constants is equivalent to pairwise

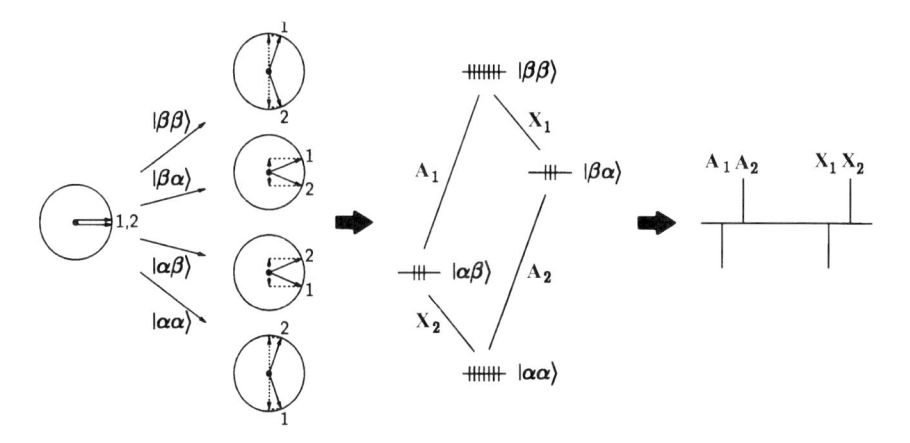

Figure 6. Explanation of a CIDNP multiplet effect in an AX spin system with vector models (left), resulting schematic population diagram (center), and NMR spectrum (right). The system is a radical pair with two protons i and j in radical 1. Parameters are triplet precursor, product of the singlet exit channel, $g_1 = g_2$, $a_i > a_j > 0$. The assignment of the transitions A_1 to X_2 assumes a positive coupling constant J_{ij}. For further details, see the text and Figure 5.

interchange of diagrams and would thus lead to the opposite polarization pattern (A/E). In contrast, taking a_i as the smaller hyperfine coupling constant would only permute the labels 1 and 2 within diagrams, and therefore leave the pattern unchanged. If the two protons were contained in different radicals, this would have the same effect as inverting the sign of one of them, as Eq. 18 shows. Finally, inverting the sign of J_{ij} would interchange the low- and high-field transitions of each of the doublets, so this would again effect a reversal of the pattern.

All these relationships are summarized by Kaptein's rule [40] for a CIDNP multiplet effect,

$$\Gamma_{ij} = \mu \times \varepsilon \times \mathrm{sgn}a_i \times \mathrm{sgn}a_j \times \sigma_{ij} \times \mathrm{sgn}J_{ij} \qquad (65)$$

$\Gamma_{ij} = +1$ corresponds to an E/A multiplet, and $\Gamma_{ij} = -1$ to an A/E multiplet. The parameter σ_{ij} is $+1$ when the two protons reside in the same radical, and -1 when they do not.

In low fields, additional phenomena such as the so-called "$n-1$ multiplets" arise, and it is possible to formulate analogous sign rules for their description [42]. However, that subject is outside the scope of this chapter.

Finally, $S-T_\pm$-type CIDNP shall be briefly treated. As the discussion of Section III.A showed, the exit channel has no influence on the polarization phase. Likewise, the sign of a is unimportant because there is no interplay with Δg. What decides the phase is the intersystem crossing pathway. If this is from $|T_-\rangle$ to $|S\rangle$, which represents the normal case (triplet precursor and negative sign of the exchange interaction J), emission results. For positive J, intersystem crossing occurs from $|T_+\rangle$ to $|S\rangle$, leading to an absorptive polarization. Reversing the intersystem crossing pathway, that is, starting from a singlet precursor, would reverse the polarization phase. It should be mentioned, however, that there is no known example of $S-T_\pm$-type CIDNP with a singlet precursor. These regularities can also be summed up by a simple sign rule for the net effect γ ($\gamma = +1$, absorption; $\gamma = -1$, emission) of $S-T_\pm$-type CIDNP,

$$\gamma = \mu \times \mathrm{sgn}J \qquad (66)$$

By drawing up population diagrams, it can easily be seen that multiplet effects are impossible in $S-T_\pm$-type CIDNP.

D. Instrumentation and Techniques

Photo-CIDNP experiments need a light source for excitation and an NMR spectrometer for observation. A pulsed and Fourier transform spectrometer is preferable; nowadays, this detection scheme is universally in use anyway. The only modification of the spectrometer concerns the probe, which must allow illumination of the sample. Probably the best way to achieve this is by off-axis insertion of a suprasil rod topped by a prism, which acts as a light guide so that the light enters the sample from the side and within the active volume of the NMR coils. Such probes are commercially available. Illumination through the bottom of the NMR tube is also feasible but less advantageous, first because a considerable part of the light is absorbed outside the active volume, where it leads to sample decomposition without detectable polarizations, and second, because the field homogeneity of a superconducting magnet is very sensitive to perturbations along the z axis close to the center of field. Illumination from the top can be realized with an optical fiber that is immersed into the sample. While the advantage of this method is that it can be done with an unmodified probe, it cannot be applied when the samples must be sealed.

Arc lamps can be used as light sources, but much greater flexibility is obtained by employing a pulsed laser, because this also allows time-resolved measurements. Steady-state experiments are performed by letting the laser work in free-running mode with a repetition interval short compared to the nuclear T_1 (or by applying continuous illumination) during acquisition of the spectra. To remove the background, that is, the normal NMR signals of molecules that have not reacted, one can subtract a spectrum acquired under identical conditions but without illumination. For time-resolved experiments, the laser is triggered by the pulser unit of the spectrometer, and the polarizations are sampled by an NMR pulse after a delay Δ; varying Δ yields a series of time-resolved spectra. With this technique, which was introduced by Ernst et al. [43] on a millisecond time scale and extended to microsecond [44a] and submicrosecond [44b] time resolution by Closs and Miller, the background signals can be easily eliminated [43b] by using a noise-modulated decoupler pulse to saturate the NMR transitions before the laser flash.

Newer methodological developments, for example, for background suppression, are discussed in Section IV.F.

E. Information Accessible by CIDNP

Because the CIDNP effect is caused by the interplay of spin dynamics, diffusional dynamics, and chemical reactivity, information about each of these fields can be obtained from CIDNP experiments. CIDNP spectroscopy has proven extremely useful for studies of the former two aspects in systems with restricted diffusion (biradicals, and, to a lesser degree, micellar systems; see Sections IV.C and IV.D). Concerning the rates and mechanisms of chemical reactions, CIDNP spectroscopy possesses several unique features, which make it a very powerful method.

1. The diamagnetic products are observed and can be characterized very well because the detection method is high-resolution NMR spectroscopy. On the other hand, the signal enhancement by the CIDNP effect mitigates the inherent low sensitivity of NMR. In consequence, even diamagnetic species that are unstable and thus present only in low concentration can be captured. An early example is the enol of acetophenone formed in the photoreaction of acetophenone with phenol [45], others are vinylamines in photoinduced hydrogen abstractions from aliphatic amines [46] (see Section V.B) and in the sensitized photoreactions of amino acids (Section V.G.1).
2. Because of the relationship between polarization intensities and hyperfine coupling constants (Section III.C), the intermediates can be identified as well, from what is called the polarization pattern (the relative polarization intensities of the different protons). This use of

CIDNP has been pioneered by Roth [46]. Applications of it are abundant (see, e.g., Sections V.A.1, and V.D.3).

3. Precursor multiplicity and exit channel of the intermediates can be determined from the overall polarization phases. The possibility of obtaining the precursor multiplicity is probably one of the most valuable assets of CIDNP, since no other kind of spectroscopy provides access to this information in such a direct way. Section V.C.2 shows examples of this.

4. The polarizations can be viewed as labels that are attached at the paramagnetic stage. This can be used in many ways to study secondary reactions qualitatively as well as quantitatively. Examples are given in Sections V.A.2, V.D.3, and V.G.1.

5. With commercially available equipment, time-resolved experiments (flash CIDNP) with a time resolution in the submicrosecond range can be performed. That this is possible at all results from the fact that the polarizations are generated during the lifetime of the paramagnetic intermediates (on the order of nanoseconds) but persist in the diamagnetic products for a time on the order of T_1 (seconds for protons). For applications of this method, see, for example, Sections IV.C, V.A.2, and V.D.1.

IV. PHYSICAL STUDIES AND METHODOLOGICAL DEVELOPMENTS

A. Spin Dynamics

Owing to their ease of use, Kaptein's rules (Eqs. 62 and 65) are the basis of most chemical applications of CIDNP. On the other hand, they represent the radical pair mechanism in a simplified form only, so the question as to the validity of this approximation is of importance. Salikhov [47a] was the first to show theoretically that these rules can be violated in systems containing more than one magnetic nucleus. This was later analyzed in more detail [47b] and verified experimentally [47b, 47c, 48].

As the simplest example, consider a radical pair containing two spin $\frac{1}{2}$ nuclei (hyperfine coupling constants a_1 and a_2) in radical 1, a triplet precursor, and a product of the singlet exit channel. The populations of the four levels of the resulting AX spin system in the product can be calculated in the same way as in Eq. 63, by using Eqs. 18, 46, and 60. By introducing the abbreviations ξ and η for the ratios a_1/a_2 and $2\Delta g\beta B_0/a_1$, respectively, and performing the decomposition into product operators described in Section III.B, one obtains

$$\mathbf{I}_{1z} \propto \sqrt{|\eta + 1 + \xi|} + \sqrt{|\eta + 1 - \xi|} - \sqrt{|\eta - 1 + \xi|} - \sqrt{|\eta - 1 - \xi|} \quad (67)$$

$$2\mathbf{I}_{1z}\mathbf{I}_{2z} \propto \sqrt{|\eta + 1 + \xi|} - \sqrt{|\eta + 1 - \xi|} - \sqrt{|\eta - 1 + \xi|} + \sqrt{|\eta - 1 - \xi|} \quad (68)$$

for the CIDNP net effect of nucleus 1 and the multiplet effect of nuclei 1 and 2. In Figure 7, these expressions have been plotted as functions of η for different values of the parameter ξ. For simplicity, only positive values of a_1, a_2, and Δg were chosen; generalization to arbitrary signs brings no new results. Kaptein's rules predict positive signs both for $\mathbf{I}_{1z}$ and for $2\mathbf{I}_{1z}\mathbf{I}_{2z}$. As far as the net effect is concerned, this is in accordance with the quantitative result of Eq. 67 as long as the nucleus with the larger hyperfine coupling constant is being considered, up to the limit of both hyperfine coupling constants possessing equal absolute values (cf. Fig. 7, curves a and b). For the nucleus with the smaller hyperfine coupling constant, however, there is a pronounced "anti-Kaptein" regime (curve c) where the net effect is opposite to that stated by Kaptein's rule; only when Δg or B_0 become large enough does the polarization phase change again to that expected from Eq. 62. In contrast, the predictions of the multiplet-effect rule (Eq. 65) are valid

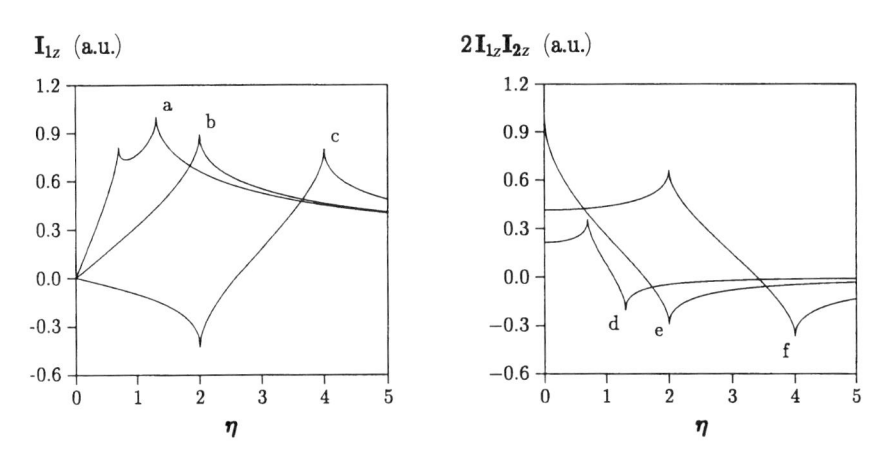

Figure 7. Plots of CIDNP net effects (Eq. 67) and multiplet effects (Eq. 68) in a radical pair with two protons as functions of the quantity η, $\eta = 2\Delta g \beta B_0 / a_1$, with the ratio of hyperfine coupling constants a_2/a_1 as parameter ξ of the curves. For a detailed explanation, see the text. (Left) net effect $\mathbf{I}_{1z}$; (right) multiplet effect $2\mathbf{I}_{1z}\mathbf{I}_{2z}$. Curves a and d, $\xi = 0.3$; curves b and e, $\xi = 1.0$; curves c and f, $\xi = 3$.

for low values of η (small Δg and/or B_0) only, and the multiplet polarizations exhibit anti-Kaptein behavior if η lies above a critical value (see curves d–f).

It is evident that the described effects are strongest when the difference between the hyperfine coupling constant of the observed nucleus 1 and the other nucleus is large (large ξ) and, at the same time, a_1 is not excessively small (otherwise its net effect is unobservable and the multiplet effect is also very small). This situation is difficult to realize when the radicals possess no other magnetic nuclei besides protons in positions of high unpaired spin density, so the only examples reported so far concern molecules containing ^{13}C at or near a radical center. For instance, Roth et al. [48] tested the above predictions on the radical pairs obtained from photocleavage of para-substituted dibenzylketones that were ^{13}CO labeled. For these systems, $\xi = a_{^{13}C} : a_{H_{benzyl}} \approx 7.5$. To vary η, experiments were performed at different fields on the one hand, and by fine-tuning Δg (without altering its sign) by the ring substituents on the other. The influence on the appearance of the CIDNP spectra was found to be pronounced, the most striking effect being that quite small modifications of Δg can lead to dramatic changes of the CIDNP patterns, by shifting the weights of net and multiplet effects. All effects could be well accounted for by the radical pair theory.

The fact that the intensity of the CIDNP net effect of a particular nucleus is influenced by the hyperfine coupling of a second nucleus (Eq. 67 and Fig. 7, left) does not in any way depend on whether or not the two nuclei are J coupled in the products. Nor is it necessary that they are contained in the same radical of the intermediate pair. Depending on the parameters, enhancement or reduction of the net polarization of the observed nucleus can result. Examples of both have been reported, again for ^{13}C labeled substrates [49]. Especially strong effects were found [49a] in the photolysis of 2-phenylcyclododecanone, where scavenging of the intermediate 1,12-biradical by CCl_4 served as the spin-independent reaction pathway (see Section III.A) required for $S-T_0$-type CIDNP. The ^{13}C CIDNP signal of the nonbenzylic α carbon atom (C^2 in the biradical) in the scavenging product was suppressed by an order of magnitude or more when ^{13}CO labeled material was used ($\xi = 2.5$). This drastic change was ascribed to a coupling of spin dynamics and chemical dynamics, the ^{13}C nucleus of the CO group with its larger hyperfine coupling constant shifting the intersystem crossing rates for the $|\alpha\rangle$ and $|\beta\rangle$ spin states of C^2 outside the kinetic window provided by the scavenging process (cf. Section IV.C).

Another matching effect can occur when molecular motion, for example, rotation of a methyl group, simultaneously modulates the hyperfine coupling constants and the chemical reactivity, the result being a violation of magnetic equivalence of the nuclei of the group. This phenomenon, which had been predicted theoretically [50a], seems to be very rare, and so far has only been observed once [50b] in low-field CIDNP experiments on an intramicellar disproportionation reaction.

B. Radical Pair Dynamics

Burri and Fischer studied the diffusion dependence of CIDNP from the radical pairs formed by triplet cleavage of di-*tert*-butyl ketone [22]. The interdiffusion coefficients D were varied by using different solvents on the one hand, and by varying the temperature on the other. It was demonstrated that for the nonviscous case, CIDNP intensities are a function of $D^{-1/2}$ (cf. Eq. 46), whereas in viscous solution a leveling-off occurs. The results were analyzed with the refined reencounter model of [32]. From the results of their fits, the authors concluded that the singlet reactivity of these radical pairs is essentially unity, that the initial distance of the radicals of a pair is only slightly larger than the encounter distance, that the value of the exchange integral at the reaction distance (5.3 Å) lies between 10^{12} and 10^{14} rad s^{-1}, and that diffusion occurs by small steps rather than by large jumps. The experimental results were also analyzed [51] by using a Green's function approach for the Laplace transform of the stochastic Liouville equation.

Diffusion without an interaction between the radicals, as in the preceding example, is well described by Eq. 42 or related formulas (cf. [1a]). In contrast, no closed-form solution is known for the conditional probability density of reencounter $f(t, d, r_0)$ in the important case of diffusion in a Coulomb potential. However, Shokhirev et al. [52a, 52b] derived an analytical approximation of the Green's function of this problem, and obtained closed-form expressions for the spin-dependent rate constants [52c]. An experimental investigation of CIDNP involving radical ion pairs in solvents of different relative permittivity ε_r was performed by Azumi et al. [53]. The authors generated radical ion pairs by photoinduced electron transfer from *trans*-stilbene to photoexcited triphenylamine in mixtures of acetonitrile-d_3 and benzene-d_6 and compared the experimental steady-state polarization intensities with numerical solutions of the stochastic Liouville equation. With decreasing ε_r, the CIDNP intensities first increase, reflecting the larger recombination probabilities, and then decrease toward zero because diffusive separation of the radical ions becomes more and more difficult. While it was possible to reproduce the shape of the experimental permittivity dependence, quantitative agreement between experiments and calculations was not very good, however.

C. Biradicals

Flexible biradicals continue to attract the attention of CIDNP spectroscopists. These intermediates differ in two respects from radical pairs. First, the exchange interaction does not vanish except for extremely long-chain biradicals. Second, there is no ready escape pathway because of the link between the radical centers. While the first factor decreases the efficiency of $S-T_0$ mixing, the second factor is the more important one for $S-T_0$-type CIDNP: Even if nuclear spin selective intersystem crossing occurs, any spin sorting is undone again because the products of the singlet and the triplet exit channels are identical. Assuming a triplet precursor, which represents the usual case as typical precursors are photoexcited cyclic ketones, those nuclear spin states that increase the intersystem crossing rate lead to faster product formation than those that decrease it. In a time-resolved experiment, the absolute CIDNP intensities thus pass through a maximum. In the absence of nuclear spin relaxation, there is complete cancelation of absorptive and emissive contributions when all biradicals have decayed to the products; with nuclear spin relaxation, some fraction of the faster component remains. The situation is very similar to that in the cyclic electron-transfer reactions discussed in Section V.A.2. This time dependence of $S-T_0$-type CIDNP from flexible biradicals was first demonstrated by Closs and Redwine [54]. A typical experimental example is shown in Figure 8.

Tsentalovich et al. [55] performed flash-CIDNP investigations of the photolysis of cycloundecanone and (see the figure) cyclododecanone (1). In the high fields of an NMR spectrometer, the contribution of $S-T_-$ polarizations is negligible for these systems. The experimental results were first analyzed [55a] with a two-position biradical model analogous to the concept of an exchange region discussed in Section II.B.5. This model assumes the spin Hamiltonian to be completely dominated either by the Δg and hyperfine terms or by the exchange interaction. While this simple approach is adequate for a description of radical pair CIDNP because the transition regime between these two regions is traversed sufficiently fast, it is less well suited to biradical CIDNP because for biradicals possessing usual chain lengths, the flexible link between the radical termini causes the time spent in the transition regime to be much longer. A more realistic treatment had already been developed in 1977 by de Kanter et al. [19]. In it, the conformational changes of the biradical are described by a diffusional motion within the precalculated distribution of distances between the radical termini; on the basis of this diffusional model, the stochastic Liouville equation is then solved numerically. By using this procedure for evaluation, Yurkovskaya et al. were able to rationalize the influence of

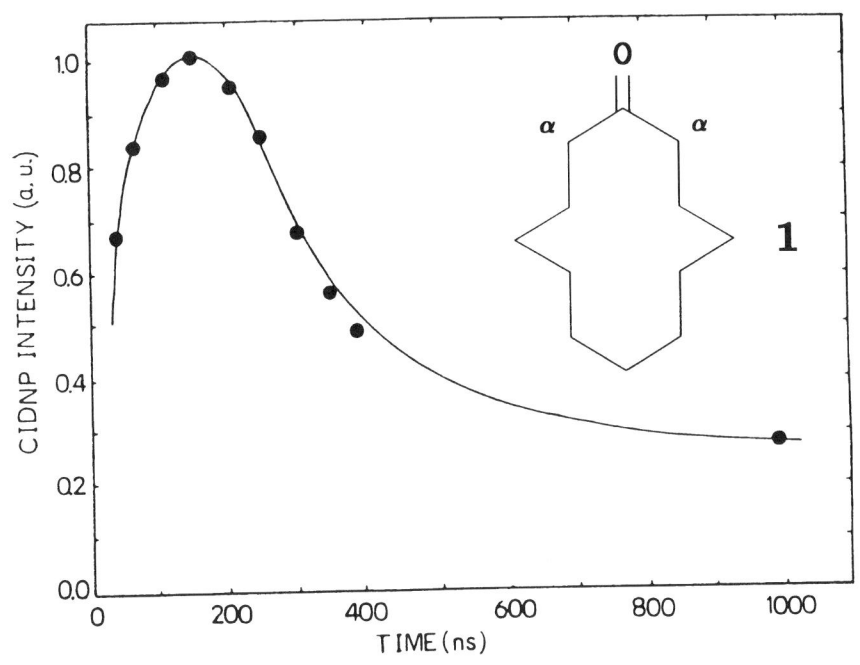

Figure 8. Time-dependent CIDNP intensities in the photolysis of cyclododecanone **1**. The symbols show the ^{1}H CIDNP signal of the α protons of the ketone, the solid line was calculated by the theoretical model of [55a]. [Adapted from ref. [55a] with permission. Copyright © 1989 Elsevier Science Publishers B.V.]

solvent viscosity and temperature on the time evolution of $S-T_0$-type CIDNP in the above-mentioned biradicals [55b], as well as the dependence of the CIDNP kinetics on scavenging reactions and on the magnetic field strength [55c].

As was first shown experimentally in 1978 by de Kanter and Kaptein [56], an escape pathway for a biradical can be provided by adding a radical scavenger **S**. Because such a scavenger reacts with one radical terminus only, the reaction is obviously independent of the electron spin multiplicity. This principle was utilized by Turro et al. [57] to make observable $S-T_0$-type CIDNP from long flexible biradicals also in steady-state experiments. The dependence of CIDNP intensities on the parameters controlling the overall rate of intersystem crossing was studied in detail in [57c]. The highest enhancement factors are obtained when the scavenging rate $k_{scav}[S]$ equals the geometric mean of the intersystem crossing rates for nuclear spin states

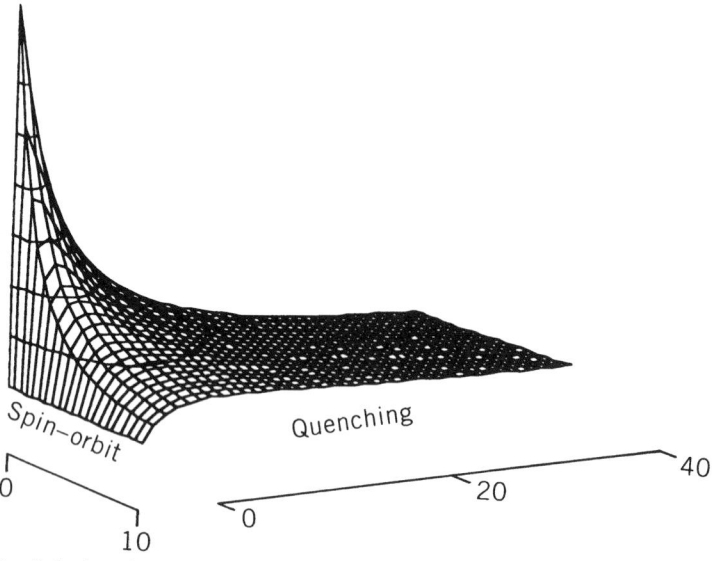

Figure 9. Calculated relative CIDNP intensities (in a.u.) of $S-T_0$-type CIDNP generated in the photolysis of phenylsubstituted (in the α position) **1** versus dimensionless rates of spin–orbit coupling and scavenging. For further explanation, see the text. [Adapted from K. C. Hwang, N. J. Turro, and C. Doubleday, *J. Am. Chem. Soc.*, **113**, 35 (1991), with permission. Copyright © 1991 American Chemical Society.]

$|\alpha\rangle$ and $|\beta\rangle$, which include spin–orbit coupling (described by a rate k_{SO}) in addition to the Δg and hyperfine terms. Thus, there exists a "kinetic window" within which polarizations can be generated; matching the rates in order to keep within that window can be conveniently accomplished by varying the rate constant k_{scav} of scavenging (by using different scavengers) or the scavenger concentration. For the acyl–alkyl biradicals derived from the cyclic ketone **1** substituted with phenyl in the α position, Figure 9 shows the dependence of relative CIDNP intensity on the reduced rates k'_{SO} and $k'_{scav}[S]$, which are taken relative to the sum of the Δg and hyperfine terms. The kinetic window is seen to become broader as the contribution of spin–orbit coupling increases.

Another escape pathway of a biradical is realized with the intermediates of the photocycloadditions between quinones and norbornadiene or quadricyclane (cf. Section V.D.3, Chart XV). In these reactions, nuclear spin dependent intersystem crossing of a 1,5-biradical competes with nuclear spin independent rearrangement to give a 1,4-biradical of the Paterno–Büchi type, which, owing to its high intersystem crossing rate, acts as a chemical sink [117b]. Consequently, very strong $S-T_0$-type CIDNP from the 1,5-

biradical is found. This result is interesting because it is a frequently stated paradigm that $S-T_0$-type CIDNP from short-chain biradicals should be impossible because the nonvanishing exchange interaction suppresses $S-T_0$ mixing. This paradigm is corroborated by numerous cases where no $S-T_0$-type CIDNP was observed for such systems. However, the occurrence of $S-T_0$-type CIDNP from the above 1,5-biradical, as well as earlier reports of such phenomena, as, for instance, from 1,6-biradicals [20, 27], from 1,5-biradicals [58], and even from a 1,4-biradical [59], seems to indicate that the influence of J on $S-T_0$-type CIDNP is far less important than is the presence or absence of an escape pathway. Still more puzzling is the strong influence of small structural changes, which was already pointed out by Doubleday [58b], for example, that $S-T_0$-type CIDNP from the 1,6-biradical derived from cyclohexanone is totally suppressed by introducing an ethano bridge between positons 4 and 6 [27]. In the light of these examples, it must be concluded that the factors giving rise to $S-T_0$-type CIDNP from short-chain biradicals are still poorly understood.

$S-T_\pm$-type CIDNP dominates in low fields. As this variety of CIDNP relies on the matching of Zeeman and exchange interactions (cf. Fig. 1), it yields information about an average value of J. The typical field dependence of the polarization intensity is a bell-shaped curve, as seen in Fig. 10. With decreasing separation of the radical termini, corresponding to increasing $|J|$, the position of the maximum is shifted toward higher field strengths. In a simple physical picture [27], the maximum should appear when Zeeman and exchange interactions match, that is, when $|2\bar{J}| = g\beta B_0$. However, as has already been shown by de Kanter et al. in 1977 [19], this model is an oversimplification, and the maxima of the field-dependent curves are not expected to coincide with $|2\bar{J}|$. Closs and Forbes [60] investigated a series of 1,7- to 1,15-biradicals derived from $\alpha, \alpha, \alpha', \alpha'$-tetramethylated cyclic ketones by variable field CIDNP, time-resolved EPR spectroscopy, and MFE measurements. Owing to the great number of spins and the fact that electron and nuclear spin states cannot be separated in the case of $S-T_\pm$ mixing, numerical solution of the stochastic Liouville equation was not feasible and the crude relation $|2\bar{J}| \approx g\beta B_0$ had to be used for evaluation of the CIDNP data. Agreement between $|\bar{J}|$ obtained from the EPR experiments and from the CIDNP experiments was found to be poor. As the authors of [60] pointed out, the EPR experiment, where the polarizations are generated by the spin-correlated radical pair mechanism [8], yields a true time-averaged J, whereas the CIDNP experiment weighs certain conformations and distances between the radical termini more heavily.

Azumi et al. carried out variable-field CIDNP experiments on the biradicals resulting from intramolecular hydrogen abstractions in polymethylene-linked xanthone and xanthene moieties [61a, 61b]. They also studied the temperature dependence of the mean exchange interaction for

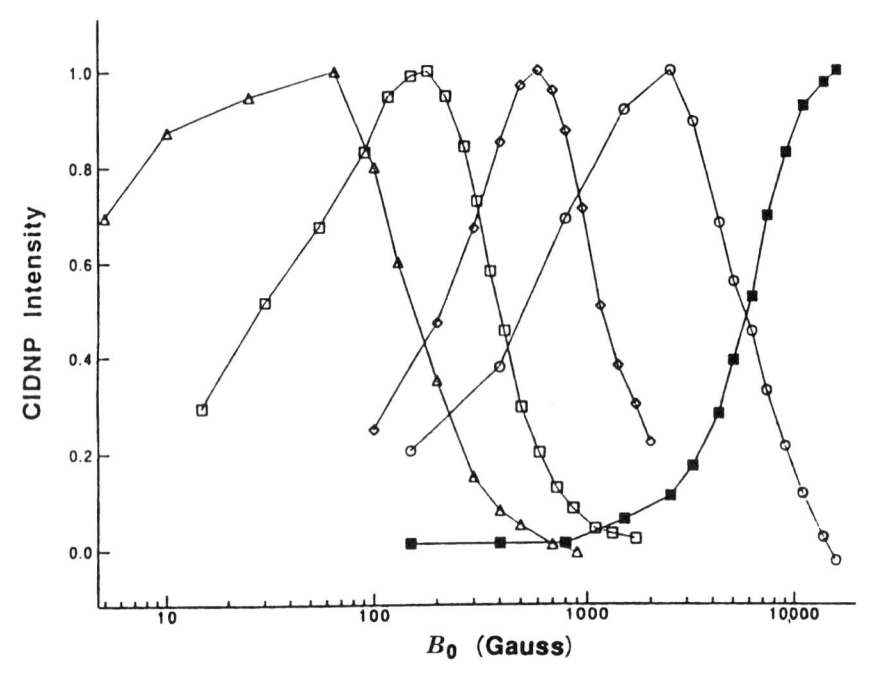

Figure 10. Field dependence of the CIDNP intensities for acyl–alkyl biradicals (open symbols) and a bisalkyl biradical (filled symbols) produced in the photoreactions of $\alpha,\alpha,\alpha',\alpha'$-tetramethylated cycloalkanones. Triangles, 1,15-biradical; open squares, 1,12-biradical; diamonds, 1,10-biradical; circles, 1,8-biradical; filled squares 1,7-biradical (resulting from decarbonylation of the 1,8 acyl–alkyl biradical). [Reproduced with permission from G. L. Closs, M. D. E. Forbes, and P. Piotrowiak, *J. Am. Chem. Soc.*, **114**, 3285 (1992). Copyright © 1992 American Chemical Society.]

biradicals derived from cyclododecanone **1** and cyclodecanone [61c]. Their experimental results, slight shifts of the CIDNP maxima to higher B_0 with rising temperature, were interpreted by a shrinking of the alkyl chain at higher temperatures.

D. Micellar Systems

After the classical studies by Turro and co-workers, who determined the exit rates of radicals from micelles from the changes in steady-state [62a] or time-resolved [62b] CIDNP spectra when an external scavenger was added, and analyzed the magnetic field dependence of the CIDNP signals [62c], activity in this field seems to have quieted down a little.

In a study of phenacylphenylsulfone photolysis, CIDNP data were taken as evidence that the primary radical pairs cannot recombine to regenerate the starting material because the micelle forces a certain orientation of the radicals [63]. From low-field ^{13}C CIDNP and SNP measurements on cleavage of benzylic ketones in sodium dodecyl sulfate micelles, it was inferred [64] that the exchange interaction in these systems is several orders of magnitude smaller ($\approx 10^{10}$ rad s^{-1} at a reduction distance of 6 Å; cf. the values in Section IV.B) and the distance dependence is much weaker ($\alpha \approx 0.5$ Å^{-1}; cf. the discussion of Eq. 10) than generally assumed for radical pairs. By numerical solutions of the stochastic Liouville equation for a model of the micelle where one of the radicals is kept fixed at the center of the micelle while the other radical is allowed to diffuse, the results of MARY experiments, ^{13}C CIDNP experiments at variable fields, and SNP experiments could be reproduced with the same set of parameters [65].

An approximative analytical treatment of $S-T_\pm$-type CIDNP of radical pairs in micelles has recently been given [66]. Comparison with numerical solutions of the stochastic Liouville equation obtained by a finite difference technique showed the accuracy of the approximate solution to be quite good.

E. Other Polarization Mechanisms

The question of whether there are other mechanisms leading to CIDNP besides the radical pair mechanism is of central importance because chemical conclusions that are drawn from CIDNP results on the basis of the latter mechanism might of course be entirely wrong with another mechanism being the source of the polarizations. There has been some evidence [67–72] that cross-relaxation in radicals, by which electron spin polarization (CIDEP) is converted into CIDNP, could provide such a mechanism. Depending on whether cross-relaxation occurs by flip–flop transitions ($\Delta m = 0$) or by double spin flips ($\Delta m = 2$), opposite or equal phases of CIDEP and CIDNP would result. Since the origin of the electron spin polarizations is usually the triplet mechanism, this cross-relaxational mechanism is sometimes referred to as the triplet mechanism of CIDNP.

Anomalous (in the sense of radical pair theory) polarization phases and magnetic field dependence were reported for substrates containing ^{19}F nuclei, and were explained by cross-relaxation [67]. Azumi and co-workers [68] investigated the photolysis of benzaldehyde. The polarizations could be accounted for by $S-T_0$ mixing at high fields and by $S-T_-$ mixing at low fields. At a field of 325 mT, however, the authors could not reconcile the CIDNP phase with the predictions of the radical pair mechanism; from additional DNP experiments, they concluded that cross-relaxation with

$\Delta m = 2$ is responsible for this polarization. The same authors studied the photolysis of benzoquinones in CDCl$_3$ [69] by CIDNP and DNP. They interpreted their results by a hydrogen abstraction from CHCl$_3$ present as an impurity in the solvent. On the basis of this chemistry, the polarization phase at low quinone concentrations agrees with Kaptein's rules, whereas the phase is reversed at high concentrations, which was again attributed to $\Delta m = 2$ cross-relaxation.

Considerable effort [70, 71] has been devoted to the photoreaction of acetone in isopropanol-d_8. This reaction proceeds via radical pairs $\overline{(CH_3)_2\dot{C}OD\ (CD_3)_2\dot{C}OD}$ that should not give rise to any CIDNP net effects owing to vanishing Δg. Observations such as an emissive net CIDNP signal that changes into absorption below $-30°C$ and an unusual kinetic behavior [70] were rationalized by $\Delta m = 0$ cross-relaxation; the same explanation was put forward [72] to account for the unusually fast polarization decay in the reactions of the radicals $(CH_3)_2\dot{C}OH$ **21·** produced by cleavage of photoexcited 1,4-dihydroxy-2,4-dimethylpentan-3-one **20** (cf. Chart XI, Section V.D.1). However, Batchelor and Fischer [71] carefully reinvestigated these systems, taking into account also the dependence on the initial radical concentration and backing up their results with CIDEP experiments. They were able to show that the reasons for the anomalies are a pronounced solvent dependence of nuclear spin relaxation in the latter case, and side reactions (e.g., two-photon cleavage of acetone) caused by the high light intensities in the former. They concluded that the contribution of the triplet mechanism in these systems is negligible at the high fields of an NMR spectrometer, that is, under usual conditions.

In the light of these results (see also [73] for an earlier example where polarizations previously attributed to the triplet mechanism [74] could be, in the words of the authors of [73] "well explained by the radical pair mechanism once the chemistry is understood"), it would seem, conversely, that the danger of falsely interpreting CIDNP spectra by applying the radical pair mechanism is probably small. The situation may be different with nuclei such as ^{19}F and in low magnetic fields, where the cross-relaxation rates could be much larger. Further studies of this important issue are certainly desirable but should be performed in systems that are thoroughly understood chemically, including side reactions.

F. New Methods and Techniques

The generation of nanosecond or shorter laser flashes presents no technical problem, so the time resolution of CIDNP experiments is limited by the width of the NMR observation pulse, which is typically in the microsecond range. This problem cannot be solved simply by using high-power pulse amplifiers because what counts is not the duration of the driving radio frequency (rf) pulse but that of the oscillating magnetic field (B_1). The high Q value of the tank circuit in the probe, which is necessary for sensitivity reasons, effects a considerable lengthening of a pulse present at the input terminals. With typical parameters, the B_1 pulse is, therefore, at least 200 ns longer than the rf pulse. This time constant constitutes the minimum width of the B_1 pulse, which cannot be improved without changes of the probe design. So far, the only feasible hardware solution seems to be to reduce Q electronically for the duration of the observation pulse, and then immediately restore the high Q value of the undamped circuit for acquisition of the free induction decay [44b].

Another way, which does not require any hardware modifications, is to stagger the observation pulses by a time interval Δ that is shorter than their duration, and apply deconvolution or iterative reconvolution techniques to the signal $S(t)$. By this, one can obtain the time-dependent CIDNP magnetization $M_{kin}(t)$ with a time resolution Δ, at the cost of a reduction in sensitivity. This method is obviously also applicable in conjunction with the former. The key relationship for this evaluation procedure is the dependence of $S(t)$ on $M_{kin}(t)$ and the shape $\omega(t)$ of the B_1 pulse. For rectangular B_1 pulses, this problem has been treated in [75]. In [76], this dependence was derived for the general case, that is, arbitrary $M_{kin}(t)$ and $\omega(t)$, and it was shown that significant deviations in the rate constants may result when the real (nonrectangular) pulse shape is not taken into account. Sampling $M_{kin}(t)$ with pulses that are not short on the time scale of the CIDNP kinetics amounts to filtering it through a low pass filter. The dependence of the transfer function of this filter on duration and shape of the B_1 pulse was analyzed in [77]; the reduction in sensitivity was also quantified there, and optimization of the tip angle and the acquisition time of flash-CIDNP experiments were discussed.

The flash-CIDNP technique can also be employed when no kinetic information is required, simply by using a delay between laser flash and acquisition pulse that is longer than the time needed for completion of the reaction; however, it has a much lower S/N than steady-state experiments. On the other hand, disadvantages of the latter are, above all, that presaturation cannot be used to eliminate the background signals, while the feasibility of subtracting "light" and "dark" spectra for that purpose may be limited by line broadening and shifts due to exchange effects. As an alternative, which combines the advantages of both time-resolved and steady-state CIDNP measurements, "pseudo-steady-state" experiments were introduced [78a] in which n laser flashes (typically $10 \cdots 20$) or gated illumination are applied during a short period (100 ⋯ 200 ms) before acquisition of the spectra in the dark. On the one hand, with this approach S/N is obviously higher by a factor of n than in a single-flash CIDNP experiment, on the other hand, the background can be removed by pulse sequences. With a very simple pulse sequence [78a], background reduction by a factor of about 150 is obtained, which is sufficient for most practical purposes; more complex sequences [78b], which, however, put some demand on spectrometer performance, lead a reduction by three orders of magnitude or more.

Owing to the dependence of the strength of a CIDNP net effect on the hyperfine coupling constants a, nuclei in the products are unpolarized when they possess negligible a in the paramagnetic intermediates. It is obvious that the absence of their signals may severely hamper product identification. Not only are the chemical shifts of the unpolarized nuclei unknown but it may also be impossible to determine whether the observable signals are due to one or more equivalent protons; for instance, when nucleus A is unpolarized and no multiplet effects occur, an AX, AX_2, or AX_3 spin system yields the same signal, a doublet for X. For coupled spin systems, this problem can be solved by coherence transfer methods [79]. If at least one nucleus 1 of the spin system is polarized (i.e., $\mathbf{I}_{1z}$ is nonzero), one can generate transversal magnetization, e.g., $\mathbf{I}_{1x}$, by an NMR pulse, let it evolve into antiphase magnetization $2\mathbf{I}_{1y}\mathbf{I}_{2z}$ under the influence of the scalar coupling with nucleus 2, and, by a second pulse, convert $2\mathbf{I}_{1y}\mathbf{I}_{2z}$ into antiphase magnetization of spin 2, $2\mathbf{I}_{1z}\mathbf{I}_{2y}$, which gives rise to a multiplet-type CIDNP signal (cf. Section III.B, Fig. 4) at the resonance frequency of the unpolarized nucleus 2. This experiment (CIDNP–COSY) is identical to the standard COSY experiment of NMR spectroscopy except that no frequency labeling is necessary. For that reason, it is most advantageous to perform it by using selective pulses. A coupling of spin 2 to a third spin can be exploited in an analogous manner (CIDNP–RCT) by a three-pulse experiment [79]. Such coherence transfer methods have also been used to simplify the spectra of proteins [80] and to separate CIDNP net and multiplet effects [81].

V. APPLICATIONS TO CHEMICAL PROBLEMS

A. Electron Transfer

1. Structures of Radical Ions. Because of the dependence of the nuclear spin polarizations on the hyperfine coupling constants, similar information about the spin density distribution, and thus structure, of radicals can be obtained by CIDNP experiments as by EPR spectroscopy (cf. Section V.A.2, Fig. 13). While EPR methods, including CIDEP, would intuitively appear to be a more direct way of investigating radicals, CIDNP possesses a number of distinct advantages over them. First, one automatically gets the signs of the individual coupling constants a_j, which cannot be obtained by standard EPR measurements. Second, while an EPR spectrum only yields a set of coupling constants a, which then must be assigned in an often tedious procedure, the CIDNP spectrum immediately establishes a correspondence between a particular value a_j and a particular nucleus j in the products. It is often straightforward to extend this correspondence to the nuclei in the radicals; in the case of radical ions, which are converted to products by back electron transfer (that is, without changes of the connectivities), this extension is usually a trivial task. Finally, despite the progress in pulsed EPR during the last years, the time scale of CIDNP—typically in the nanosecond range—is still somewhat shorter than that of EPR, and detection is not hampered by fast secondary reactions at the paramagnetic or diamagnetic stage. For instance, the radical cation $2^{\cdot +}$ of quadricyclane, which until recently [82a] has eluded all attempts of observation by EPR, has been shown by CIDNP more than a decade ago [82b] to exist and to be a different minimum on the $C_7H_8^+$ potential hypersurface than the radical cation $3^{\cdot +}$ of norbornadiene (see Chart II).

Another example [83] is the radical cation of hexamethylprismane **4** (cf. Chart III). Because the CIDNP spectra in the photoreactions of the sensitizer anthraquinone with **4** and with hexamethyldewarbenzene **5** are different, it was concluded that $4^{\cdot +}$ and $5^{\cdot +}$ are chemically distinct species. In the reaction of **4**, weak polarizations were observed for the regenerated starting material, and stronger polarizations of the same overall phase for the two products **5** and hexamethylbenzene **6** (not shown in the chart), which must both be due to radical cation rearrangement followed by back electron transfer. From this, it was inferred that $4^{\cdot +}$ is a very short-lived species, the rearrangement to $5^{\cdot +}$ and $6^{\cdot +}$ occurring within the lifetime of the geminate radical ion pair, that is, on a nanosecond to subnanosecond

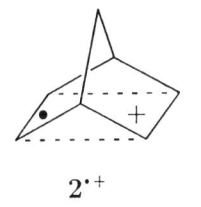

$2^{\cdot +}$ $3^{\cdot +}$

Chart II

4   5

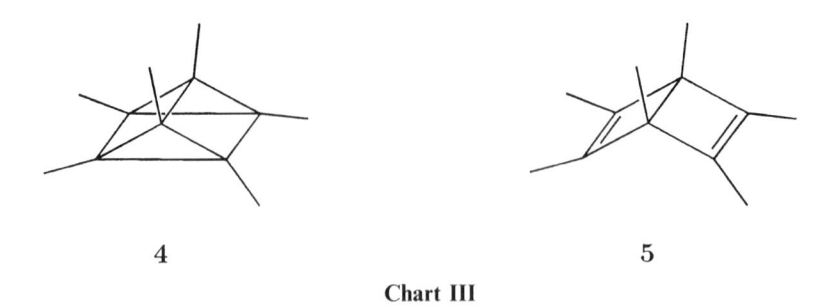

Chart III

time scale. The polarization pattern is in accordance with the $^2B_1$ state of $4^{\cdot+}$.

Homoconjugative interactions between a cyclopropane ring and a double bond were studied for the three bicyclic compounds **7–9**, each possessing a locked vinylcyclopropane fragment [84]. The polarization patterns in the chloranil-sensitized photoreactions of these substrates were used to probe the spin density distribution in the radical cations. A strong dependence of electronic structure of the radical cations on steric effects was found (see Chart IV).

The radical cation $7^{\cdot+}$ derived from bicyclo[3.1.0]hex-2-ene is best described by a weakened lateral cyclopropane bond. When the size of the larger ring is increased by one CH_2 unit, as in the radical cation $8^{\cdot+}$ of norcarene, the internal bond of the cyclopropyl moiety is weakened instead. Attaching methyl groups to the unsubstituted position of the three-membered ring ($9^{\cdot+}$) again reverses the effect. While the latter change of electronic structure can be explained by a hyperconjugative interaction between the double bond and methyl, the former was rationalized by the higher rigidity of the bicyclohexene system, which only allows orbital overlap of the lateral cyclopropane bond with the p orbitals of the olefin moiety; with the more flexible bicycloheptene system delocalization into the

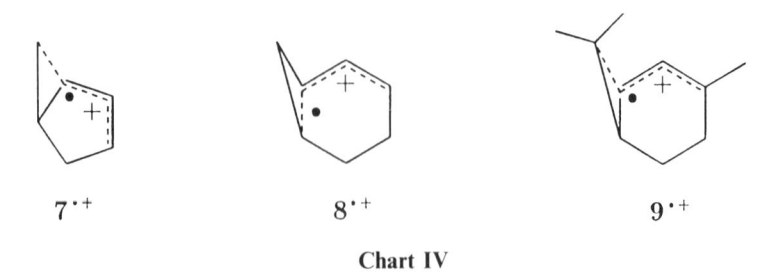

$7^{\cdot+}$ $8^{\cdot+}$ $9^{\cdot+}$

Chart IV

Chart V

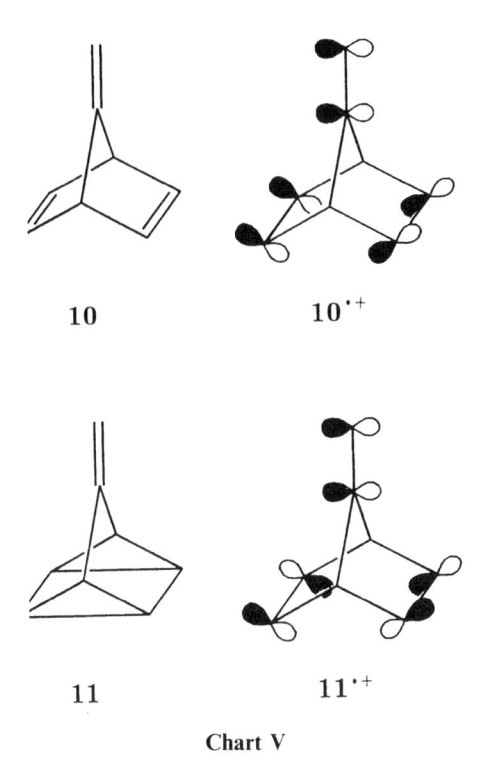

other cyclopropane bond is also possible, and is preferred because this is the more highly substituted bond.

Investigations along these lines were also carried out to elucidate the interactions of an exocyclic double bond or cyclopropane ring with the olefinic system of norbornadiene or with the cyclopropane moieties of quadricyclane [85]. With the radical cation of 7-methylenenorbornadiene **10** (Chart V), a substantial negative hyperfine coupling constant was found for the exocyclic protons, indicating a hyperconjugative interaction of the singly occupied molecular orbital (SOMO) of the norbornadiene moiety with the frontier orbitals of the exocyclic double bond, which possess compatible symmetry. In contrast, the hyperfine coupling constant of these protons in the radical cation **11**$^{\cdot+}$ of 7-methylenequadricyclane is large and positive. This result was explained by an exchange interaction mechanism of π, π spin polarization involving the SOMO of the quadricyclane part and the frontier orbitals of the olefinic moiety, which are orthogonal to the former. Similar, but weaker effects were observed for the corresponding compounds with a cyclopropane ring instead of the exocyclic double bond.

2. Kinetic and Mechanistic Studies. The simplest electron transfer conceivable is that between a radical ion and its parent compound, for example,

$$D^{\cdot+} + D \rightleftharpoons D + D^{\cdot+} \tag{69}$$

Although these so-called "self-exchange" or "degenerate electron-transfer" reactions do not lead to new products, they are very important nevertheless, because their activation barriers are key parameters in all current theories of electron transfer in solution. Despite the absence of net chemical change, the rates of self-exchange reactions can be measured if the reactants are labeled (#) as to lift the degeneracy of the two sides of Eq. 69 without influencing the energetics or kinetics of the reaction,

$$\#D^{\cdot+} + D \rightleftharpoons \#D + D^{\cdot+} \tag{70}$$

For obvious reasons, nuclear spins or electron spins are extremely well suited for a labeling under these constraints. If stationary reactant concentrations are used, one can evaluate the line broadening in NMR or EPR spectra. If, on the other hand, radicals $D^{\cdot+}$ are created as transients, any kind of nonequilibrium magnetization in the system, for instance, as is generated by the CIDNP effect, will with time distribute evenly between $D^{\cdot+}$ and D by the self-exchange; hence, the progress of this reaction can be monitored by a time-resolved experiment. An advantage of this technique compared to line-broadening methods is that rather unstable systems can be investigated as well.

A typical mechanism (photoreaction of an excited triplet acceptor $^3A^*$ with a donor D in a system where back electron transfer in the triplet state is not feasible thermodynamically) for the explanation of the CIDNP kinetics is shown in Chart VI; Fig. 11 displays the resulting time-resolved CIDNP signals in the system anthraquinone (**12**)/*N,N*-dimethylaniline (**13**) [86a]. Quenching of $^3A^*$ by D (rate constant k_q) yields a triplet radical ion pair. With the magnetic parameters of **12**$^{\cdot-}$ and **13**$^{\cdot+}$, the radical pair mechanism leads to nuclear spin polarizations of the aliphatic protons of the donor that are emissive in the singlet recombination product D and absorptive in the free radicals $D^{\cdot+}$, as symbolized by ↓ and ↑ in the chart. By the self-exchange (rate constant k_{ex}), the latter polarizations are also transferred to D, but more slowly, and gradually compensate the former polarizations. This exchange cancellation of CIDNP in cyclic reactions was first recognized by Closs [87]. In the absence of nuclear spin relaxation in the free radicals, the compensation would be perfect. In practice, relaxation (rate constant $1/T_1^{rad}$) cannot be neglected, and residual polarizations remain at long times, which would be emissive in this example. Thus, the

APPLICATIONS TO CHEMICAL PROBLEMS 119

$$^3A^* + D \xrightarrow{k_q} \overline{^3A^{\cdot -} + \uparrow D^{\cdot +}} \rightleftarrows \overline{^1A^{\cdot -} + \downarrow D^{\cdot +}} \longrightarrow A + \downarrow D$$

$$A + \uparrow D^{\cdot +} \xrightarrow{\quad k'_{ex} \quad} \uparrow D$$
$$\phantom{A + \uparrow D^{\cdot +}} \quad D \nearrow \quad \searrow D^{\cdot +}$$

$$\downarrow 1/T_1^{rad}$$

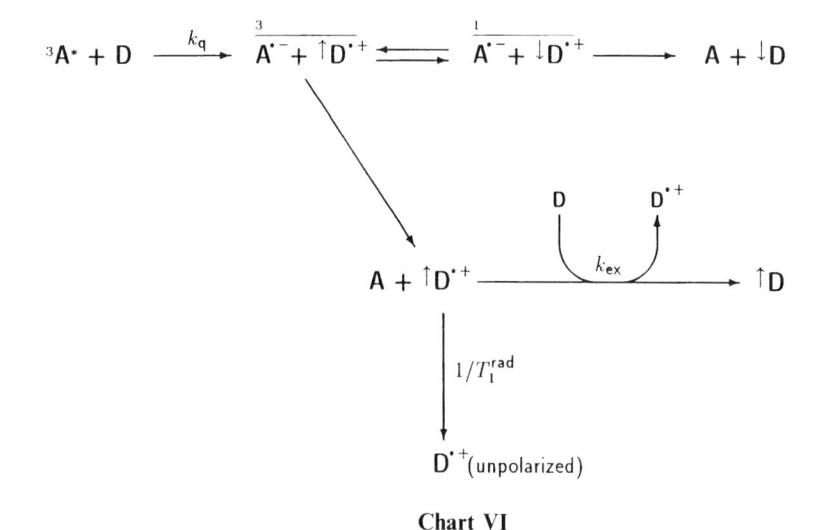

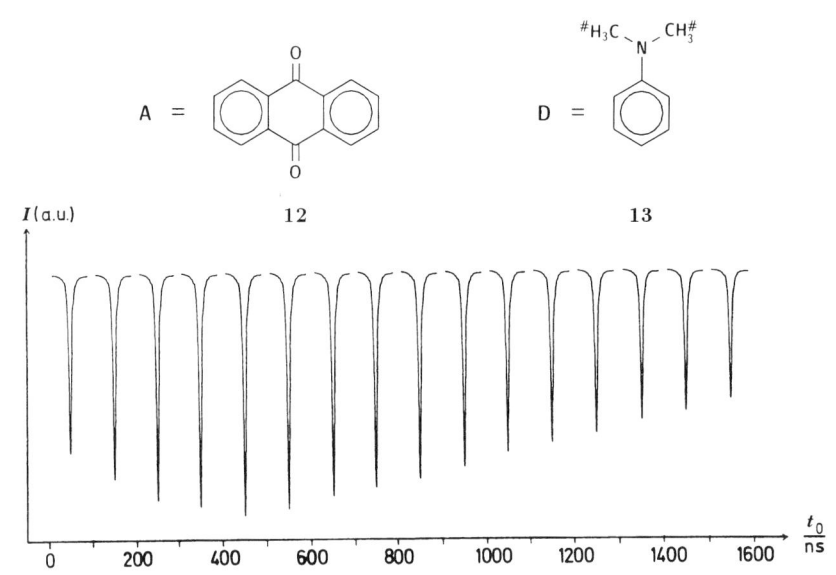

$$D^{\cdot +} \text{(unpolarized)}$$

Chart VI

A = [anthraquinone structure]   D = [N,N-dimethylaniline structure with #H₃C, CH₃#]

12   13

Figure 11. Time-resolved CIDNP spectra in a system described by the mechanism of Chart VI: photoreaction of the acceptor anthraquinone **12** (8×10^{-4} M) with the donor N,N-dimethylaniline **13** (3.2×10^{-4} M) in acetonitrile-d_3. Experimental parameters $T = 257$ K, excitation wavelength 343 nm. Shown is the dependence of the signal I of the dimethylamino protons (marked with # in the formula) on the delay time t_0 between laser flash and acquisition pulse. [Adapted from ref. [86a] with permission. Copyright © 1990 Elsevier Science Publishers B.V.]

quenching rate is reflected by the rise of the signals in Fig. 11, the decay of the signals yields the self-exchange rate, and from the residual signal one obtains the relaxation rate.

For a number of organic radical cations derived from para-substituted N,N-dimethylanilines [86b] and phenol ethers $C_6H_{6-n}(OCH_3)_n$ [86d], rate constants and activation parameters of the self-exchange reactions were determined in this way by time-resolved CIDNP spectroscopy. Within the two series, isokinetic relationships were found. The self-exchange rate constants agreed well with theoretical predictions of a model [86c] based on the Marcus theory, where all molecular parameters (geometries, force constants, and charge densities at the individual atoms) the model needs as input were obtained by semiempirical molecular orbital (MO) calculations with the AM1 [88] Hamiltonian.

By the same experimental technique, the temperature dependence of the nuclear spin relaxation rates was investigated for the radical cations of dimethoxy- and trimethoxybenzenes [89]. The rates of these processes do not appear to be accessible by other methods. As was shown, $1/T_1^{rad}$ of an aromatic proton in these radicals is proportional to the square of its hyperfine coupling constant. This result could be explained qualitatively by a simple MO model. Relaxation predominantly occurs by the dipolar interaction between the proton and the unpaired spin density in the p_z orbital of the carbon atom the proton is attached to. Calculations on the basis of this model were performed with the density matrix formalism of MO theory and gave an agreement of experimental and predicted relaxation rates within a factor of 2.

Becker et al. [90] focused on another aspect of electron transfer, namely the rates of electron return. They studied a series of para-substituted arene diazonium salts 14^+, with rubrene 15 being used as (singlet) sensitizer. The relevant steps of the mechanism are displayed in Chart VII. The aryldiazo radicals 14˙ produced by electron-transfer quenching of excited 15 are σ radicals possessing a negligible interaction between the para substituent and the unpaired electron. Consequently, the magnetic parameters of the radical pairs may be taken as constant within the series. The same holds for the diffusional dynamics because the solvent is not changed and the molecular sizes of the quenchers are practically identical. The relative CIDNP enhancement factor per radical pair is thus directly related to the pair lifetime, that is, to the rate constant k_{-e} of electron return. Plotting the experimental enhancement factors of ^{13}C CIDNP against the free enthalpy of charge recombination $\Delta G^°_{-e}$ which was calculated from electrochemical data, gave a Marcus curve, with the para-dimethylamino compound, which possesses a much higher value of $\Delta G^°_{-e}$ than the other diazonium salts, clearly falling into the inverted region.

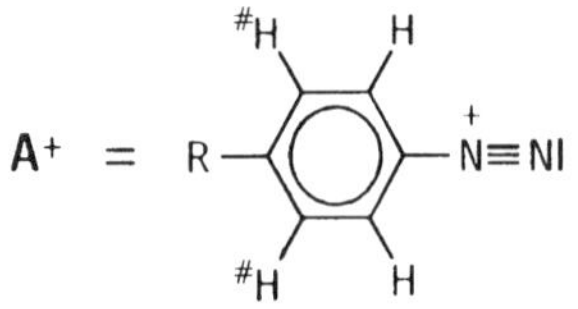

APPLICATIONS TO CHEMICAL PROBLEMS 121

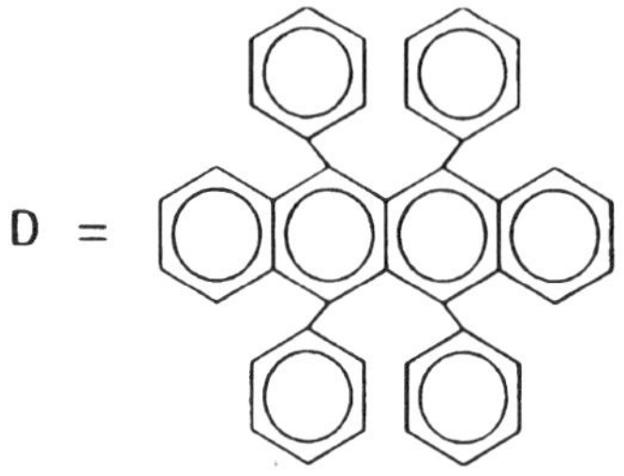

$$A^+ = R\!-\!\underset{\#H\ \ H}{\overset{\#H\ \ H}{\bigcirc}}\!-\!N\!\equiv\!N|^+ \qquad D = [\text{structure 15}]$$

$$14^+ \qquad\qquad 15$$

$$A^+ + {}^1D^* \longrightarrow \overset{1}{{}^{\#}A^{\cdot} + D^{\cdot+}} \rightleftharpoons \overset{3}{{}^{\#}A^{\cdot} + D^{\cdot+}}$$

$$\downarrow k_{-e}$$

$${}^{\#}A^+ + D \qquad\qquad\qquad \text{escape products}$$

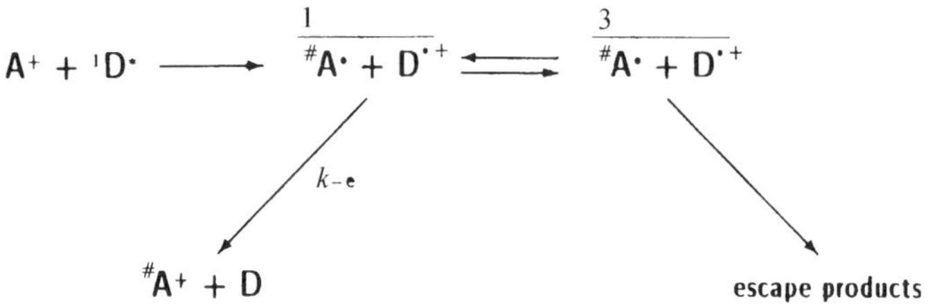

Chart VII

The CIDNP phases allow a clear distinction between back electron transfer of singlet pairs and of triplet pairs. This has been put to use by Fischer and Schaffner [91]. In a system such as (cf. Chart VIII) *N,N*-dimethyl-l-naphthylamine (**16**)/benzonitrile (**17**), charge recombination is not only possible in the singlet state but also in the triplet state because the triplet energy of **16** is lower than the energy of the radical ion pair. However, the triplet species cannot be observed by NMR, so the polarizations are hidden in it for a duration on the order of its triplet lifetime. Consequently, only polarizations from the singlet exit channel are visible at early times even if electron return is faster for triplet pairs; the subsequent deactivation triplet product → ground-state product leads to time-dependent CIDNP effects similar as those described above. Addition of triplet scavengers reduces the triplet lifetime and, therefore, increases the rate of appearance of the polarizations from the triplet exit channel in the CIDNP spectrum. Depending on the scavenging rate, that is, the scavenger concentration, the CIDNP signals thus undergo a phase inversion some time after the laser flash, or — with very efficient scavenging — their phase from the beginning is opposite to that in the experiment without scavenger (see Fig. 12). The actual CIDNP kinetics are complicated by nuclear spin relaxation in the triplet species and the above-mentioned polarization transfer by electron

A = ⟨phenyl⟩—CN

17

D = ⟨N,N-dimethylaminonaphthalene with $^\#H_3C$, $CH_3^\#$⟩

16

Chart VIII

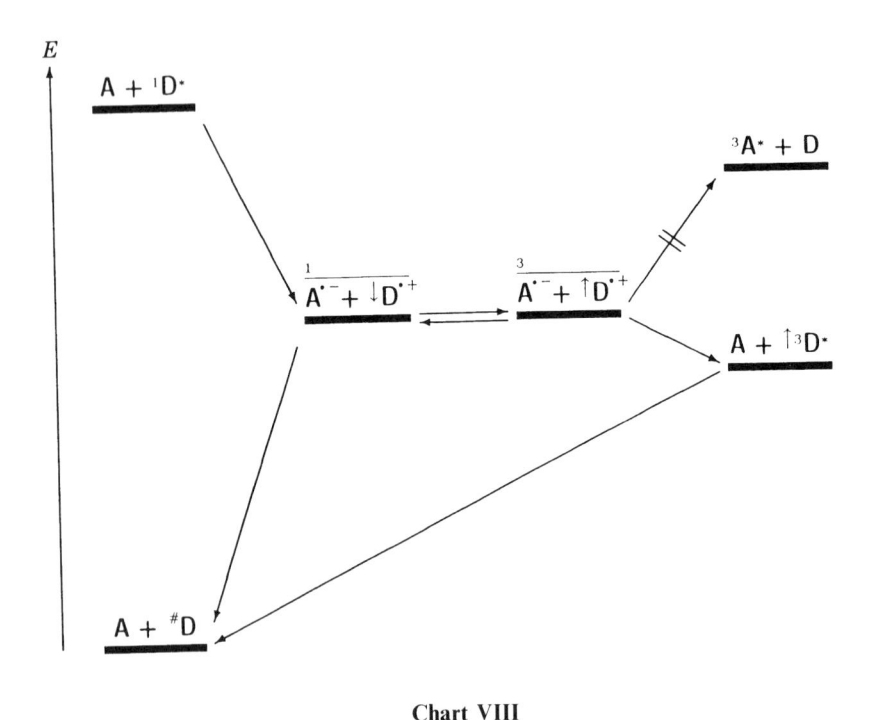

self-exchange. Nevertheless, Fischer and Schaffner [91b] were able to separate singlet and triplet back electron transfer by a global analysis of time dependence and scavenger dependence and determining as many of the kinetic parameters as possible from independent photophysical and photochemical experiments. Their results show a Marcus-type behavior for the rate constants, with the singlet recombination (higher energy gap) in the inverted region and the triplet recombination (lower energy gap) in the normal region.

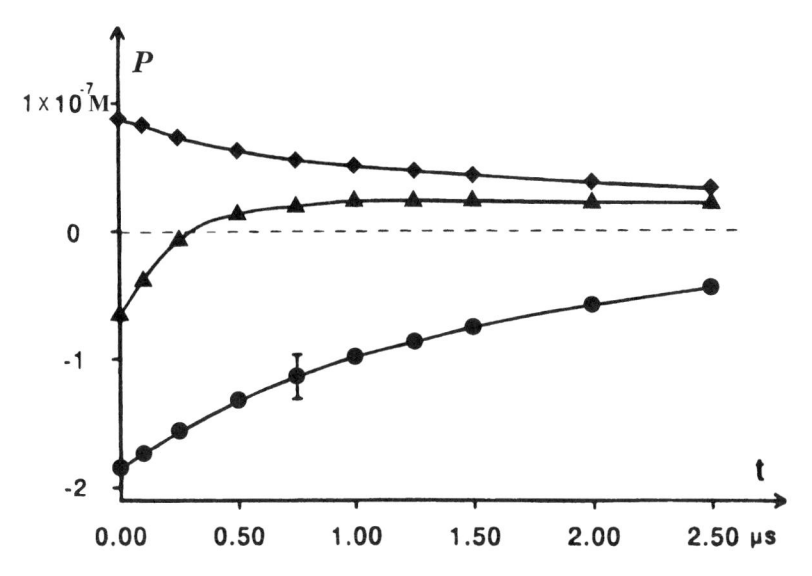

Figure 12. Time dependence of the CIDNP intensities P in the system of Chart VIII for different concentrations of the triplet quencher 1,3-cyclohexadiene (diamonds, 9.4 mM; triangles, 0.94 mM; circles, without quencher). [Adapted from E. Schaffner and H. Fischer, *J. Phys. Chem.*, **99**, 102 (1995) with permission. Copyright © 1995 American Chemical Society.]

Finally, CIDNP results were taken as evidence for benzoquinone-mediated photoinduced electron transfer in a donor–acceptor system [92], and the rates of electron self-exchange and nuclear spin relaxation of the radical anion of C_{60} have been measured by time-resolved CIDNP [93].

B. Hydrogen Abstractions

Hydrogen abstractions in ketone/amine and quinone/amine systems continue to attract the attention of CIDNP spectroscopists [94] despite the facts that the application of CIDNP to these reactions dates back to 1974 [95] and that the basic mechanism—electron transfer from the amine **DH** followed by deprotonation of the resulting aminium cation **DH**$^{\cdot+}$ to give an α-aminoalkyl radical **D**$^{\cdot}$–has already been cleared up in those early investigations [46]. CIDNP spectroscopy is very well suited to probe the microscopic details of such reactions that involve more than one radical intermediate. Polarizations can arise in both **DH**$^{\cdot+}$ and **D**$^{\cdot}$, but the spin density distributions of these two radicals differ strongly. Hence, the polar-

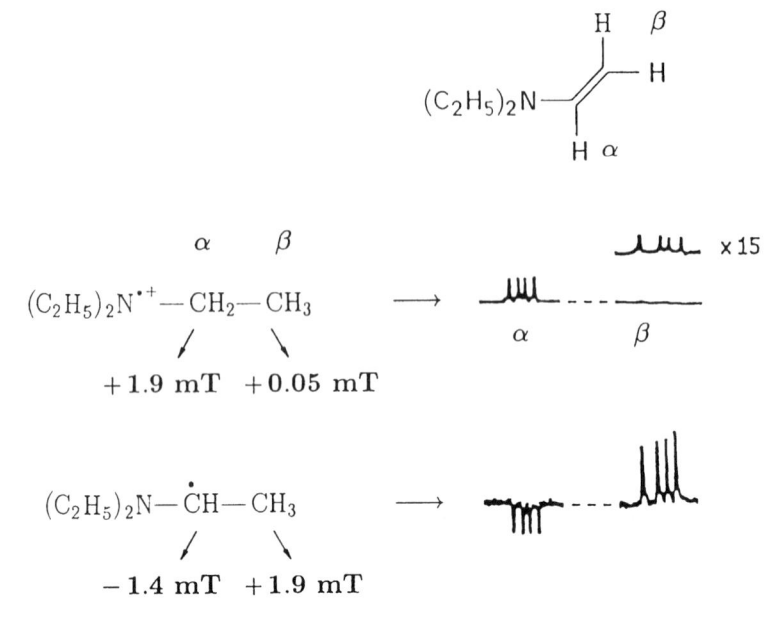

Figure 13. (Left) Hyperfine coupling constants [46] of the α and β protons in the aminium cation **DH**$^{·+}$ and the α-aminoalkyl radical **D**$^{·}$ derived from triethylamine. (Right) Resulting polarization pattern of these protons in the reaction product N,N-diethylvinylamine (see formula at the top) of the photoinduced hydrogen abstraction by anthraquinone **12**. [Adapted from ref. [94e] with permission. Copyright © 1994 VCH Verlagsgesellschaft Weinheim.]

ization pattern (cf. Fig. 13) allows conclusions as to where the polarizations stem from, which in turn yields mechanistic and kinetic information.

In a mechanistic study [94c], it was found that two independent pathways exist for the deprotonation of aminium cations (see Chart IX). The proton is either abstracted within the cage, by the radical anion **A**$^{·-}$ of the sensitizer, or outside the cage, the base in the latter case being another amine molecule. Both pathways lead to the same products but can be distinguished by the polarization patterns, which are symbolized by $^{\#}$ and § in the chart, corresponding to the upper and lower pattern in Fig. 13, respectively. Which pathway is taken is determined by the competition of in-cage deprotonation with intersystem crossing and escape of the radical ions from the cage. The standard free enthalpy $\Delta G°$ of both deprotonation reactions was obtained from thermodynamic cycles. These thermodynamic calculations explained why amines such as N,N-dimethylaniline and diazabicyclo[2.2.2]octane are not deprotonated in the photoreactions with

Chart IX

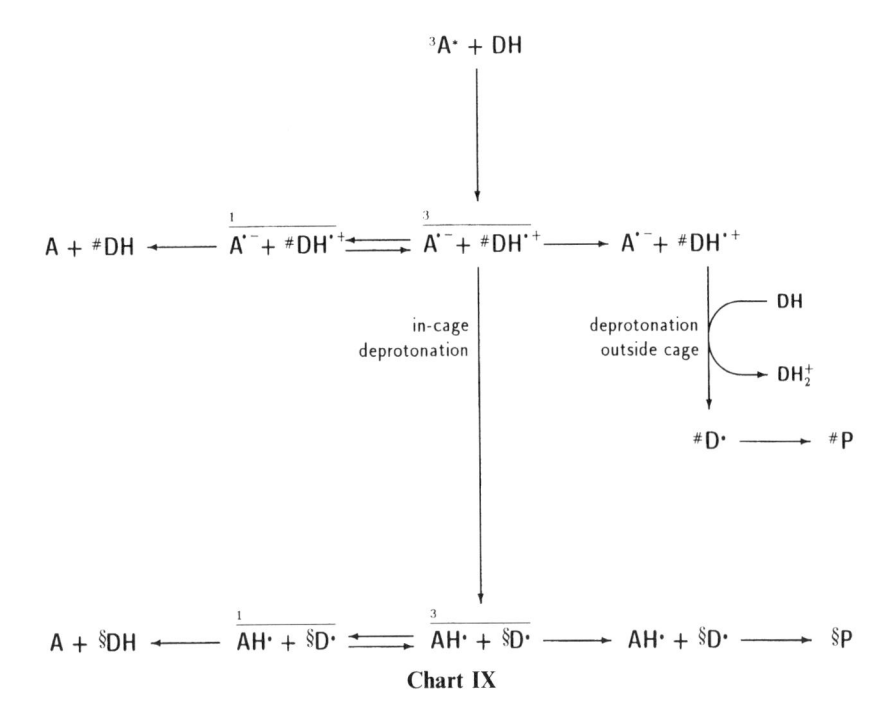

```
                        ³A• + DH
                           |
                           ↓
                    1            3
A + #DH  ←—  A•⁻ + #DH•⁺ ⇌ A•⁻ + #DH•⁺  ⟶  A•⁻ + #DH•⁺
                           |                    |  ⎡— DH
                     in-cage            deprotonation |
                   deprotonation         outside cage |
                                                      ⎣→ DH₂⁺
                                                 #D•  ⟶  #P
                           |
                           ↓
                    1           3
A + §DH  ←—   AH• + §D•  ⇌  AH• + §D•  ⟶  AH• + §D•  ⟶  §P
```

anthraquinone, so that CIDNP investigations of electron self-exchange are possible in these systems (cf. Section V.A.2). The dependence of the rate of in-cage deprotonation on the driving force $\Delta G^\circ_{\text{in-cage}}$ was also studied. On the one hand, $\Delta G^\circ_{\text{in-cage}}$ was varied by variation of the sensitizer [94c], on the other hand by variation of the solvent [94e]. An interesting threshold behavior was found (see Fig. 14): for increasing exergonicity, the polarization pattern, and thus the deprotonation pathway, changes abruptly within a small range of about 25 kJ mol^{-1}. Hence, $\Delta G^\circ_{\text{in-cage}}$ determines both sensitizer dependence and solvent dependence of these reactions.

The contributions of radical ions $\text{DH}^{•+}$ and neutral radicals $\text{D}^•$ to the product polarizations were also separated by variable-field stationary CIDNP [94a], time-resolved CIDNP, and SNP [94b]. From an evaluation of the CIDNP memory effect, the rate of in-cage proton transfer from the radical cation of triethylamine to the radical anion of *trans*-stilbene was estimated to be $3 \times 10^7 \text{ s}^{-1}$ [94f].

With unsymmetrically substituted tertiary aliphatic amines, deprotonation can occur at different sites, and CIDNP spectroscopy was employed to measure relative group reactivities [94d]. For these complex reaction mechanisms, where two deprotonation routes can lead to the same products,

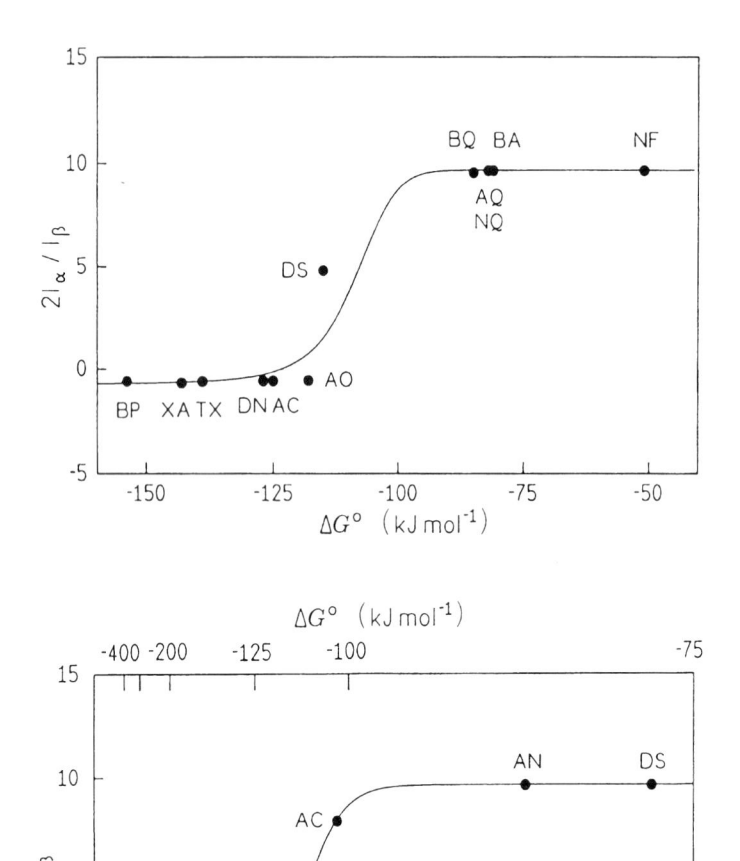

Figure 14. Dependence on $\Delta G°_{\text{in-cage}}$ of the deprotonation pathway (see Chart IX) of the aminium cation derived from triethylamine. The ratio $2I_\alpha/I_\beta$ of polarizations of the olefinic α and β protons of diethylvinylamine (cf. Fig. 13) is shown as function of $\Delta G°_{\text{in-cage}}$; a value of about -1 indicates complete in-cage deprotonation, a value of about $+9$ exclusive deprotonation outside the cage. (Top) Variation of $\Delta G°_{\text{in-cage}}$ by variation of the sensitizer. (Bottom) By variation of the solvent. The solid lines are a global best fit to both data sets. The labels (see [94e]) at the curves denote sensitizer and solvent, respectively. [Reproduced from ref. [94e] with permission. Copyright © 1994 VCH Verlagsgesellschaft Weinheim.]

this approach appears to possess an important advantage over chemical studies of the final product distribution, namely that CIDNP detection allows one to conclude whether or not the yields obtained refer to the same deprotonation pathway. As was shown, this is not always the case. For example, with the system *N*-ethyldiallylamine/anthraquinone in acetonitrile, ethyl is only deprotonated outside the cage, whereas allyl is also deprotonated within the cage.

Hydrogen abstractions by deoxybenzoin from the solvent, undeuterated and perdeuterated cyclohexane, were studied by using stationary ^{13}C CIDNP [96]. As was inferred from the polarizations, the products are formed to a large degree by secondary encounters of escaped radicals. These reactions are accompanied by α-cleavage of the ketone, which is of comparable rate as deuterium abstraction but significantly slower than hydrogen abstraction.

Two time-resolved CIDNP investigations of hydrogen transfer were reported, in which the rates of hydrogen exchange between carbonyl compounds (benzaldehyde and benzophenone [97a]; benzoquinone [97b]) and their ketyl radicals were measured. The experiments also yielded the homogeneous recombination rate of the radicals.

Other studies have dealt with hydrogen abstraction from the solvent by the photosensitizing drug nalidixic acid [98a], hydrogen transfer in anthraquinone/xanthene systems [98b], photoreductions of quinones by alcohols [98c] and of acetylenic ketones by various hydrogen donors [98d], the oxidation of NADH analogues [98e–98g], and the reaction of 4-methyl-2-quinolinecarbonitrile with optically active phenylpropionic acid [98h].

C. Fragmentations

1. α-Cleavage of Carbonyl Compounds. A long-standing paradigm of ketone photochemistry, namely that α-cleavage takes place exclusively from the triplet state, has recently been challenged by Azumi and co-workers [99]. They report stationary CIDNP experiments on dibenzyl ketone **18** in different solvents. By comparing experimental polarization intensities with theoretical calculations they arrive at the conclusion that in deuterochloroform the contribution of α-cleavage from the excited *singlet* state of **18** is as high as 75–85%, depending on temperature, while in toluene it lies between 25 and 40%. (Earlier CIDNP results by others [100] on aliphatic ketones in CDCl$_3$ had also been regarded as evidence for α-cleavage via a singlet pathway involving an exciplex with the solvent.) Fischer et al. [101] pointed out a number of shortcomings of that study, above all that in [99] a variant of radical pair theory was used that is known to be grossly inaccurate for singlet pairs. They carefully repeated and refined these experiments using time-resolved CIDNP spectroscopy and testing the radical pair model used for their evaluation on another ketone, pivalophenone, for which cleavage has been established to occur from a long-lived triplet state. According to their results, the contribution of an excited singlet state of **18** to α-cleavage is negligible. However, as Stern–Volmer experiments again gave evidence for some (20–30%) participation of such a state in this fragmentation [102], the issue cannot be regarded as settled conclusively, and CIDNP spectroscopy may play an important role in deciding it.

Flash-CIDNP experiments have been used to distinguish between one- and two-photon processes in ketone chemistry [103]. Several strategies have been followed. With the first, the concentration of radical pairs is probed through the CIDNP signal P_{gem} of the geminate recombination product, that is, the starting ketone **K**. Because the lifetime of higher excited triplet states $^3K^{**}$ is shorter than that of the first excited triplet state $^3K^*$, only the latter can be quenched by addition of a triplet quencher. Hence, Stern–Volmer plots obtained from P_{gem} become more and more curved with increasing intensity of the exciting laser pulse (cf. Fig. 15). By an appropriate fitting procedure, it was possible to incorporate the fact that cleavage of $^3K^*$ and of $^3K^{**}$ may yield different radicals, and rate constants and triplet extinction coefficients could be determined in this way. Second, for an unsymmetrical ketone such as methylethylketone the relative polarizations of the different protons in the recombination product were monitored as

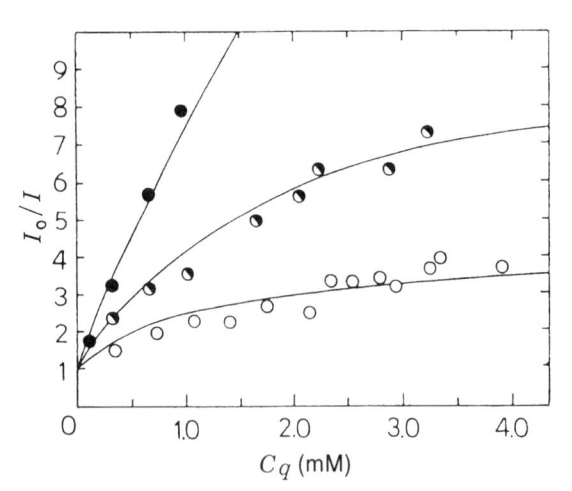

Figure 15. Stern–Volmer plot obtained from the nuclear polarizations observed in α-cleavage of methylethylketone; quencher *cis*-piperylene (concentration C_9). Filled circles, laser energy (308 nm) 3 mJ; half-filled circles, 7.4 mJ; open circles, 18 mJ. [Adapted from ref. [103] with permission. Copyright © 1993 Elsevier Sequoia.]

functions of the laser intensity. With this ketone, cleavage of $^3K^*$ produces the pair $\overline{\text{CH}_3\dot{\text{C}}\text{O} \; \dot{\text{C}}\text{H}_2\text{CH}_3}$ while cleavage of $^3K^{**}$ is less selective and also yields the pair $\overline{\text{CH}_2\text{CH}_3\dot{\text{C}}\text{O} \; \dot{\text{C}}\text{H}_3}$. Because high-field CIDNP intensities are proportional to the hyperfine coupling constants, the first pair leads to weak polarization of the methyl substituent in the geminate product and to strong polarization of the methylene protons, whereas polarization intensities from the second pair are just the other way round. With increasing laser intensity, the polarization ratio of methylene to methyl protons thus decreases. Third, when $^3K^*$ and $^3K^{**}$ react to different products, the CIDNP intensities of these products can be evaluated. This is the case with acetone in isopropanol, where $^3K^*$ reacts with the solvent to give, among other products, an enol, while $^3K^{**}$ fragments, the escaping methyl radicals leading to methane. The ratio of polarizations of enol and methane was found to be a linear function of the laser intensity with vanishing intercept, thus indicating that α-cleavage of acetone in isopropanol exclusively occurs through a two-photon mechanism.

2. Other Cleavage Reactions. Considerable effort has been devoted [104–106] to the photodissociations of carbon–heteroatom bonds in compounds of the general structure Ar—X—R, where X = O, N, or S (see Chart X). CIDNP spectroscopy played a central part in clearing up the mechanisms. The findings led to a unified description [105] of reactions that until then had been regarded as unrelated, namely β-cleavage of aromatic ketones, photo-Fries and photo-Claisen rearrangements. Especially with respect to β-cleavage of aromatic carbonyl compounds (R = CH$_2$C(O)R′), the results caused a revision of mechanistic interpretations previously advanced in the literature.

By evaluation of the polarization phases, it could be demonstrated unambiguously [104a, 104b] that β-cleavage of phenoxyacetophenones occurs from an excited singlet state. As these reactions had been classified as reactions of the carbonyl chromophor, this appeared to be a violation of a well-established paradigm of ketone photochemistry (cf. Section V.B.1). However, from a comparison of the reactivity of a series of differently substituted phenoxyacetophenones [104b], it was concluded that β-cleavage is not a carbonyl reaction but a reaction of the Ar—X moiety, which bears resemblance to the photo-Claisen rearrangement.

In a study of the β-cleavage of aryl substituted acetones [104c], it was found that this process takes place both from the singlet state ($n\pi^*$) and from the triplet state ($\pi\pi^*$), with comparable quantum yields. This was established by combining CIDNP detection with sensitization experiments: Upon addition of a triplet sensitizer, the CIDNP signals of all products that are due to β-cleavage become inverted while their absolute intensities

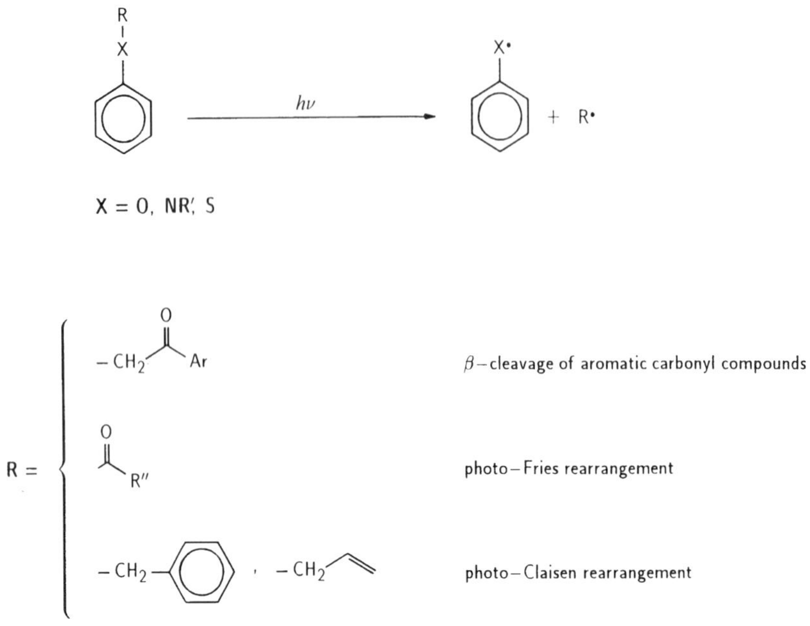

Chart X

remain essentially constant. Temperature-dependent CIDNP spectra further show that in the direct photoreaction the precursor multiplicity changes with decreasing temperature (at high temperature, singlet; at low temperature, triplet), which manifests itself by an inversion of the CIDNP signals (see Fig. 16). Such a behavior had only been described once in the literature [107]. It can be explained by a competition between thermally activated bond scission in the singlet state with temperature-independent intersystem crossing. From an analysis of the intensities of the CIDNP signals of Fig. 16, it was inferred that the singlet reaction possesses the lower activation energy.

All experimental observations with respect to these reactions, including the dependence on neither orbital character nor spin multiplicity of the excited state (which is in striking contrast to α-cleavage of carbonyl compounds) and the influence of substituents of the aryl moiety on the reaction rate, could be rationalized by correlation diagrams that were obtained on the basis of semiempirical MO calculations [105]. β-Cleavage of phenoxyketones as well as photo-Fries and photo-Claisen rearrangements were characterized to be $\pi\sigma^*$ photodissociations.

For further corroboration, the photodissociations of arylethers [104d, 104e] and arylthioethers [104e] (photo-Claisen reactions) were investigated.

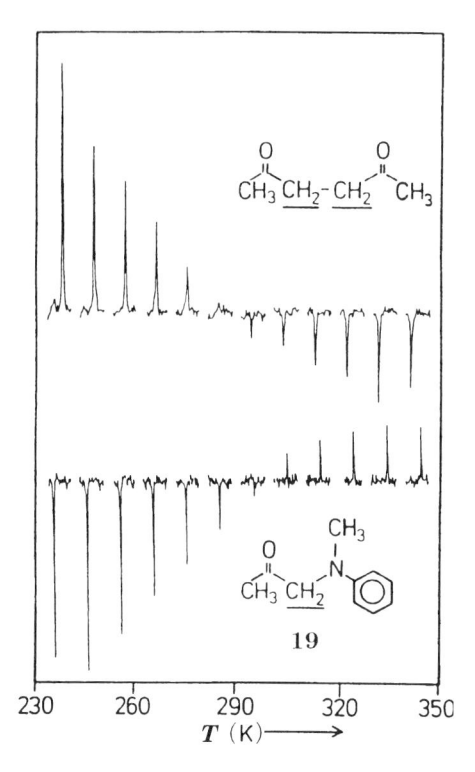

Figure 16. Photocleavage of (*N*-methylanilino)acetone **19** in acetonitrile-d_3. The plot shows the temperature dependence of the CIDNP signals of the CH_2-protons (underlined in the formulas given) in the geminate reaction product **19** (bottom) and in the escape product 2,5-hexadione (top). [Adapted from ref. [104c] with permission. Copyright © 1992 Elsevier Sequoia.]

With the aid of CIDNP spectroscopy it was shown that the primary step of these reactions is cleavage of the X—R bond, which upon direct excitation occurs both from the singlet and the triplet state. By combining the results of CIDNP experiments and photophysical measurements, rate constants of singlet and triplet reactions could be determined. In these systems, cleavage is again faster from the singlet state.

Cleavage from the singlet state is also the dominant pathway in the photo-Fries rearrangement of 1-naphthylacetate [106], as was inferred from laser flash photolysis and measurements of quantum yields. In that study, however, the authors concluded that the CIDNP signals mainly derive from the triplet precursor. This was explained by the involvement of a higher excited triplet state that is populated by a two-photon process and cleaves efficiently, while the first excited triplet state decays mostly by triplet–triplet annihilation.

The photochemical decomposition of 9,10-diphenylanthracene endoperoxide was investigated by ^{1}H and ^{13}C CIDNP [108]. Depending on the excitation wavelength (λ_{exc}), the first step of this reaction is fragmentation of a carbon–oxygen bond ($\lambda_{exc} \leq 290$ nm) or of the oxygen–oxygen bond ($\lambda_{exc} \geq 350$ nm). The precursor multiplicity is singlet in both cases. The subsequent steps involve biradicals and radical pairs, and the CIDNP experiments have been able to shed some light on the complex reaction mechanism.

D. Secondary Reactions of Radicals and Biradicals

1. Radical Additions. Batchelor and Fischer [109] measured absolute rate constants k_{add} for addition of 2-hydroxy-2-propyl radicals **21·** to alkenes by a new method, using time-resolved CIDNP. Photoexcited 2,4-dihydroxy-2,4-dimethylpentan-3-one **20** served as a source of these radicals. Chart XI summarizes the relevant steps of the underlying reaction scheme in a simplified manner. Their key idea was to evaluate not the net polarizations, which may be strongly distorted by nuclear spin relaxation in the radicals, but the multiplet polarizations, which had previously [110] been shown to be uninfluenced by relaxation at the paramagnetic stage provided that the

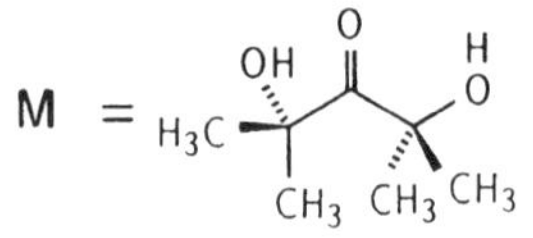

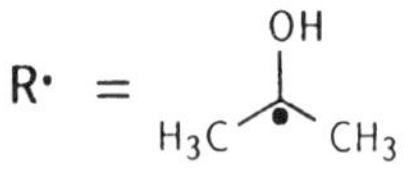

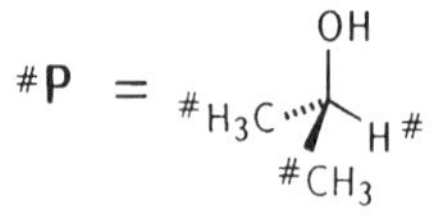

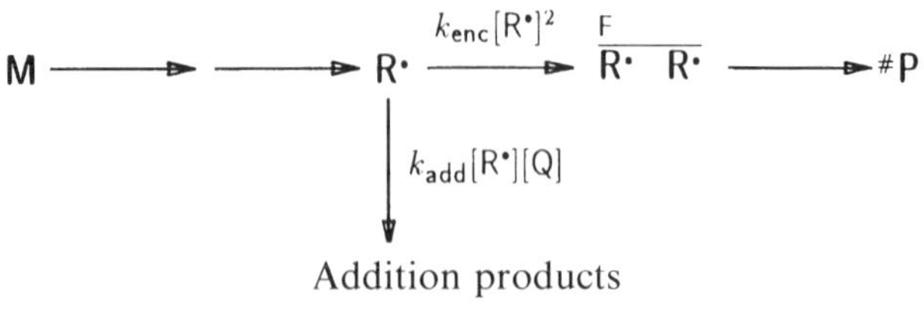

Chart XI

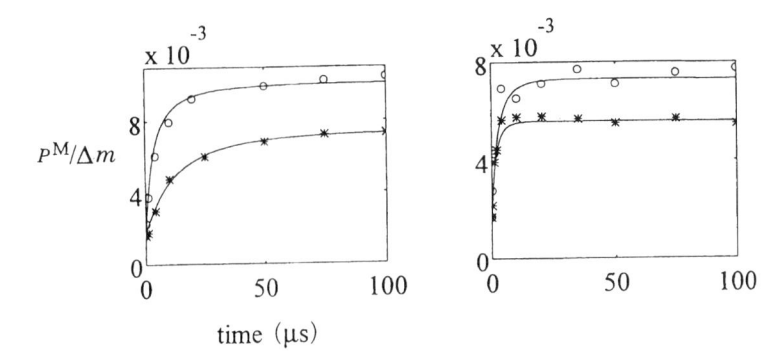

Figure 17. Time dependence of the multiplet polarization $P^M/\Delta M$ in the product **22** upon quenching of the radical **R·** (**21·**) by an olefin **Q**; solvent isopropanol. For further explanation, see Chart XI and the text. (Left) Constant concentration (3.3×10^{-3} M) of the quencher $CH_2=CPh_2$, variable initial radical concentration (open circles, [**R·**] = 11.6×10^{-5} M; stars, [**R·**] = 2.1×10^{-5} M). (Right) Variable concentration of the quencher $CH_2=CHCN$ (open circles, [**Q**] = 3.4×10^{-4} M; stars, [**Q**] = 7.5×10^{-4} M), constant initial radical concentration ([**R·**] = 10.9×10^{-5} M). [Adapted from S. N. Batchelor and H. Fischer, *J. Phys. Chem.*, **100**, 9794 (1996) with permission. Copyright © 1996 American Chemical Society.]

coupled protons exhibiting the multiplet effect in a product stem from different radicals of the pairs (i.e., that $\sigma_{ij} = -1$). The latter condition is fulfilled for the product they observe, isopropanol **22**. As the CIDNP multiplet effect arises in F pairs (cf. Section II.B.2) **21·21·**, which are formed by bimolecular encounters of the radicals, the multiplet polarization of **22** in the absence of scavengers thus follows pure second-order kinetics. In the presence of an alkene **Q**, **21·** is scavenged by the addition reaction. This leads to a shortening of the rise time and a decrease of the final value of the multiplet polarization in **22**. Figure 17 shows experimental examples. It was found that negligible amounts of **22** are formed by cross-reactions. Hence, the rate of addition of **21·** to an alkene can be determined by a one-parameter fit if one knows the rate constant of self-termination of **21·** and the initial radical pair concentration. The former is constant in a given solvent and can be measured by an experiment without scavenger, the latter can be obtained by a single control experiment with a different concentration of **Q** (Fig. 17, right) or a different initial concentration of **21·** (Fig. 17, left).

To keep within the kinetic window of time-resolved CIDNP, the reaction rate is controlled by the alkene concentration. The method was reported to be applicable for k_{add} being in the range $10^5 \cdots 10^9 \, M^{-1} \, s^{-1}$, the limits being dictated by possible interference of the alkene with the initial steps of the reaction at very high concentration, and nonnegligible consumption of the alkene during a measurement at very low concentration. Its main advantage compared to other methods for the determination of k_{add} is speed. By this technique, a series of alkenes was investigated. A good correlation between k_{add} and alkene electron affinity was found, the rate constant varying over eight orders of magnitude in this series, which shows the radical 21˙ to be very nucleophilic. Solvent effects and steric effects on k_{add} were also studied.

Other examples of the application of CIDNP spectroscopy to radical additions and substitutions include the self-substitution of quinones in the presence of tertiary aliphatic amines [111], the photoreactions between hexamethyldisilane and quinones [112], and the allylation of quinones via photoinduced electron transfer from allylstannanes [113] (see also Section V.E). Cycloadditions via radical ions are treated in a separate section (V.D.3).

2. Radical Fragmentations. The electron-transfer induced dediazatation of the cyclic azo derivatives **23** and **24** (see Chart XII) of quadricyclane **2** and norbornadiene **3** were investigated [114]. No polarizations were observed when singlet sensitizers were employed, which was explained by fast back electron transfer of singlet radical ion pairs. With the triplet sensitizer chloranil (**25**), **23** gave a CIDNP spectrum in which **3** was the only product exhibiting substantial polarizations. The polarization pattern of **3** was virtually identical to that observed [82b] in its photoreaction with **25**, showing the radical pair $\overline{25^{\cdot -} \; 3^{\cdot +}}$ to be the source of the polarizations. In contrast, when excited **25** was quenched by electron transfer from **24**, the CIDNP spectrum displayed polarizations of **2** and **3**, the polarization patterns indicating CIDNP to arise both in $\overline{25^{\cdot -} \; 2^{\cdot +}}$ and in $\overline{25^{\cdot -} \; 3^{\cdot +}}$. No polarizations could be attributed to pairs containing 23˙⁺ or 24˙⁺. From this, it was concluded that these radical cations are short lived on the CIDNP time scale; furthermore, dediazatation of 23˙⁺ obviously occurs via a single pathway, affording 2˙⁺ only, whereas two competing pathways exist for dediazatation of 24˙⁺, which lead to 2˙⁺ and 3˙⁺.

CIDNP spectroscopy was further employed to study the cleavage of aliphatic ditertiary ethers following photoionization in the vacuum UV [115]. Other examples from the field of radical fragmentations are treated elsewhere, see Sections V.F (fragmentation of the neutral radicals derived from onium salts) and V.G.1 (decarboxylation of sulfur-centered radical cations of cysteines). Strictly speaking, deprotonations of aminium cations, that is, two-step hydrogen abstractions from amines (Section V.B), also belong in this category.

APPLICATIONS TO CHEMICAL PROBLEMS   135

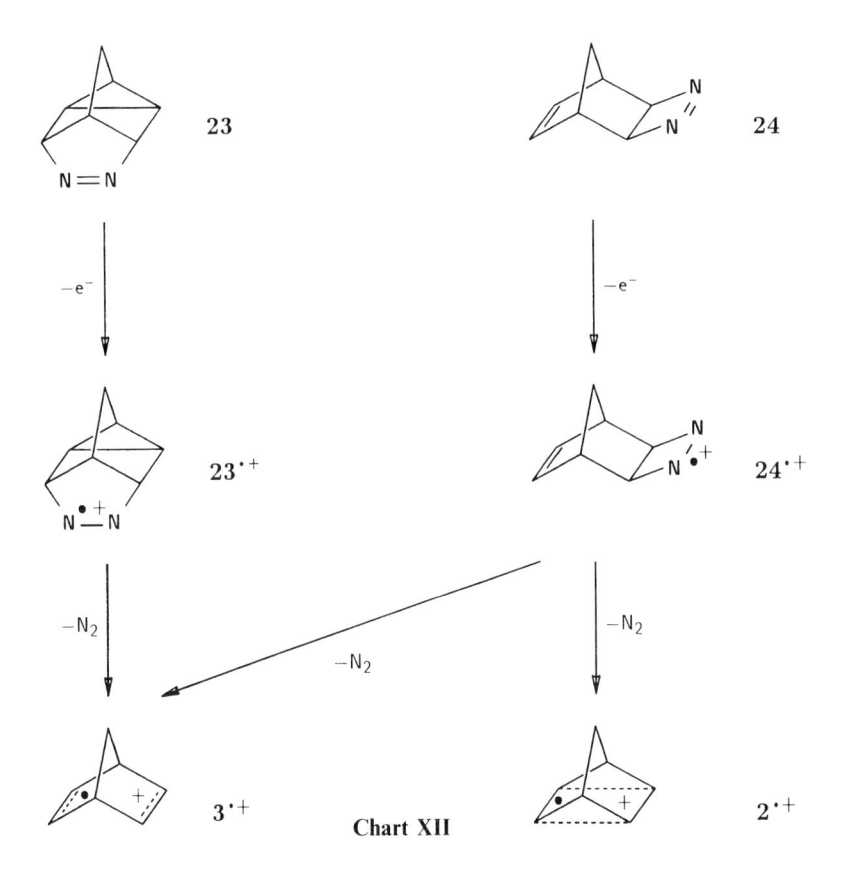

Chart XII

3. Cycloadditions, Cycloreversions, and Isomerizations. Because these processes are interrelated (e.g., cycloadditions are frequently accompanied by isomerization of the substrates via the same intermediates) they are not discussed in separate sections.

Roth and Hutton [116] investigated the cycloreversion of the radical cation of the dimethylindene dimer **26** (cf. Chart XIII), which they generated by photoinduced electron transfer from **26** to triplet chloranil. As they showed by calculating spin density distributions for the radical cations that can be expected as intermediates, the polarization pattern observed both for regenerated **26** and for the monomer **27**, which is characterized by equal signal phase and comparable intensities for α and β protons, cannot be reconciled with CIDNP arising in a single type of radical pair. They were able to explain the polarization patterns by assuming that CIDNP is generated in two subsequent radical pairs. From an analysis with den

Chart XIII

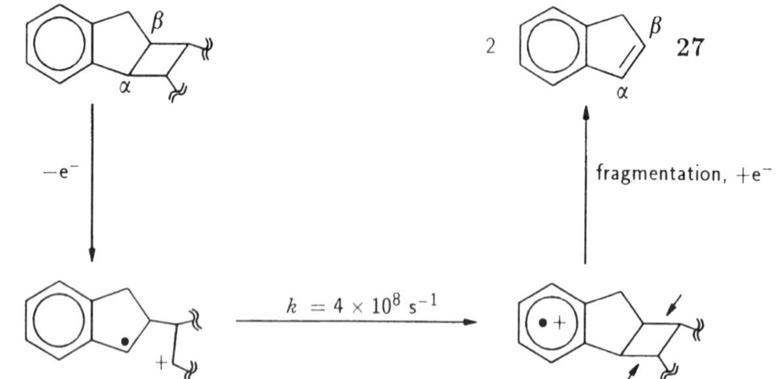

Hollander's theory [41] of pair substitution, they concluded that an open chain or extended radical cation **28**$^{\cdot+}$ must precede a closed radical cation **29**$^{\cdot+}$ in which spin and charge are localized in an aromatic moiety; with the reverse sequence of intermediates, which would perhaps appear more natural from the point of view of chemical intuition, the experimental polarization pattern could not be simulated. Based on their analysis, they obtained a rate constant of $4 \times 10^8 \, \text{s}^{-1}$ for the transformation **28**$^{\cdot+} \to$ **29**$^{\cdot+}$. Related cycloreversions of biological significance (photosensitized splitting of pyrimidine dimers) are dealt with in Section V.G.2.

All these processes are retrodimerizations. The characteristic of the corresponding cycloadditions is that the sensitizer, while being essential for activating the reactants (by photoinduced electron transfer) and for making the products persist (by back electron transfer), takes no part in the cycloaddition proper, the key step of which is the combination of a radical ion with its parent molecule. For cycloadditions of unlike species, another mechanistic pathway is open, namely, via direct combination of radical ion *pairs*, as was reported in several recent publications [117].

By a detailed CIDNP investigation [117a] of the Paterno–Büchi reactions of anetholes **31** with quinones **30** in polar medium earlier mechanistic hypotheses were disproved. Stationary and time-resolved experiments showed the mechanism to have the following novel features (cf. Chart XIV): Spin-correlated radical ion pairs (i.e., $\overline{\mathbf{30}^{\cdot-} \, \mathbf{31}^{\cdot+}}$) are key intermediates for cycloadduct formation; free radical ions do not play a significant role. In the singlet state, these pairs undergo back electron transfer; geminate reaction of triplet pairs leads to triplet biradicals, which are the precursors to the photoproducts.

Mechanistic conclusions drawn from CIDNP results are frequently doubted on the grounds that CIDNP is blind to pathways other than via

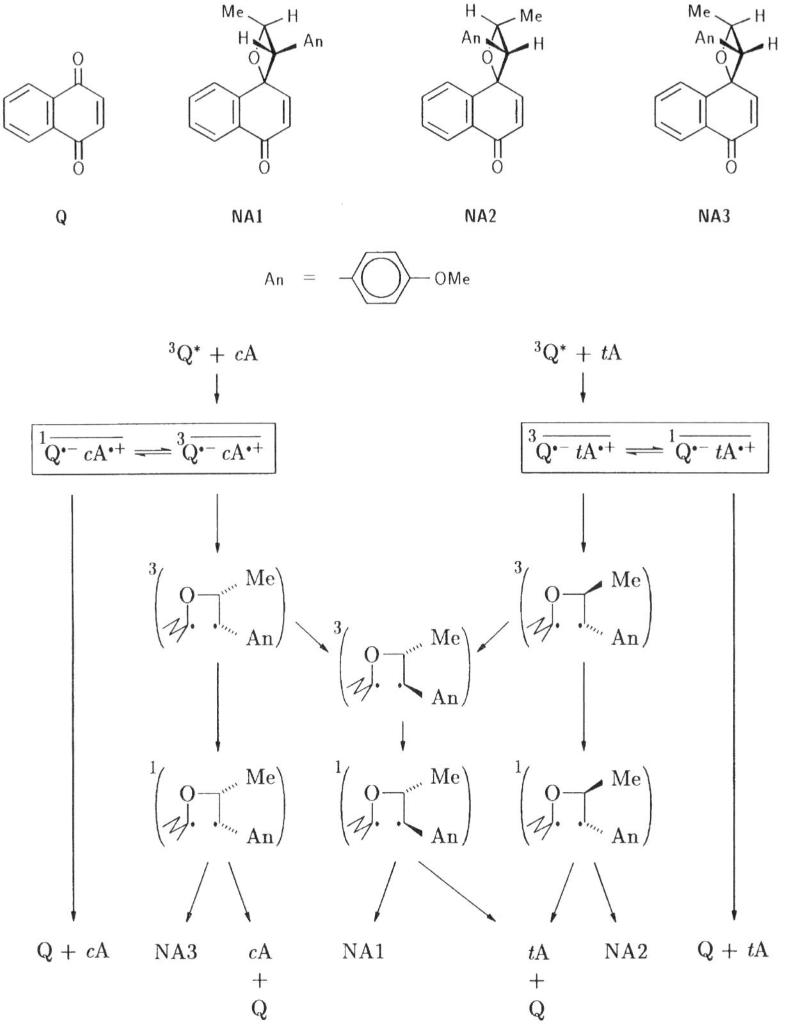

APPLICATIONS TO CHEMICAL PROBLEMS

Chart XIV

radical pairs; hence, the objection is raised that the observed polarizations might not be caused by the main reaction but might just stem from a minor side channel. This problem was also addressed in [117a]. For the systems studied there, exciplexes $*(30^{\delta-} \cdots 31^{\delta+})$ could conceivably lead to cycloaddition but would not give rise to any CIDNP effects. To rule out this pathway, photoinduced electron transfer (PET) sensitization was employed, by which the radical ion pairs $30^{\cdot -}\ 31^{\cdot +}$ were prepared indirectly without the possibility of formation of *($30^{\delta-} \cdots 31^{\delta+}$) (see Fig. 18). To this end, 9-cyanoanthracene **CNA** was used as auxiliary (singlet) sensitizer, which could be selectively excited and was quenched by *cis*-anethole *c***A**. In this two-component system, no CIDNP arises because for the radical pairs **CNA**$^{\cdot-}$ *c***A**$^{\cdot+}$ back electron transfer is equally feasible in the singlet and in the triplet state, so the nuclear spin polarizations from both exit channels cancel. However, adding benzoquinone **BQ** as a third component in high concentration (0.1 M) leads to quantitative interception of **CNA**$^{\cdot-}$ by the quinone during diffusive excursions of the primary pairs **CNA**$^{\cdot-}$ *c***A**$^{\cdot+}$, and, as the top trace of Fig. 18 shows, to strong CIDNP signals. In this three-component system, the desired radical ion pairs **BQ**$^{\cdot-}$ *c***A**$^{\cdot+}$ are obtained as secondary species. For energetic reasons, neither photoexcited **BQ** nor photoexcited *c***A** can be formed in this process. Hence, an exciplex between **BQ** and *c***A** is also impossible. The CIDNP spectrum in this PET-sensitized experiment and the CIDNP spectrum of the direct photoreaction between ^{3}**BQ** and *c***A** (bottom trace of Fig. 18) are mirror images of one another. The inversion of overall signal phase is due to the change in precursor multiplicity (triplet in the direct reaction, singlet in the PET-sensitized reaction) and is thus proof of the interception process. As conditions were chosen such that the same number of radical pairs **BQ**$^{\cdot-}$ *c***A**$^{\cdot+}$ was formed in both experiments, the equal absolute intensities in the two CIDNP spectra shows that the pathway via spin-correlated radical ion pairs is indeed the main pathway to cycloaddition.

Another aspect of CIDNP that was put to use in these investigations [117a] is the facile characterization, both qualitatively (structures) and quantitatively (relative yields), of the diamagnetic products. While this would of course have been equally possible by standard methods of product separation and analysis, the signal enhancement by the polarizations causes the CIDNP method to be advantageous in two respects. On the one hand, the product distribution can be sampled at low consumption of the reactants and low concentration of accumulated products, so, for instance, the danger of being misled by a secondary reaction of a product is minimized. On the other hand, the method is very fast because product separation is unnecessary and the spectrum from which the information is to be extracted is present within minutes after starting the illumination of the sample. In this way, a very detailed picture of the reaction was obtained from the CIDNP results. The primary electron transfer was found to play an important role for the structure of the cycloadducts. On the one hand, it decides the regiochemistry, because the atoms bearing the largest positive charge density in the radical cations and those bearing the largest negative charge density in the radical anions are connected in biradical formation. On the other hand, it is partly responsible for the stereochemistry (see Chart XIV).

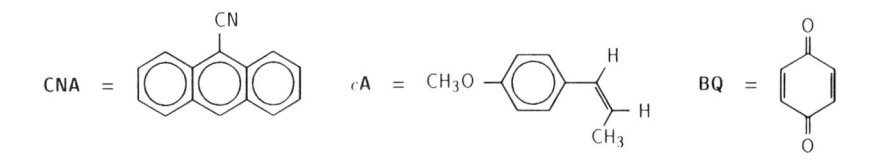

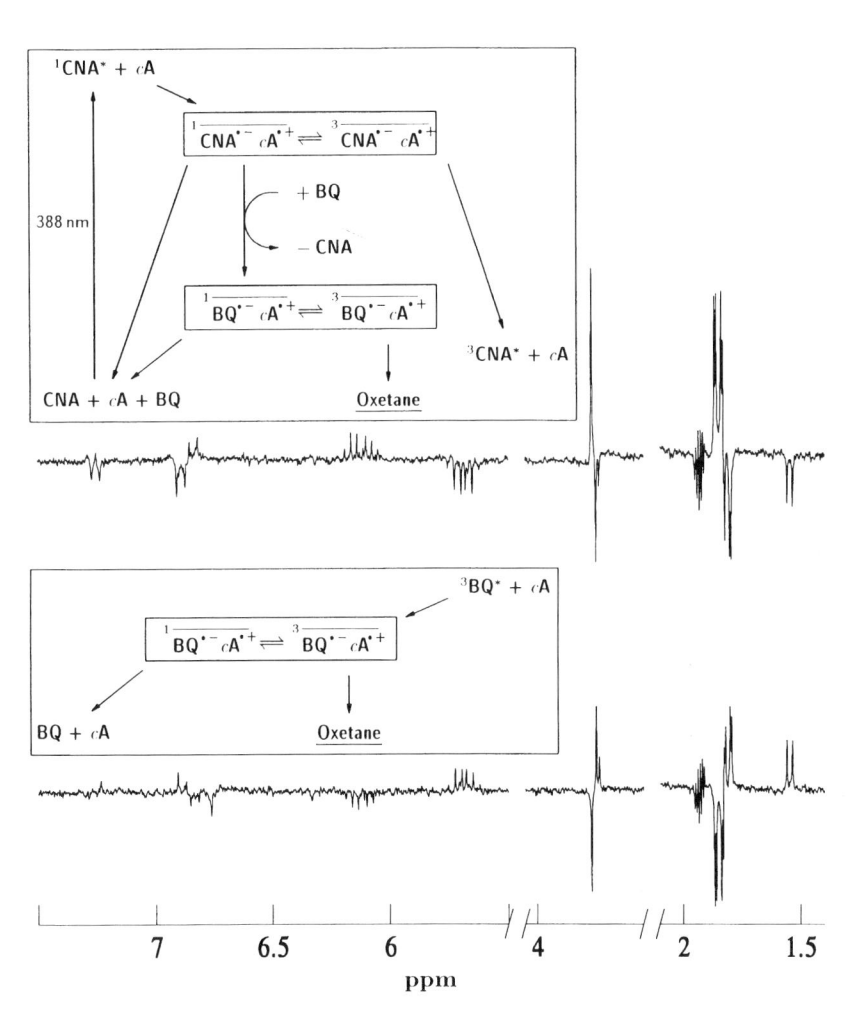

Figure 18. Cycloaddition of benzoquinone **BQ** with *cis*-anethole *c*A. The lower CIDNP spectrum is observed in the direct photoreaction, the upper one in the PET-sensitized reaction (sensitizer 9-cyanoanthracene **CNA**). The pertaining reaction mechanisms are given above the respective trace; the chemical formulas of the reactants are shown at the top. For further explanation, see the text.

As the CIDNP spectra show, with naphthoquinone **Q** at low temperatures, one predominantly obtains the product possessing the same configuration of the anethole moiety as the starting olefin (**NA2** from *trans*-anethole *t***A**, **NA3** from *c***A**); in contrast, at room temperature the thermodynamically most stable cycloadduct **NA1** is formed, which possesses a trans configuration of the anethole moiety regardless of the configuration of the starting olefin. This product distribution and its temperature dependence, as well as the observation that cycloaddition is accompanied by one-way cis–trans isomerization of the olefin, the amount of which increases with temperature, are explained by the mechanism shown in Chart XIV. The biradicals are first formed in a geometry allowing maximum Coulombic stabilization of the transition state, hence from a sandwich-like arrangement of the radical ions. After intersystem crossing of these biradicals, ring closure yields the respective low-temperature oxetane **NA2** or **NA3**, and biradical scission regains the starting olefin. In the biradicals, the Coulombic stabilization is no longer present, so an energetically more favorable conformation can be reached by rotations around single bonds; this conformation is the same for both anetholes. Ring closure of the relaxed biradical leads to the high-temperature oxetane **NA1**, and scission to *trans*-anethole *t***A**, that is, to one-way isomerization. Bond rotation is thermally activated and competes with temperature-independent intersystem crossing of the biradicals. This competition was studied by CIDNP. From the temperature dependence of the distribution of the polarizations among the different products, the activation energies for bond rotation, the triplet lifetimes of the initially formed biradicals, and the scission probability of the relaxed biradical were obtained. Information from the CIDNP experiments thus extended not only to the stage of the radical ion pair but also to the biradical stage of the reaction.

Another example is provided by the photocycloadditions of quinones with quadricyclane **2** or norbornadiene **3** to give oxolanes **32** and oxetanes **33** (Chart XV), where CIDNP experiments again yielded detailed insight into the mechanism [117b]. In these systems, polarizations to some extent arise in the radical ion pairs but to the greatest part in the biradicals. The overall CIDNP phases gave evidence that singlet biradicals lead to the oxolanes and triplet biradicals to the oxetanes. Furthermore, by scavenging experiments it was shown that (triplet) radical ion pairs precede triplet biradicals. Hence, the same cycloaddition mechanism as in the quinone/anethole systems is realized, but in this case the CIDNP effects predominantly stem from the second paramagnetic stage of the reaction, namely the biradicals.

Both combination of the radical ion pairs to give biradicals and oxetane formation from the biradicals are accompanied by skeleton rearrangements. By using the polarizations as labels, these rearrangements could be traced

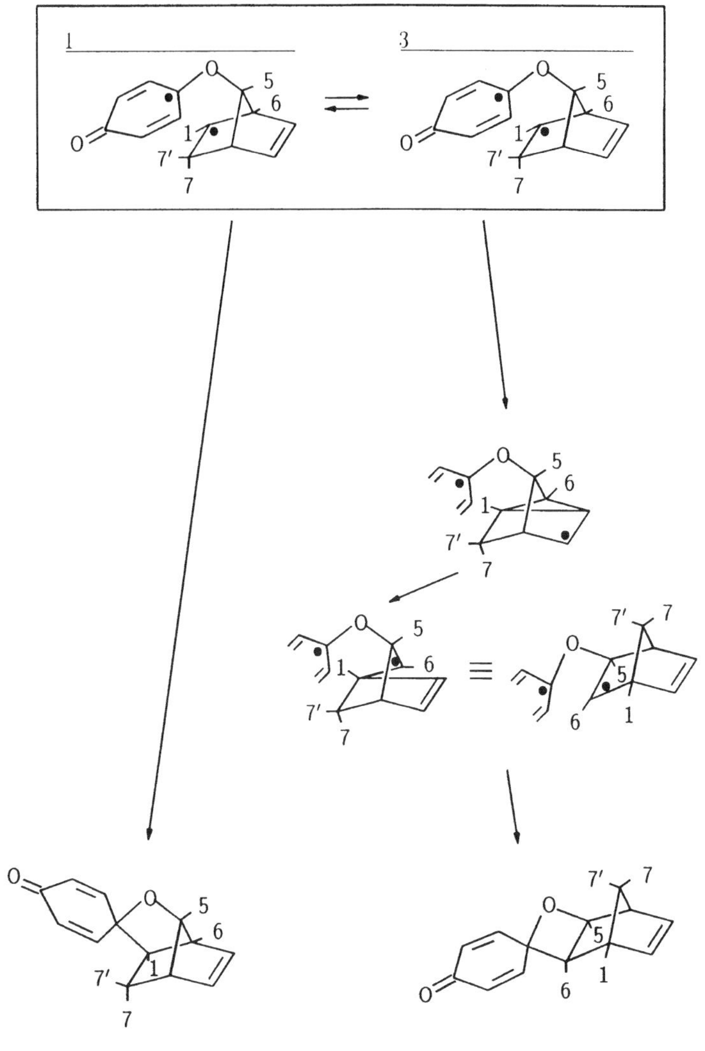

Chart XV

[117b]. The results of these experiments are summarized in Chart XV. The protons labeled with numbers in the chart can be unambiguously distinguished by their widely differing hyperfine coupling constants in the biradical. Of interest in this respect is that CIDNP generation at the preceding paramagnetic stage, in the radical ion pair, effects an additional labeling (not

shown in Chart XV). Such a two-step labeling would obviously be impossible to achieve by any chemical labeling procedure.

Finally, the same basic cycloaddition mechanism, via formation of triplet biradicals by geminate reaction of spin-correlated radical ion pairs, was also found for the mixed [2 + 2] photocycloadditions of anetholes 31 with fumarodinitrile 34 in polar solvents [117c]. While the regiochemistry in these reactions is again determined by the Coulombic interactions of the radical ions, the stereochemistry is uninfluenced by the primary electron transfer because product formation exclusively occurs from the relaxed conformation of the biradicals owing to their longer lifetime (lower intersystem crossing rate) compared to the Paterno–Büchi biradicals. In these systems, back electron transfer is also feasible for radical ion pairs in triplet states, in contrast to the previously discussed cases. From an analysis of the CIDNP data, it was concluded that this process indeed occurs and manifests itself by bidirectional isomerization of the reactants. However, it is only about half as efficient as formation of a chemical bond between $31^{\cdot+}$ and $34^{\cdot-}$, which leads to the biradicals and subsequently to cycloaddition as well as unidirectional cis–trans isomerization of the anetholes. This at first glance surprising difference in reactivity was explained by the small exergonicity of back electron transfer in the triplet state and the much larger exergonicity of biradical formation. Contrary to mechanistic hypotheses put forward for analogous systems [118], the pathways via triplet anethole or triplet fumarodinitrile do not lead to cycloaddition, as follows from the CIDNP results.

These examples seem to indicate that the described mechanism of photocycloadditions via radical ion pairs possesses some generality: It covers several, at least formally different, categories and electronic configurations (heterocycloadditions: [$2\pi + 2\pi$] in the Paterno–Büchi reactions of quinones with anetholes or norbornadiene, [$2\pi + 4\sigma$] in the corresponding reactions of quinones with quadricyclane; homocycloadditions: [$2\pi + 2\pi$] in cyclobutane formation from anetholes and fumarodinitrile), is independent of whether the acceptor is excited (in the first two examples) or the donor (in the last example), and is independent of the electron spin multiplicity of the excited state (triplet in the former two instances, singlet in the latter). The last case is interesting insofar as it shows that the triplet radical ion pair → triplet biradical pathway is preferred to other cycloaddition pathways even if it involves two intersystem crossing steps (singlet–triplet in the radical ion pairs, and triplet–singlet in the biradicals).

CIDNP experiments have also been performed to gain insight into the mechanisms of PET induced [2 + 4] cyclizations of quinones with allenes [119], as well as of the photocyloadditions of chloranil with furan derivatives [120] and with N-vinylpyrrolin-2-one [121]. The sensitized valence isomerizations of norbornadiene and quadricyclane have received some attention [122], and CIDNP spectroscopy was applied to the rearrangement of spirofluorenebicyclo[6.1.0]nonatriene to spirofluorenebarbaralane [123]. Information about the rearrangements of the radical cations of other strained hydrocarbons was also obtained in several of the studies discussed in Section V.A.1.

E. Inorganic and Metal Organic Substrates

Kaptein and co-workers [124] addressed the question to what degree $S-T_\pm$-type CIDNP is influenced by a heavy-atom effect. To this end, they performed variable-field (5 ⋯ 800 mT) CIDNP investigations on the radical pairs $\overline{Ph-CH_2^\cdot\, Me_3Sn^\cdot}$, which were generated by photoinduced cleavage of dibenzyl ketone followed by hydrogen abstraction from ^{117}Sn-enriched trimethyltin hydride and F-pair formation. The absence of magnetic isotope effects (MIE) in a number of reactions involving tin radicals had been taken as evidence that nuclear spin dependent intersystem crossing cannot compete successfully with nuclear spin independent intersystem crossing via the spin–orbit coupling pathway [125]; on the other hand, strong ^{119}Sn CIDNP effects had been reported [126], which would be incompatible with this interpretation. For their system, the authors of [124] found that the radical pair mechanism can account completely for the results of their variable-field CIDNP experiments. As an explanation of the failure to observe MIE, they suggested isotope scrambling during the reaction. Tin-centered radicals are also intermediates in the allylation of quinones via photoinduced electron transfer from allylstannanes [113].

Lehnig et al. [127] used ^{1}H CIDNP to investigate the insertions of photochemically generated heavy carbene analogues R_2M, with M being Si, Ge, or Sn, and R being methyl or *tert*-butyl, into C—Cl and C—Br bonds. The aim of their study was to distinguish between concerted and stepwise insertion. The former is CIDNP-inactive while the latter proceeds via intermediate radical pairs formed by halogen abstraction from the substrate by the carbene, and can thus give rise to CIDNP. This diagnostic criterion had already been applied to carbene chemistry by Closs [128] soon after the discovery of the CIDNP effect. As the results of [127] show, insertion of dimethylsilylene occurs by the stepwise mechanism when the substrate is benzyl chloride, the precursor multiplicity being singlet. In contrast, in the case of allyl chloride the concerted mechanism is realized, which is not only indicated by the absence of CIDNP signals but is also corroborated by the absence of radical dimerization products. With simple, unstabilized alkyl halides, no reaction takes place. Dimethylgermylene is less reactive than dimethylsilylene. Finally, although the formal insertion product EtSnBu$_2$Br is formed when $(Bu_2Sn)_6$ is photolysed in the presence of ethyl bromide, the reaction does not seem to proceed via a stannylene; it was concluded that this product results from a radical substitution that does not involve radical pairs.

CIDNP spectroscopy has been employed to unravel the reactions following photoexcitation of $PtMe_4$(bipyridyl) [129]. From the polarization, it was inferred that the precursor multiplicity is triplet and that the primary photochemical step is cleavage of a Pt—Me-bond. Escape products are formed by deuterium abstraction from the solvent and geminate products by disproportionation.

In the reactions of excited quinones with hexamethyldisilane, CIDNP spectroscopy provided evidence that the polarizations arise in radical ion pairs. Furthermore, it was concluded that product formation involves scission of the Si—Si bond not concomitantly with but after geminate combination of radical cation and radical anion [112].

Buchachenko et al. [130] explored the analogy between the photoreactions of uranyl and carbonyl compounds, both groups being capable of inducing a hydrogen abstraction from a donor **DH** by a two-step pathway (cf. Section V.B),

$$>C=O + DH \rightarrow >\dot{C}-O^- + D^{\cdot} + H^+$$
$$=U=O^{2+} + DH \rightarrow =\dot{U}=O^+ + D^{\cdot} + H^+$$

The polarizations observed for the geminate disproportionation product benzophenone in the photoreactions of UO_2^{2+} with Ph_2CHOH or $Ph_2C(OH)COOH$ are consistent with a triplet precursor multiplicity and a g value of $\dot{U}O_2^+$ that is smaller than that of $Ph_2\dot{C}OH$, which is also in accordance with EPR data. Rykov et al. [131] studied the photodecompositions of uranyl carboxylates by CIDNP spectroscopy. With monocarboxylates, the spectra were dominated by CIDNP effects stemming from secondary radical pairs that did not contain uranoyl radicals [131a]; with the dicarboxylate uranyl succinate, the CIDNP signals were interpreted by $S-T_-$ mixing in an intermediate biradical, probably $O_2\dot{U}(OOC-CH_2-CH_2^{\cdot})$ [131b].

F. Polymerization Initiators

A detailed investigation of the photochemistry of the important photocuring agent 2,2-dimethoxy-2-phenylacetophenone **35** has appeared [132]. The reactions relevant in the present context are displayed in Chart XVI.

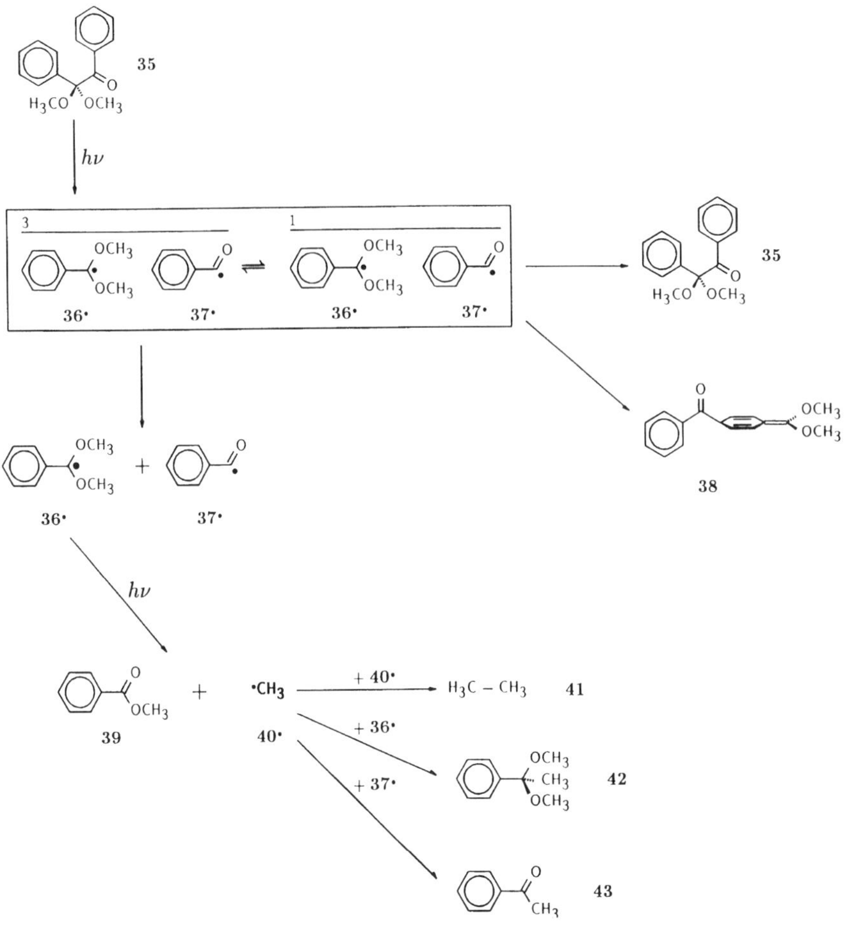

Chart XVI

A key step is cleavage of the primarily formed dimethoxybenzyl radical **36·** to give methylbenzoate **39** and methyl radical **40·**. Rate and mechanism of this process had been a point of controversy. By using CIDNP, Fischer et al. [132] were able to show that the demethylation occurs via electronic excitation of **36·** and subsequent fast fragmentation of the photoexcited radical. This follows unambiguously from the dependence of CIDNP intensities on the intensity I of the excitation light: Plotting net nuclear polarizations per initial radical pair $\overline{36\cdot 37\cdot}$, $|P/\Delta M|$, as functions of I yields constants (see Fig. 19, top, curves a and b) for the geminate products **35** and

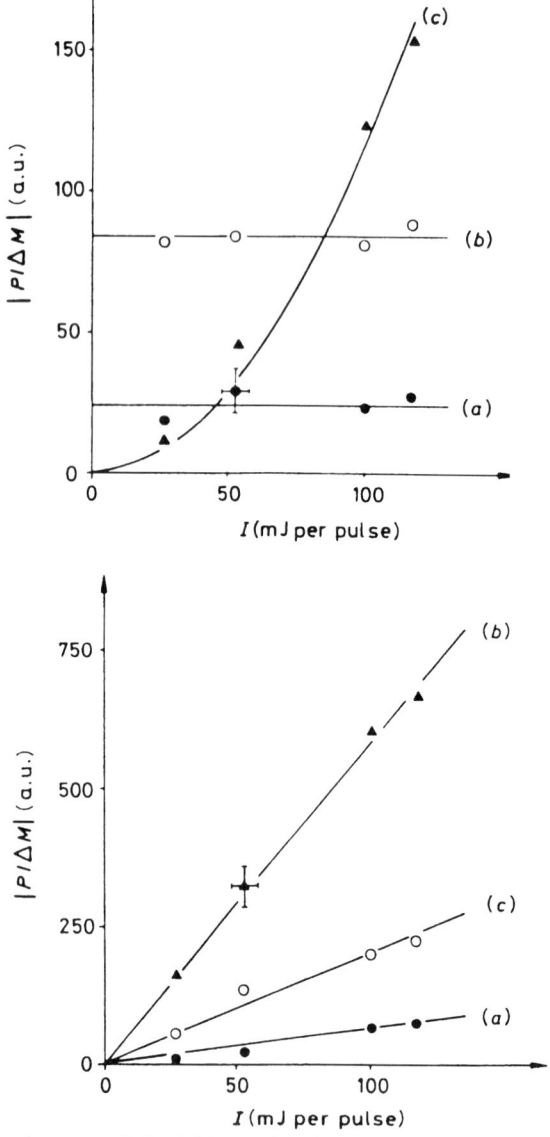

Figure 19. Photocleavage of the initiator 2,2-dimethoxy-2-phenylacetophenone **35**. Net nuclear polarizations per initial radical pair $|P/\Delta M|$ are plotted as functions of the light intensity I. (Top) Curves for the geminate products **35** (a) and **38** (b) that are formed without participation of secondary methyl, and for the combination product **41** (c) of two methyl radicals. (Bottom) Curves for the products **39** (a), **43** (b), and **42** (c), the formation of which involves one methyl radical. For further information, see the text and Chart XVI. [Reproduced from ref. [132] with permission. Copyright © 1990, The Royal Society of Chemistry.]

APPLICATIONS TO CHEMICAL PROBLEMS 147

38 of the initial pairs, that is, products that are formed *without* participation of methyl radicals. In contrast, $|P/\Delta M|$ increases linearly with I (Fig. 19, bottom) both for the diamagnetic product of the radical cleavage, **39**, and for the products **42** and **43**, the formation of which involves *one* methyl radical. Finally, $|P/\Delta M|$ depends quadratically on I (Fig. 19, top, curve c) in the case of ethane **41**, the combination product of *two* methyl radicals. On the one hand, this experiment demonstrates very elegantly that cleavage of the radical **36·** relies on photoexcitation. On the other hand, the fact that these relations are also observed for **42** and **43** implies that fragmentation of excited **36·** is fast compared with the lifetimes of the radicals **36·** and **37·**.

Chart XVII

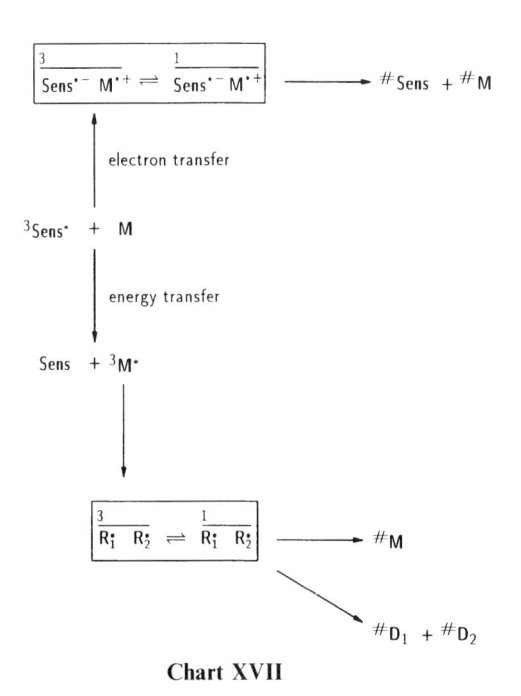

The distinction between energy-transfer and electron-transfer sensitization has been the topic of a study [133] on xanthone/α-aminoketone systems **Sens/M**, which are of major importance for UV curing of pigmented coatings. The basic idea is that while a decision between these alternatives might be impossible on the basis of the products if the reaction mechanism is complex, it should be possible on the basis of the intermediate radicals because (cf. Chart XVII) energy transfer leads to two radicals derived from the initiator, $\mathbf{R}_1^{\cdot}$ and $\mathbf{R}_2^{\cdot}$, and electron transfer to one sensitizer-based and one initiator-based radical, **Sens**$^{\cdot-}$ and $\mathbf{M}^{\cdot+}$. Consequently, any polarization of the sensitizer observed in the bimolecular reaction cannot be due to energy transfer, whereas geminate polarizations in a disproportionation product $\mathbf{D}_1$ or $\mathbf{D}_2$ of the initiator are clear evidence for it. This method proved to be very successful for the systems in question.

CIDNP spectroscopy has been applied to both direct and sensitized photoreactions of onium salts (diaryliodonium and triarylsulfonium salts) [134]; see Chart XVIII. In a recent investigation [134f] ambiguities of earlier studies on sensitized photolysis of iodonium salts [134a, 134b], where $S-T_{\pm}$-type polarizations had to be postulated to rationalize the polariz-

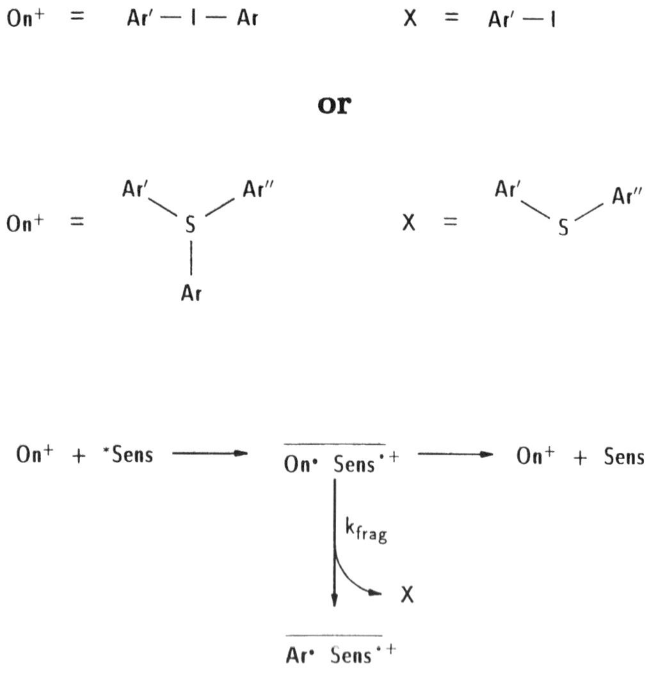

$$On^+ = Ar' - I - Ar \qquad X = Ar' - I$$

or

$$On^+ = \begin{array}{c} Ar' \diagdown \diagup Ar'' \\ S \\ | \\ Ar \end{array} \qquad X = \begin{array}{c} Ar' \diagdown \diagup Ar'' \\ S \end{array}$$

$$On^+ + {}^{\cdot}Sens \longrightarrow \overline{On^{\cdot} \; Sens^{\cdot +}} \longrightarrow On^+ + Sens$$

$$\downarrow k_{frag}$$

$$\longrightarrow X$$

$$\overline{Ar^{\cdot} \; Sens^{\cdot +}}$$

Chart XVIII

ations, could be removed by carefully choosing systems for which thermodynamic and photophysical data as well as control experiments exclude energy transfer and unambiguously fix precursor multiplicity μ and exit channel ε. As was shown in this way, normal radical pair theory ($S-T_0$ mixing) is capable of explaining all the observed polarizations.

These onium salts **On**$^+$ are of interest not only because they are important initiators of cationic as well as radical photopolymerizations but also from the point of view of CIDNP spectroscopy itself. First, their reductions by electron-transfer sensitizers **Sens** do not involve charge separation but only charge shifts. As there is no Coulombic attraction between the radicals of the resulting pairs **On˙Sens˙**$^+$, the pair dynamics should essentially be uninfluenced by the solvent polarity (cf. the polarity dependence of radical ion pair CIDNP, Section IV.B). This was verified experimentally [134f]. Second, the primarily formed iodonium or sulfonium radicals **On˙** fragment to give iodoarene or diarylsulfide **X**, respectively, and an aryl radical **Ar˙**; escaping **Ar˙** terminates predominantly by abstraction of deuterium from the solvent. Contrary to previous studies [134a, 134b], it was found [134f] that the main source of the polarizations of the deuterated arene is the secondary radical pair **Ar˙Sens˙**$^+$. Furthermore, it was shown that the amount of polarizations from the primary and secondary pairs crucially depends on the fragmentation rate k_{frag} of **On˙**. This key step is faster with sulfonium than with iodonium salts, and its rate increases with decreasing solvent polarity. On the basis of these findings, a unified description of electron-transfer sensitized photolysis of both types of onium salts could be given.

G. Biologically Relevant Molecules

CIDNP studies of biopolymers and their building units have been covered up to 1992 in a comprehensive review [135]. The following sections therefore deal with more recent investigations only.

1. Amino Acids, Peptides, and Proteins. The determination of the accessibility of amino acid residues is the "standard" application of CIDNP to proteins and larger peptides, the key idea being that only amino acids exposed to the surface can react with a photoexcited dye. The photoreactions must be reversible to avoid unwanted structural changes of the biopolymer that are induced by the experiment itself. This can be realized with cyclic electron-transfer (cf. Section V.A.2, Chart VI) or hydrogen-transfer reactions. Because of the photochemistry of amino acids, the only

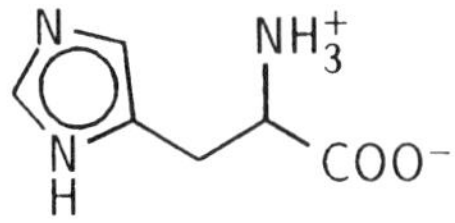

Chart XIX

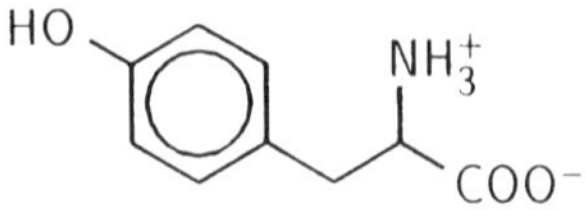

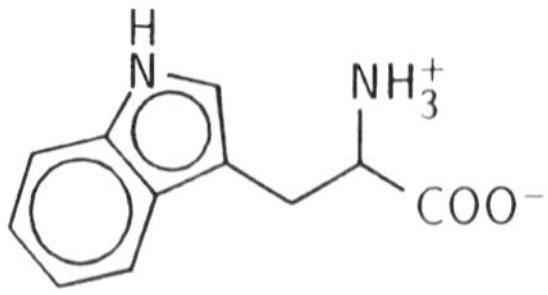

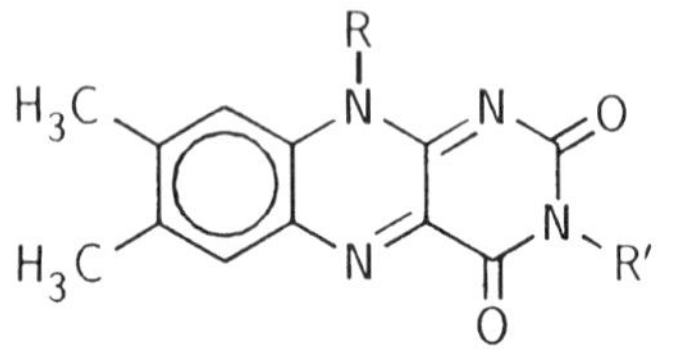

commonly occurring ones amenable to this technique are histidine **44**, tyrosine **45**, and tryptophan **46**, which all react at their aromatic side chains. As dyes, flavins **47** are used almost exclusively.

A fundamental problem of this method, which so far appears to have been ignored in all these investigations, has been addressed by Hore [136]: The polarization intensity of an amino acid residue is not simply a constant that is specific for the radical pair resulting with the dye but also comprises a Stern–Volmer term describing the competition with the other amino acids in the polymer for the excited dye molecules. Hence, an undetectable CIDNP signal from a particular residue does not guarantee inaccessibility of the latter; predominant quenching of the dye by other amino acids must be excluded first. Likewise, a decrease of a particular CIDNP signal during a structural change does not necessarily mean a decrease of accessibility of the corresponding residue; it could also be due to increased accessibility of other, more efficiently quenching residues. The problem is most severe with histidine because its Stern–Volmer constants for flavin quenching are six and eight times lower than those of tyrosine and tryptophan. As no dye is known at present that specifically polarizes histidine, the use of mixtures of amino acids as model systems seems to be the only reliable, though tedious, solution.

The (irreversible) photoreactions of cysteine derivatives **48** with 4-carboxybenzophenone **49** in D_2O have been investigated [137], the interest in these reactions being due to their model character for the damage of cell components. CIDNP spectroscopy allowed a detailed analysis of the rather complex mechanism (see Chart XX).

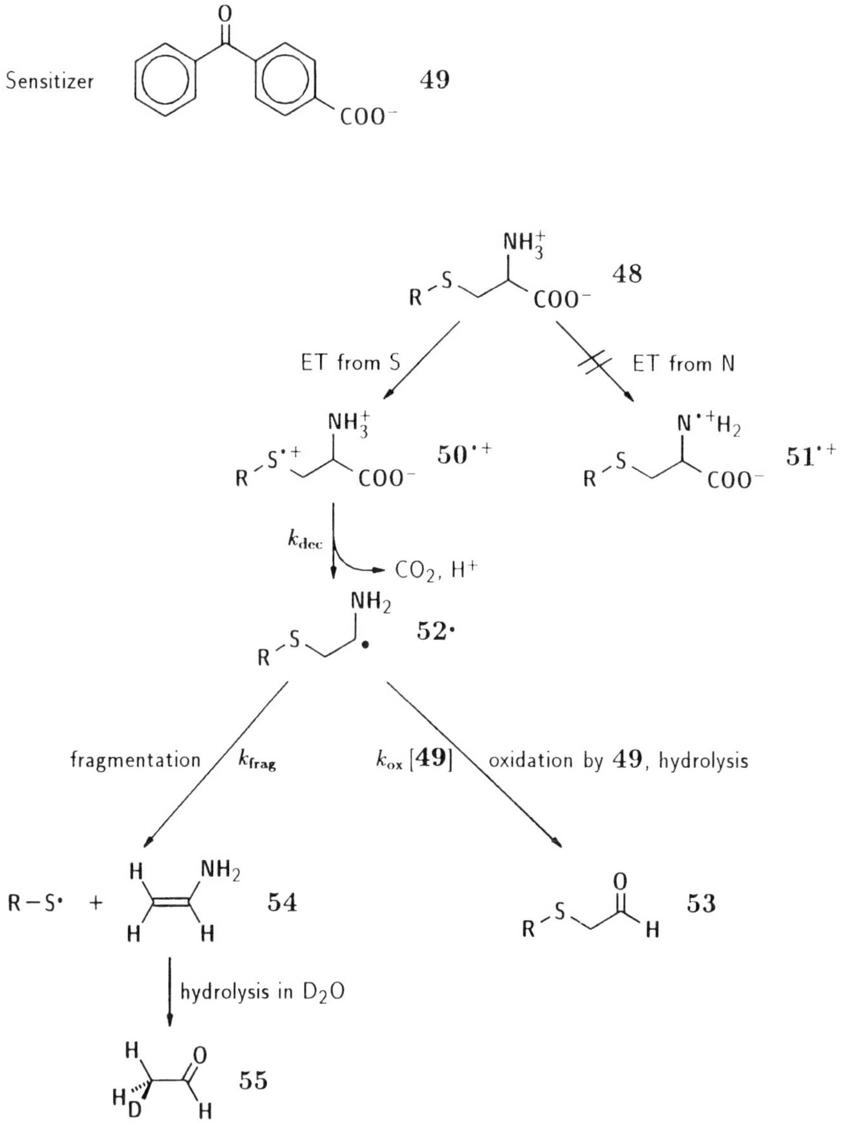

APPLICATIONS TO CHEMICAL PROBLEMS

Chart XX

At pH above pK_{a2}, where the amino group is deprotonated, these substrates contain two possible donor sites, sulfur and nitrogen. From the polarization patterns, it could be concluded that **48** quenches the sensitizer triplet by electron transfer from sulfur only. The CIDNP effects arise in two successive radical pairs because the sulfur-centered radical cations **50**$^{\cdot+}$

rapidly lose CO_2 to give α-aminoalkyl radicals **52·**, the counterradical is **49·⁻** in both cases. As the rate k_{dec} of this decarboxylation is comparable to the rate of $S-T_0$ mixing, strong cooperative effects occur (cf. Section III.C). Evaluation of these effects provided estimates of the decarboxylation rate, some 10^9 s^{-1} at pH below pK_{a2} and about an order of magnitude faster above pK_{a2}. The escaping α-aminoalkyl radicals decay by two competing pathways. One is oxidation by the sensitizer in its ground state, which finally produces a sulfur-containing aldehyde **53** via an as yet uncharacterized intermediate, probably an imine. The other is β-fragmentation to give thiyl radicals R—S·, which add to the sensitizer, and vinylamine **54**, which in acidic medium yields monodeuterated acetaldehyde **55**. The polarizations of the β protons in the α-aminoalkyl radicals were used as labels to investigate these secondary reactions. Relative rates of cleavage and oxidation were obtained from the distribution of these polarizations among the different products, and evaluation of the temperature dependence gave the activation energy of the β-fragmentation ($\sim 55 \text{ kJ mol}^{-1}$).

CIDNP spectroscopy was further applied to the photoreactions between tyrosine and synthetic water-soluble prophyrins [138]. As concerns biopolymers, photo-CIDNP experiments were used to investigate an intact glycoprotein in solution [139a], to test the molten-globule hypothesis for a monomeric insulin analogue [139b], to study the interaction of T4 endonuclease V with DNA [139c] and the aromatic side-chain interactions in myotoxin A [139d], as well as to determine the DNA-binding site of the Gal4 protein [139e].

2. Nucleic Acids. Efforts in this area have focused almost exclusively on the photosensitized splitting of pyrimidine dimers [140] (see Chart XXI, which shows the dimer **D** and the monomer **M** of dimethyluracil). Motivation of these studies has been to gain insight into the mechanism of photorepair of DNA by DNA photolyase. The operating principle of this enzyme is photoinduced electron transfer. However, the key question whether excited DNA photolyase donates or accepts the electron is still unresolved.

With the model systems used in these investigations, dimer splitting can be induced both by reduction and by oxidation, depending on the sensitizer. While it was inferred from the existence of dimer radical anions **D·⁻** that splitting via the former route occurs in two steps, it has been debated for some time whether this also holds for the latter route, that is, whether dimer oxidation and cleavage are concerted or successive. CIDNP spectroscopy is particularly well suited to answer such questions because the intermediates leave their EPR spectrum (their polarization pattern) in the products. Thus, not only can the intermediates be identified by this signature, even if they are rather short lived, but the occurrence of their polarizations in a product

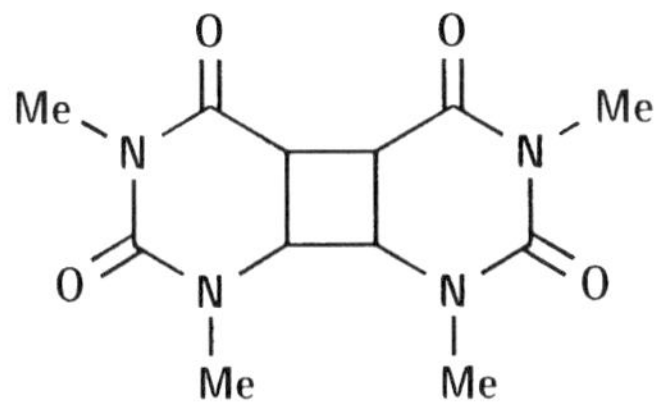

Chart XXI

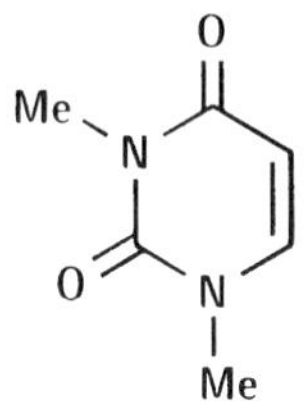

proves that they are its precursors. By virtue of this, it could be shown [140a] that dimer radical cations $\mathbf{D}^{·+}$ indeed exist and are intermediates in pyrimidine dimer splitting via the oxidative route. Similarly, the intermediacy of $\mathbf{D}^{·-}$ in reductive splitting could be confirmed both in aqueous [140b] and nonaqueous [140c] medium.

During the correlated life of the radicals, CIDNP can also arise in the monomer radical cations or anions $\mathbf{M}^{·+}$ produced by splitting of the dimer radical ions $\mathbf{D}^{·\pm}$. The rate of $S-T_0$ mixing in the pairs provides an inner clock for the reactions. For long-lived $\mathbf{D}^{·\pm}$, polarizations stem from the pairs containing $\mathbf{D}^{·\pm}$, and for short-lived $\mathbf{D}^{·\pm}$ from those containing $\mathbf{M}^{·\pm}$. Since the resulting polarization patterns differ, these two limits can be easily distinguished. As was found, the lifetime of $\mathbf{D}^{·+}$ is strongly influenced by substituents; furthermore, it is increased by a bridge between the two pyrimidine rings [140e]. In the case of $\mathbf{D}^{·-}$, the solvent polarity also plays a significant role [140c, 140f]. From the occurrence of polarizations from $\mathbf{M}^{·\pm}$ in the diamagnetic dimers, it was concluded that dimer splitting is reversible at least for bridged pyrimidines [140d–f].

As recent investigations [140e, 140f] of a larger number of pyrimidine monomers and dimers showed, the polarization patterns generated in the anion radicals $\mathbf{D}^{·-}$ and $\mathbf{M}^{·-}$ are totally different from those due to the cation radicals $\mathbf{D}^{·+}$ and $\mathbf{M}^{·+}$. This opens up the possibility of studying systems consisting of DNA photolyase and DNA that contains pyrimidine dimers with the aim of deciding the question of oxidative or reductive dimer splitting.

Purine bases have received much less interest from CIDNP spectroscopists [135]. The photoreaction of guanine with synthetic water-soluble porphyrins was addressed in [141]. Besides an earlier study of the single-stranded region of tRNA [142], the only reported examples of CIDNP from the corresponding biopolymers are two investigations of oligonucleotide duplexes [143].

3. Photosynthesis. Gust et al. [144] applied photo-CIDNP techniques to photoinduced electron transfer in porphyrin–quinone model systems of photosynthesis. The CIDNP intensities in aromatic solvents were found to be strongly dependent on the concentration of added trifluoroacetic acid through a protonation preequilibrium. Maruyama et al. used CIDNP spectroscopy to study the radical anions of pheophytin *a* derivatives [145a] and the photoinduced cross-coupling with benzoquinone of covalently linked tyrosine- and tryptophan-pyropheophorbide *a* compounds [145b, 145c].

The most fascinating development in this field of CIDNP within the last years has been the observation, by Zysmilich and McDermott [146], of nuclear spin polarized (solid state) ^{15}N NMR spectra from photosynthetic reaction centers in which the forward electron transfer from the primary charge-separated state to the accepting quinone was blocked. The all-emissive polarizations were proposed to be due to a radical pair mechanism, though many of the details are still not very clear. The reaction scheme is virtually identical to that of Chart VIII (Section V.A.2), the donor **D** being the special pair and the acceptor **A** the pheophytin. As in that example, the polarizations from the triplet exit channel are hidden in the triplet product ^{3}**D*** for the lifetime of the latter. This feature, in combination with the fact that nuclear spin relaxation in the molecular triplet localized on the special pair is relatively fast, serves to avoid the cancellation of CIDNP that would occur otherwise because the products from both exit channels are identical.

ACKNOWLEDGMENTS

Special thanks are due to Prof. Dr. U. E. Steiner, University Konstanz, for his participation in the planning of this chapter, critical readings of the manuscript, and helpful discussions. Gratitude is expressed to Prof. Dr. H. Fischer, University Zurich, to Dr. M. D. E. Forbes, Argonne National Laboratory, to Dr. S. Grimme, University Bonn, to Dr. Yu. Tsentalovich, Institute of Chemical Kinetics and Combustion, Novosibirsk, and to Dr. N. Turro, Columbia University, New York, for providing me with copies of original figures from their works.

REFERENCES

1. The whole field of spin chemistry is reviewed in (*a*) U. E. Steiner and T. Ulrich, *Chem. Rev.*, **89**, 51 (1989). (b) A. L. Buchachenko and E. L. Frankevich, *Chemical Generation and Reception of Radio- and Microwaves*, VCH, New York, 1994.

REFERENCES 155

2. For reviews of magnetic field effects besides [1], see (a) P. W. Atkins and T. P. Lambert, *Annu. Rep. Prog. Chem.*, **72A**, 67 (1975). (b) A. L. Buchachenko, *Russ. Chem. Rev. (Engl. Transl.)*, **45**, 761 (1976). (c) R. Z. Sagdeev, K. M. Salikhov and Y. M. Molin, *Russ. Chem. Rev. (Engl. Transl.)*, **46**, 569 (1977). (d) Y. M. Molin, R. Z. Sagdeev, and K. M. Salikhov, *Sov. Sci. Rev., Sect. B, Chem. Rev.*, **1**, 67 (1979). (e) N. J. Turro and B. Kräutler, *Acc. Chem. Res.*, **13**, 369 (1980). (f) K. A. McLauchlan, *Sci. Prog.*, **67**, 509 (1981). (g) N. J. Turro, *Pure Appl. Chem.*, **53**, 259 (1981). (h) S. G. Boxer, C. E. D. Chidsey, and M. G. Roelofs, *Annu. Rev. Phys. Chem.*, **34**, 389 (1983). (i) K. M. Salikhov, Y. M. Molin, R. Z. Sagdeev, and A. L. Buchachenko, *Spin Polarization and Magnetic Effects in Radical Reactions*, Elsevier, Amsterdam, 1984. (j) I. R. Gould, N. J. Turro, and M. B. Zimmt, *Adv. Phys. Org. Chem.*, **20**, 1 (1984). (k) A. J. Hoff, *Photochem. Photobiol.*, **43**, 727 (1986). (l) U. E. Steiner and H.-J. Wolff, in J. F. Rabek, Ed., *Photochemistry and Photophysics, Vol. IV*, CRC Press, Boca Raton, FL, 1991, pp. 1–130.

3. For reviews of magnetic isotope effects, see (a) [2e]. (b) Reference [2g]. (c) A. L. Buchachenko, *Prog. React. Kinet.*, **13**, 163 (1984). (d) Reference [2j]. (e) N. J. Turro and B. Kräutler, *Isotopes Org. Chem.*, **6**, 107 (1984). (f) Reference [1b].

4. RYDMR is reviewed in W. Lersch and M. E. Michel-Beyerele, in A. J. Hoff, Ed., *Advanced EPR in Biology and Biochemistry*, Elsevier, Amsterdam, 1989, pp. 685–705, in [1a] and [2l], and, very detailed, in [1b].

5. For a review of RIMIE, see [1b].

6. For reviews on CIDNP, see (a) H. Fischer, *Fortschr. Chem. Forsch.*, **24**, 1 (1971). (b) A. L. Buchachenko and F. M. Zhidomirov, *Russ. Chem. Rev. (Engl. Transl.)*, **40**, 801 (1971). (c) H. R. Ward, *Acc. Chem. Res.*, **5**, 18 (1972). (d) A. R. Lepley and G. L. Closs, Eds., *Chemically Induced Magnetic Polarization*, Wiley, London, 1978. (e) R. G. Lawler, *Prog. Nucl. Magn. Reson. Spectrosc.*, **9**, 145 (1973). (f) C. Richard and P. Granger, *Chemically Induced Dynamic Nuclear and Electron Polarizations—CIDNP and CIDEP*, Springer, Berlin, 1974. (g) R. Kaptein, *Adv. Free Radical Chem.*, **5**, 319 (1975). (h) L. T. Muus, P. W. Atkins, K. A. McLauchlan, and J. B. Pedersen, Eds., *Chemically Induced Magnetic Polarization*, Reidel, Dordrecht, 1977. (i) J. K. S. Wan, *Adv. Photochem.*, **12**, 283 (1980). (j) Ref. [2i]. (k) P. J. Hore, in R. V. Bensasson, G. Jori, and E. J. Land, Eds., NATO ASI Ser. A 85, *Primary Photoprocesses in Biology and Medicine*, Plenum, New York, 1985, pp. 111–122. (l) P. J. Hore and R. W. Broadhurst, *Prog. Nucl. Magn. Reson.*, **25**, 345 (1993). (m) M. Goez, *Concepts Magn. Reson.*, **7**, 69 (1995).

7. CIDEP is reviewed in (a) Reference [6b]. (b) Reference [6d]. (c) Reference [6h]. (d) P. J. Hore and K. A. McLauchlan, *Rev. Chem. Intermed.*, **3**, 89 (1979). (e) Ref. [7e]. (f) A. J. Hoff, *Q. Rev. Biophys.*, **14**, 599 (1981). (g) Reference [2i]. (h) C. D. Buckley and K. A. McLauchlan, *Mol. Phys.*, **54**, 1 (1985). (i) Reference [2k].

8. (a) G. L. Closs, M. D. E. Forbes, and J. R. Norris, Jr., *J. Phys. Chem.*, **91**, 3592 (1987). (b) C. D. Buckley, D. A. Hunter, P. J. Hore, and K. A. McLauchlan, *Chem. Phys. Lett.*, **135**, 307 (1987).

9. For reviews on SNP, see (a) R. Z. Sagdeev and E. G. Bagryanskaya, *Pure Appl. Chem.*, **62**, 1547 (1990). (b) E. G. Bagryanskaya and R. Z. Sagdeev, *Z. Phys. Chem. Neue Folge*, **180**, 111 (1993). (c) Ref. [1b].
10. (a) R. G. Lawler, *Acc. Chem. Res.*, **5** 25 (1972). (b) J. H. Freed and J. B. Pedersen, *Adv. Magn. Reson.*, **8**, 1 (1976). (c) P. W. Atkins and G. T. Evans, *Adv. Chem. Phys.*, **35**, 1 (1976). (d) F. J. Adrian, *Rev. Chem. Intermed.*, **3**, 3 (1979). (e) Ref. [1a]. (f) Ref. [2l]. (g) K. A. McLauchlan, U. E. Steiner, *Mol. Phys.*, **73**, 241 (1991). (h) F. J. Adrian, *Res. Chem. Intermed.*, **16**, 99 (1991). (i) M. Goez, *Concepts Magn. Reson.*, **7**, 137 (1995).
11. (a) J. B. Pedersen, *Theories of Chemically Induced Magnetic Polarization*, Odense University Press, Odense, 1979. (b) Reference [6f]. (c) Reference [6h]. (d) Reference [2i]. (e) Reference [1b].
12. (a) A. L. Buchachenko, L. V. Ruban, E. N. Step, and N. J. Turro, *Chem. Phys. Lett.*, **233**, 315 (1995). (b) A. L. Buchachenko and V. L. Berdinsky, *Chem. Phys. Lett.*, **242**, 43 (1995).
13. W. S. Jenks and N. J. Turro, *Res. Chem. Intermed.*, **13**, 237 (1990).
14. (a) R. M. Noyes, *J. Am. Chem. Soc.*, **77**, 2042 (1954). (b) R. M. Noyes, *J. Chem. Phys.*, **22**, 1349 (1954). (c) R. M. Noyes, *J. Am. Chem. Soc.*, **78**, 5486 (1956).
15. (a) J. Bargon, *J. Am. Chem. Soc.*, **99**, 8350 (1977). (b) G. L. Closs and M. S. Czeropski, *J. Am. Chem. Soc.*, **99**, 6127 (1977), and references cited therein.
16. (a) F. J. Adrian, *J. Chem. Phys.*, **53**, 3374 (1970). (b) F. J. Adrian, *J. Chem. Phys.*, **54**, 3912 (1971). (c) F. J. Adrian, *J. Chem. Phys.*, **54**, 3918 (1971).
17. F. J. Adrian, *Chem. Phys.*, **57**, 5107 (1972).
18. Compare [10g], p. 244, and pertaining references.
19. F. J. J. de Kanter, J. A. den Hollander, A. H. Huizer, and R. Kaptein, *Mol. Phys.*, **34**, 857 (1977).
20. F. J. J. de Kanter and R. Kaptein, *J. Am. Chem. Soc.*, **104**, 4759 (1982).
21. K. Schulten and R. Bittl, *J. Chem. Phys.*, **84**, 5155 (1986).
22. J. Burri and H. Fischer, *Chem. Phys.*, **139**, 497 (1989).
23. N. J. Turro and B. Kräutler, in *Diradicals*, W. T. Borden, Ed., Wiley, New York, 1982, p. 259.
24. (a) P. W. Atkins, *Chem. Phys. Lett.* **66**, 403 (1979). (b) H. A. Syage, *Chem. Phys. Lett.*, **91**, 378 (1982).
25. G. P. Zientara and J. H. Freed, *J. Chem. Phys.*, **70**, 1359 (1979).
26. R. Kaptein, *J. Am. Chem. Soc.*, **94**, 6251 (1972).
27. G. L. Closs, in [6h], p. 225.
28. (a) C. P. Slichter, *Principles of Magnetic Resonance*, 3rd ed., Springer, Berlin, 1989. (b) T. C. Farrar, *Concepts Magn. Reson.*, **2**, 1 (1990).
29. L. Monchick and F. J. Adrian, *J. Chem. Phys.*, **68**, 4376 (1978).
30. (a) J. B. Pedersen, *J. Chem. Phys.*, **67**, 4097 (1977). (b) J. B. Pedersen, in [6h], p. 297.
31. G. T. Evans, P. D. Fleming, and R. G. Lawler, *J. Chem. Phys.*, **58**, 2071 (1973).

32. J.-K. Vollenweider and H. Fischer, *Chem. Phys.*, **124**, 333 (1988).
33. L. Monchick, *J. Chem. Phys.*, **24**, 381 (1956).
34. (a) F. J. Adrian and L. Monchick, *J. Chem. Phys.*, **71**, 2600 (1979). (b) F. J. Adrian and L. Monchick, *J. Chem. Phys.*, **72**, 5786 (1980).
35. (a) J. Bargon, H. Fischer, and U. Johnsen, *Z. Naturforsch.*, **22a**, 1551 (1967). (b) H. R. Ward and R. G. Lawler, *J. Am. Chem. Soc.*, **89**, 5518 (1967).
36. (a) R. Kaptein and L. J. Oosterhoff, *Chem. Phys. Lett.*, **4**, 195 (1969). (b) G. L. Closs, *J. Am. Chem. Soc.*, **91**, 4552 (1969).
37. S. Schäublin, A. Höhener, and R. R. Ernst, *J. Magn. Reson.*, **13**, 196 (1974).
38. R. Hany, J.-K. Vollenweider, and H. Fischer, *Chem. Phys.*, **120**, 169 (1988).
39. M. Goez, *Concepts Magn. Reson.*, **7**, 263 (1995).
40. R. Kaptein, *J. Chem. Soc. Chem. Commun.*, 732 (1971).
41. (a) J. A. den Hollander, *J. Chem. Soc. Chem. Commun.*, 352 (1975). (b) J. A. den Hollander, *Chem. Phys.*, **10**, 167 (1975). (c) J. A. den Hollander and R. Kaptein, *Chem. Phys. Lett.*, **41**, 257 (1976).
42. R. Kaptein and J. A. den Hollander, *J. Am. Chem. Soc.*, **94**, 6269 (1972).
43. (a) S. Schäublin, A. Wokaun, and R. R. Ernst, *Chem. Phys.*, **14**, 285 (1976). (b) S. Schäublin, A. Wokaun, and R. R. Ernst, *J. Magn. Reson.*, **27**, 273 (1977).
44. (a) G. L. Closs and R. J. Miller, *J. Am. Chem. Soc.*, **101**, 1639 (1979). (b) R. J. Miller and G. L. Closs, *Rev. Sci. Instrum.*, **52**, 1876 (1981).
45. S. M. Rosenfeld, R. G. Lawler, and H. R. Ward, *J. Am. Chem. Soc.*, **95**, 946 (1973).
46. H. D. Roth and M. L. Manion, *J. Am. Chem. Soc.*, **97**, 6886 (1975).
47. (a) K. M. Salikhov, *Chem. Phys.*, **64**, 371 (1982). (b) P. J. Hore, S. Stob, J. Kemmink, and R. Kaptein, *Chem. Phys. Lett.*, **98**, 409 (1983). (c) R. S. Hutton, H. D. Roth, and S. H. Bertz, *J. Am. Chem. Soc.*, **105**, 6371 (1983).
48. H. D. Roth, K. C. Hwang, R. S. Hutton, N. J. Turro, and K. M. Welsh, *J. Phys. Chem.*, **93**, 5697 (1989).
49. (a) K. C. Hwang, N. J. Turro, H. D. Roth, and C. Doubleday, *J. Phys. Chem.*, **95**, 63 (1991). (b) K. C. Hwang, H. D. Roth, N. J. Turro, and K. M. Welsh, *J. Phys. Org. Chem.*, **5**, 209 (1992).
50. (a) I. A. Shkrob, *Chem. Phys. Lett.*, **210**, 432 (1993). (b) V. F. Tarasov and I. A. Shkrob, *J. Magn. Reson. A*, **109**, 65 (1994).
51. A. I. Shushin, *Chem. Phys.*, **152**, 133 (1991).
52. (a) A. A. Zharikov and N. V. Shokhirev, *Z. Phys. Chem. Neue Folge*, **177**, 37 (1992). (b) N. V. Shokhirev, A. A. Zharikov, and E. B. Krissinel, *J. Chem. Phys.*, **99**, 2643 (1993). (c) N. V. Shokhirev and E. B. Krissinel, *Chem. Phys. Lett.*, **236**, 247 (1995).
53. T. Aizawa, T. Sakata, S. Itoh, K. Maeda, and T. Azumi, *Chem. Phys. Lett.*, **195**, 16 (1992).
54. G. L. Closs and O. D. Redwine, *J. Am. Chem. Soc.*, **107**, 6131 (1985).

55. (a) Y. P. Tsentalovich, A. A. Shargorodsky, A. A. Obynochny, R. Z. Sagdeev, P. A. Purtov, and A. V. Yurkovskaya, *Chem. Phys.*, **139**, 307 (1989). (b) A. V. Yurkovskaya, Y. P. Tsentalovich, N. N. Lukzen, and R. Z. Sagdeev, *Res. Chem. Intermed.*, **17**, 145 (1992). (c) A. V. Yurkovskaya, O. B. Morozova, R. Z. Sagdeev, S. V. Dvinskih, G. Buntkowsky, and H. M. Vieth, *Chem. Phys.*, **197**, 157 (1995).

56. F. J. J. de Kanter and R. Kaptein, *Chem. Phys. Lett.*, **58**, 340 (1978).

57. (a) N. J. Turro, K. C. Hwang, V. P. Rao, and C. Doubleday, *J. Phys. Chem.*, **95**, 1872 (1991). (b) N. H. Han, K. C. Hwang, X. G. Lei, and N. J. Turro, *J. Photochem. Photobiol. A*, **61**, 35 (1991). (c) K. C. Hwang, N. J. Turro, and C. Doubleday, *J. Am. Chem. Soc.*, **113**, 2850 (1991).

58. (a) C. Doubleday Jr., *Chem. Phys. Lett.*, **64**, 67 (1979). (b) C. Doubleday Jr., *Chem. Phys. Lett.*, **94**, 375 (1981).

59. R. Kaptein, J. J. de Kanter, and G. H. Rist, *J. Chem. Soc., Chem. Commun.*, 499 (1981).

60. G. L. Closs, M. D. E. Forbes, and P. Piotrowiak, *J. Am. Chem. Soc.*, **114**, 3285 (1992).

61. (a) K. Maeda, M. Terazima, T. Azumi, and Y. Tanimoto, *J. Phys. Chem.*, **95**, 197 (1991). (b) K. Maeda, Q. X. Meng, T. Aizawa, M. Terazima, T. Azumi, and Y. Tanimoto, *J. Phys. Chem.*, **96**, 4884 (1992). (c) K. Maeda, N. Suzuki, and T. Azumi, *J. Phys. Chem.*, **97**, 9562 (1993).

62. (a) G. F. Lehr and N. J. Turro, *Tetrahedron*, **37**, 3411 (1981). (b) N. J. Turro, M. B. Zimmt, and I. R. Gould, *J. Am. Chem. Soc.*, **105**, 6347 (1983). (c) M. B. Zimmt, C. Doubleday, and N. J. Turro, *J. Am. Chem. Soc.*, **106**, 3363 (1984).

63. E. N. Step, A. L. Buchachenko, and N. J. Turro, *Chem. Phys.*, **162**, 189 (1992).

64. I. A. Shkrob, V. F. Tarasov, and E. G. Bagryanskaya, *Chem. Phys.*, **153**, 427 (1991).

65. V. F. Tarasov, N. D. Ghatlia, N. I. Avdievich, and N. J. Turro, *Z. Phys. Chem. Neue Folge*, **182**, 227 (1993).

66. A. I. Shushin, J. B. Pedersen, and L. I. Lolle, *Chem. Phys.*, **188**, 1 (1994).

67. (a) F. J. Adrian, *Chem. Phys. Lett.* **26**, 437 (1974). (b) F. J. Adrian, H. M. Vyas, and J. K. S. Wan, *Chem. Phys.*, **65**, 1454 (1976). (c) H. D. Roth, R. S. Hutton, and M. L. M. Schilling, *Rev. Chem. Intermed.*, **3**, 169 (1979).

68. (a) Y. Yamakage, Q. X. Meng, K. Maeda, and T. Azumi, *Chem. Phys. Lett.*, **204**, 411 (1993). (b) Y. Yamakage, Q. X. Meng, S. S. Ali, K. Maeda, and T. Azumi, *Proc. Indian Acad. Sci.*, **105**, 629 (1993).

69. (a) Q. X. Meng, K. Suzuki, K. Maeda, M. Terazima, and T. Azumi, *J. Phys. Chem.*, **97**, 1265 (1993). (b) Q. Meng, Y. Yamakage, K. Maeda, and T. Azumi, *Z. Phys. Chem. Neue Folge*, **180**, 95 (1993).

70. (a) A. V. Yurkovskaya, Y. P. Tsentalovich, and R. Z. Sagdeev, *Chem. Phys. Lett.*, **171**, 406 (1990). (b) Y. P. Tsentalovich, A. A. Frantsev, A. B. Doktorov, A. V. Yurkovskaya, and R. Z. Sagdeev, *J. Phys. Chem.*, **97**, 8900 (1993). (c) Y. P. Tsentalovich, A. V. Yurkovskaya, A. A. Frantsev, A. B. Doctorov, and R. Z.

Sagdeev, *Z. Phys. Chem. Neue Folge*, **182**, 119 (1993). (d) O. B. Morozova, Y. P. Tsentalovich, A. V. Yurkovskaya, and R. Z. Sagdeev, *Chem. Phys. Lett.*, **246**, 499 (1995).
71. S. N. Batchelor and H. Fischer, *J. Phys. Chem.*, **100**, 556 (1996).
72. M. Salzmann, Y. P. Tsentalovich, and H. Fischer, *J. Chem. Soc. Perkin Trans.* 2, 2119 (1994).
73. G. L. Closs and R. J. Miller, *J. Am. Chem. Soc.*, **100**, 3483 (1978).
74. K. Y. Choo and J. K. S. Wan, *J. Am. Chem. Soc.*, **97**, 7127 (1975).
75. M. Läufer, *Chem. Phys. Lett.*, **127**, 136 (1986).
76. M. Goez, *Chem. Phys. Lett.*, **165**, 11 (1990).
77. M. Goez, *J. Magn. Reson.*, **123**, 161 (1996).
78. (a) M. Goez, *Chem. Phys. Lett.*, **188**, 451 (1992). (b) M. Goez, *Appl. Magn. Reson.*, **5**, 113 (1993).
79. M. Goez, *J. Magn. Reson. A*, **102**, 144 (1993).
80. (a) R. M. Scheek, S. Stob, R. Boelens, K. Dijkstra, and R. Kaptein, *Faraday Discuss. Chem. Soc.*, **78**, 245 (1984). (b) R. M. Scheek, S. Stob, R. Boelens, K. Dijkstra, and R. Kaptein, *J. Am. Chem. Soc.*, **107**, 705 (1985). (c) S. Stob, R. M. Scheek, R. Boelens, K. Dijkstra, and R. Kaptein, *Isr. J. Chem.*, **28**, 319 (1988).
81. R. Boelens, A. Podoplelov, and R. Kaptein, *J. Magn. Reson.*, **69**, 116 (1986).
82. (a) K. Ishiguro, I. V. Khudyakov, P. F. McGarry, N. J. Turro, and H. D. Roth, *J. Am. Chem. Soc.*, **116**, 6933 (1994). (b) H. D. Roth, M. L. M. Schilling, and G. Jones II, *J. Am. Chem. Soc.*, **103**, 1246 (1981).
83. K. Raghavachari and H. D. Roth, *J. Am. Chem. Soc.*, **111**, 7132 (1989).
84. H. D. Roth and T. Herbertz, *J. Am. Chem. Soc.*, **115**, 9804 (1993).
85. H. D. Roth, X. M. Du, H. X. Weng, P. S. Lakkaraju, and C. J. Abelt, *J. Am. Chem. Soc.*, **116**, 7744 (1994).
86. (a) M. Goez, *Chem. Phys.*, **147**, 143 (1990). (b) M. Goez, *Z. Phys. Chem. Neue Folge*, **169**, 123 (1990). (c) M. Goez, *Z. Phys. Chem. Neue Folge*, **169**, 133 (1990). (d) M. Goez and G. Eckert, *Ber. Bunsenges. Phys. Chem.*, **95**, 1179 (1991).
87. G. L. Closs, *Chem. Phys. Lett.*, **32**, 277 (1975).
88. M. J. S. Dewar, E. G. Zoebisch, E. F. Healy, and J. J. P. Stewart, *J. Am. Chem. Soc.*, **107**, 3902 (1985).
89. M. Goez and G. Eckert, *Z. Phys. Chem. Neue Folge*, **182**, 131 (1993).
90. H. G. O. Becker, K. Urban, and D. Pfeifer, *J. Chem. Soc. Faraday Trans. 2*, **85**, 1765 (1989).
91. (a) E. Schaffner and H. Fischer, *J. Phys. Chem.*, **99**, 102 (1995). (b) E. Schaffner and H. Fischer, *J. Phys. Chem.*, **100**, 1657 (1996).
92. Y. C. Liu, Z. L. Liu, M. X. Zhang, and L. Yang, *J. Photochem. Photobiol. A*, **67**, 279 (1992).
93. E. Schaffner and H. Fischer, *J. Phys. Chem.*, **97**, 13149 (1993).
94. (a) N. E. Polyakov and T. V. Leshina, *J. Photochem. Photobiol. A*, **55**, 43 (1990).

(b) Y. P. Tsentalovich, E. G. Bagryanskaya, Y. A. Grishin, A. A. Obynochny, R. Z. Sagdeev, and H. K. Roth, *Chem. Phys.*, **142**, 75 (1990). (c) M. Goez and I. Sartorius, *J. Am. Chem. Soc.*, **115**, 11123 (1993). (d) M. Goez and I. Frisch, *J. Photochem. Photobiol. A*, **84**, 1 (1994). (e) M. Goez and I. Sartorius, *Chem. Ber.*, **127**, 2273 (1994). (f) H. Klaukien and M. Lehnig, *J. Photochem. Photobiol. A*, **84**, 221 (1994).

95. H. D. Roth and A. A. Lamola, *J. Am. Chem. Soc.*, **96**, 6270 (1974).

96. K. C. Hwang, N. J. Turro, and H. D. Roth, *J. Org. Chem.*, **59**, 1102 (1994).

97. (a) Y. P. Tsentalovich, R. Z. Sagdeev, and A. A. Obynochny, *Chem. Phys.*, **139**, 301, 306 (1989). (b) T. Aizawa, Y. Araki, K. Shindo, K. Maeda, and T. Azumi, *Spectrochim. Acta A*, **50**, 1443 (1994).

98. (a) G. Vermeersch, A. Filali, A. Couture, J. P. Catteau, and J. Marko, *J. Photochem. Photobiol. B*, **4**, 85 (1989). (b) M. Terazima, K. Maeda, M. Sugawara, S. Takahashi, and T. Azumi, *J. Chem. Soc. Faraday Trans.*, **86**, 253 (1990). (c) T. V. Leshina and N. E. Polyakov, *J. Phys. Chem.*, **94**, 4379 (1990). (d) V. I. Maryasova, A. S. Zanina, A. I. Kruppa, and T. V. Leshina, *J. Photochem. Photobiol. A*, **61**, 201 (1991). (e) M. B. Taraban, A. I. Kruppa, N. E. Polyakov, T. V. Leshina, V. Lusis, D. Muceniece, and G. Duburs, *J. Photochem. Photobiol. A*, **73**, 151 (1993). (f) A. I. Kruppa, M. B. Taraban, N. E. Polyakov, T. V. Leshina, V. Lusis, D. Muceniece, and G. Duburs, *J. Photochem. Photobiol. A*, **73**, 159 (1993). (g) N. E. Polyakov, M. B. Taraban, A. I. Kruppa, N. I. Avdievich, V. V. Mokrushin, P. V. Schastnev, T. V. Leshina, V. Lusis, D. Muceniece, and G. Duburs, *J. Photochem. Photobiol. A*, **74**, 75 (1993). (h) S. S. Ali, K. Maeda, and T. Azumi, *Chem. Lett.* 227 (1995).

99. T. Sakata, S. Takahashi, M. Terazima, and T. Azumi, *J. Phys. Chem.*, **95**, 8671 (1991).

100. (a) M. X. Zhang, L. M. Wu, Y. C. Liu, and Z. L. Liu, *Magn. Reson. Chem.*, **27**, 451 (1989). (b) M. Zhang, L. Yang, Z. Liu, and Y. Liu, *Chin. J. Chem.*, **9**, 54 (1991).

101. R. Hany and H. Fischer, *Chem. Phys.*, **172**, 131 (1993).

102. T. Noh, E. Step, and N. J. Turro, *J. Photochem. Photobiol. A*, **72**, 133 (1993).

103. Y. P. Tsentalovich, A. V. Yurkovskaya, and R. Z. Sagdeev, *J. Photochem. Photobiol. A*, **70**, 9 (1993).

104. (a) W. U. Palm, H. Dreeskamp, *J. Photochem. Photobiol. A*, **52**, 439 (1990). (b) W. U. Palm, H. Dreeskamp, H. Bouaslaurent, and A. Castellan, *Ber. Bunsenges. Phys. Chem.*, **96**, 50 (1992). (c) S. Grimme and H. Dreeskamp, *J. Photochem. Photobiol. A*, **65**, 371 (1992). (d) G. Pohlers, S. Grimme, and H. Dreeskamp, *J. Photochem. Photobiol. A*, **79**, 153 (1994). (e) G. Pohlers, H. Dreeskamp, and S. Grimme, *J. Photochem. Photobiol. A*, **95**, 41 (1996).

105. S. Grimme, *Chem. Phys.*, **163**, 313 (1992).

106. N. P. Gritsan, Y. P. Tsentalovich, A. V. Yurkovskaya, and R. Z. Sagdeev, *J. Phys. Chem.*, **100**, 4448 (1996).

107. (a) A. Henne, N. P. Y. Siew, and K. Schaffner, *Helv. Chim. Acta*, **62**, 1952 (1979). (b) A. Henne, N. P. Y. Siew, and K. Schaffner, *J. Am. Chem. Soc.*, **101**, 3671 (1979).
108. F. Carissimorietsch, J. Marko, G. Vermeersch, and J. M. Aubry, *J. Photochem. Photobiol. A*, **69**, 175 (1992).
109. S. N. Batchelor and H. Fischer, *J. Phys. Chem.*, **100**, 9794 (1996).
110. J.-K. Vollenweider and H. Fischer, *Chem. Phys.*, **108**, 365 (1986).
111. Y. Liu, M. Zhang, L. Yang, and Z. Liu, *J. Chem. Soc. Perkin Trans. 2*, 1919 (1992).
112. M. Igarashi, T. Ueda, M. Wakasa, and Y. Sakaguchi, *J. Organomet. Chem.*, **421**, 9 (1991).
113. K. Maruyama and H. Imahori, *Bull. Chem. Soc. Jpn.*, **62**, 816 (1989).
114. G. W. Sluggett, N. J. Turro, and H. D. Roth, *J. Am. Chem. Soc.*, **117**, 9982 (1995).
115. K. Hildenbrand, C. von Sonntag, and H. P. Schuchmann, *J. Photochem. Photobiol. A*, **48**, 219 (1989).
116. H. D. Roth and R. S. Hutton, *J. Phys. Org. Chem.*, **3**, 119 (1990).
117. (a) G. Eckert and M. Goez, *J. Am. Chem. Soc.*, **116**, 11999 (1994). (b) M. Goez and I. Frisch, *J. Am. Chem. Soc.*, **117**, 10486 (1995). (c) M. Goez and G. Eckert, *J. Am. Chem. Soc.*, **118**, 140 (1996). (d) G. Eckert and M. Goez, *J. Inf. Rec. Mats.*, **22**, 561 (1996).
118. (a) P. C. Wong and D. R. Arnold, *Can. J. Chem.*, **57**, 1037 (1979). (b) K. A. Brown-Wensley, S. L. Mattes, and S. Farid, *J. Am. Chem. Soc.*, **100**, 4162 (1978).
119. K. Maruyama and H. Imahori, *J. Org. Chem.*, **54**, 2692 (1989).
120. J. Xu, J. Xu, B. Yan, and H. Yuan, *Chin. Chem. Lett.*, **2**, 831 (1991).
121. H. C. Yuan and B. Z. Yan, *Chin. Chem. Lett.*, **3**, 25 (1992).
122. (a) Z. Liu, M. Zhang, L. Yang, Y. Liu, Y. L. Chow, and C. I. Johansson, *J. Chem. Soc. Perkin Trans. 2*, 585 (1994). (b) Q. H. Wu, B. W. Zhang, Y. F. Ming, and Y. Cao, *J. Photochem. Photobiol. A*, **61**, 53 (1991). (c) L. Yang, M. X. Zhang, Y. C. Liu, Z. L. Liu, and Y. L. Chow, *J. Chem. Soc. Chem. Commun.*, 1055 (1995).
123. T. Miyashi, H. D. Roth, M. L. M. Schilling, C. J. Abelt, T. Mukai, Y. Takahashi, and A. Konno, *J. Org. Chem.*, **54**, 1445 (1989).
124. A. V. Podoplelov, R. Kaptein, and S. Stob, *Chem. Phys. Lett.*, **160**, 233 (1989).
125. A. V. Podoplelov, Sen Chen Su, R. Z. Sagdeev, M. S. Shtein, V. M. Moralev, V. I. Gol'danskii, and Yu. N. Molin, *Izv. Akad. Nauk SSSR Ser. Khim.*, **10**, 2207 (1985).
126. M. Lehnig, *Chem. Phys.*, **54**, 323 (1981).
127. M. Lehnig, H. Klaukien, and F. Reininghaus, *Ber. Bunsenges. Phys. Chem.*, **94**, 1411 (1990).
128. (a) G. L. Closs and L. E. Closs, *J. Am. Chem. Soc.*, **91**, 4549 (1969). (b) G. L. Closs and A. D. Trifunac, *J. Am. Chem. Soc.*, **91**, 4554 (1969).

129. J. E. Hux and R. J. Puddephatt, *J. Organomet. Chem.*, **437**, 251 (1992).

130. (a) A. L. Buchachenko, A. Z. Yankelevitch, E. S. Klimtchuk, I. V. Khudyakov, and L. A. Margulis, *J. Photochem. Photobiol. A*, **46**, 281 (1989). (b) A. L. Buchachenko and I. V. Khudyakov, *Acc. Chem. Res.*, **24**, 177 (1991).

131. (a) S. V. Rykov, I. V. Khudyakov, E. D. Skakovsky, H. D. Burrows, S. J. Formosinho, and M. D. M. Miguel, *J. Chem. Soc. Perkin Trans. 2*, 835 (1991). (b) S. V. Rykov, I. V. Khudyakov, E. D. Skakovskii, L. Y. Tychinskaya, and M. M. Ogorodnikova, *J. Photochem. Photobiol. A*, **66**, 127 (1992).

132. H. Fischer, R. Baer, R. Hany, I. Verhoolen, and M. Walbiner, *J. Chem. Soc. Perkin Trans. 2*, 787 (1990).

133. G. Rist, A. Borer, K. Dietliker, V. Desobry, J. P. Fouassier, and D. Ruhlmann, *Macromol.*, **25**, 4182 (1992).

134. (a) R. J. Devoe, M. R. V. Sahyun, and E. Schmidt, *Can. J. Chem.*, **66**, 319 (1988). (b) R. J. Devoe, M. R. V. Sahyun, and E. Schmidt, *J. Imag. Sci.*, **33**, 39 (1989). (c) N. P. Hacker, D. C. Hofer, and K. M. Welsh, *J. Photopolym. Sci. Technol.*, **5**, 35 (1992). (d) K. M. Welsh, J. L. Dektar, N. P. Hacker, and N. J. Turro, *Polym. Mater. Sci. Eng.*, **61**, 181 (1989). (e) K. M. Welsh, J. L. Dektar, M. A. Garciagaribaya, N. P. Hacker, and N. J. Turro, *J. Org. Chem.*, **57**, 4179 (1992). (f) G. Eckert, M. Goez, B. Maiwald, and U. Mueller, *Ber. Bunsenges. Phys. Chem.*, **100**, 1191 (1996).

135. P. J. Hore and R. W. Broadhurst, *Prog. Nucl. Magn. Reson.*, **25**, 345 (1993).

136. S. L. Winder, R. W. Broadhurst, and P. J. Hore, *Spectrochim. Acta A*, **51**, 1753 (1995).

137. M. Goez, J. Rozwadowski, and B. Marciniak, *J. Am. Chem. Soc.*, **118**, 2882 (1996).

138. D. Lenouen, J. Marko, G. Vermeersch, N. Febvaygarot, A. L. Combier, M. Perreefauvet, and A. Gaudemer, *J. Phys. Org. Chem.*, **3**, 69 (1990).

139. (a) K. Hard, J. P. Kamerling, and J. F. G. Vliegenthart, *Carbohydr. Res.*, **236**, 315 (1992). (b) Q. X. Hua, J. E. Ladbury, and M. A. Weiss, *Biochemistry*, **32**, 1433 (1993). (c) B. J. Lee, H. Sakashita, T. Ohkubo, M. Ikehara, T. Doi, K. Morikawa, Y. Kyogoku, T. Osafune, S. Iwai, and E. Ohtsuka, *Biochemistry*, **33**, 57 (1994). (d) K. A. Muszkat, V. Preygerzon, and A. T. Tu, *J. Protein Chem.*, **13**, 333, 337 (1994). (e) Y. Serikawa, M. Shirakawa, and Y. Kyogoku, *FEBS Lett.*, **299**, 205 (1992).

140. (a) T. Young, R. Nieman, and S. D. Rose, *Photochem. Photobiol.*, **52**, 661 (1990). (b) R. F. Hartman, S. D. Rose, P. J. W. Pouwels, and R. Kaptein, *Photochem. Photobiol.*, **56**, 305 (1992). (c) R. R. Rustandi and H. Fischer, *J. Am. Chem. Soc.*, **115**, 2537 (1993). (d) P. J. W. Pouwels, R. F. Hartman, S. D. Rose, and R. Kaptein, *J. Am. Chem. Soc.*, **116**, 6967 (1994). (e) P. J. W. Pouwels, R. F. Hartman, S. D. Rose, and R. Kaptein, *Photochem. Photobiol.*, **61**, 563 (1995). (f) P. J. W. Pouwels, R. F. Hartman, S. D. Rose, and R. Kaptein, *Photochem. Photobiol.*, **61**, 575 (1995).

141. D. Lenouen, G. Vermeersch, J. Marko, A. Lablachecombier, N. Febvaygarot, A. Gaudemer, and M. Perreefauvet, *Photochem. Photobiol.*, **49**, 7 (1989).
142. E. F. McCord, K. M. Morden, A. Pardi, I. Tinoco Jr., and S. G. Boxer, *Biochemistry*, **23**, 1926 (1984).
143. (a) M. Katahira, R. Sakaguchikatahira, F. Hayashi, S. Uesugi, and Y. Kyogoku, *J. Am. Chem. Soc.*, **113**, 8647 (1991). (b) M. Katahira, H. Sato, K. Mishima, S. Uesugi, and S. Fujii, *Nucl. Acids Res.*, **21**, 5418 (1993).
144. D. Gust, T. A. Moore, A. L. Moore, X. C. C. Ma, R. A. Nieman, G. R. Seely, R. E. Belford, and J. E. Lewis, *J. Phys. Chem.*, **95**, 4442 (1991).
145. (a) K. Maruyama, H. Yamada, and A. Osuka, *Bull. Chem. Soc. Jpn.*, **63**, 3462 (1990). (b) K. Maruyama, H. Yamada, and A. Osuka, *Chem. Lett.*, 833 (1989). (c) K. Maruyama, H. Yamada, and A. Osuka, *Photochem. Photobiol.*, **53**, 617 (1991).
146. (a) M. G. Zysmilich and A. McDermott, *J. Am. Chem. Soc.*, **116**, 8362 (1994). (b) M. G. Zysmilich and A. McDermott, *J. Am. Chem. Soc.*, **118**, 5867 (1996).

PHOTOPHYSICS OF GASEOUS AROMATIC MOLECULES: EXCESS VIBRATIONAL ENERGY DEPENDENCE OF RADIATIONLESS PROCESSES

Edward C. Lim*

Department of Chemistry, The University of Akron, Akron, OH 44325

CONTENTS

I. Introduction, 166
II. Intramolecular character of S_n ($n \geq 2$) → S_1 IC and spectral characteristics of dispersed fluorescence in collision-free molecules, 168
III. Excess energy dependence of radiationless transitions: theoretical considerations, 171
IV. Excess energy dependence of radiationless transitions in molecules with nearby $n\pi^*$ and $\pi\pi^*$ states: theoretical considerations, 180
V. Excess energy dependence of radiationless transitions in aromatic hydrocarbons: experimental results, 186

*Holder of the Goodyear Chair in Chemistry at The University of Akron.

Advances in Photochemistry, Volume 23, Edited by Douglas C. Neckers, David H. Volman, and Günther von Bünau
ISBN 0-471-19289-9 © 1997 by John Wiley & Sons, Inc.

VI. Excess energy dependence of radiationless transitions in nitrogen heterocyclic compounds: experimental results, 194
VII. Channel three decay in benzene, 202
VIII. Concluding remarks, 206
Acknowledgments, 208
References, 208

I. INTRODUCTION

Radiationless transition, a process whereby a molecule makes a transition from one electronic state to another of lower energy without emission of radiation, plays a central role in photophysics and photochemistry of polyatomic molecules. More specifically, the competitions among different nonradiative decay channels and with radiative decay determine the fate of the electronically excited molecules, and hence influence their photophysical and photochemical properties. In the condensed phase, where vibrational relaxation is fast on the time scale of electronic relaxation, radiationless transition originates from the vibrationless level of the initial electronic state. The photophysics of organic molecules with low levels of vibrational excitation is a very mature field in which several important generalizations exist concerning the factors that influence radiationless transitions [1]. Perhaps the best known of these is the "energy-gap law," which states that the rate constant for radiationless transition increases very rapidly with decreasing energy gap between the electronic states involved in the transition. Since the electronic energy gaps are much smaller between excited singlet states than between the lowest excited singlet state (S_1) and the ground state (S_0), the energy-gap law readily accounts for the well-known experimental observation that the fluorescence almost always originates from the lowest excited singlet state, Kasha's rule [2], independent of the electronic state to which the molecule is initially excited. The $S_n(n > 2) \to S_{n-1}$ radiationless transitions with small electronic energy gaps are usually so fast that fluorescence from the higher excited singlet state cannot compete with the ultrafast internal conversion (radiationless transition between electronic states of the same spin multiplicity) between the excited states. Once the molecule is in its S_1 state, however, it can either fluoresce or pass nonradiatively to the triplet (T) manifold by intersystem crossing (radiationless transition between electronic states of different spin multiplicity), due to the inefficient $S_1 \to S_0$ internal conversion (IC) caused by the large S_1–S_0 electronic energy gap. For most aromatic hydrocarbons in solution, the sum of the quantum yields of fluorescence and of singlet–triplet intersystem crossing (ISC) is near unity [3].

A question of considerable interest concerning the above generalization of the unimportance of $S_1 \to S_0$ IC is whether it applies only to vibrationally relaxed molecules, or applies also to collision-free gaseous molecules with substantial vibrational excitation. Although the answer to this question has been known for over two decades [4], monographs on photophysics and photochemistry of organic molecules do not adequately address this topic, presumably due to their primary focus on condense-phase phenomena. This is rather unfortunate since the vibrational energy dependence of electronic relaxation is a topic directly relevant to photochemical reactions in gaseous systems.

In this chapter, we describe the excess vibrational energy dependence of $S_1 \to S_0$ IC and $S_1 \to T$ ISC in collision-free aromatic molecules. It will be shown that the IC rate increases much more rapidly with increasing vibrational excitation than does the ISC rate. Because of the very different energy dependence of the two radiationless transitions, internal conversion to the ground state dominates over the intersystemm crossing for molecules with large excess energies in S_1. This effect of vibrational excitation is shown to be especially dramatic in nitrogen heterocyclic compounds that possess close-lying $n\pi^*$ and $\pi\pi^*$ singlet states. We also show how the excitation energy dependence of the nonradiative decay rate can be related to the question of the efficiency of intramolecular vibrational redistribution (IVR). The organization of this chapter is as follows. We begin with a discussion of the spectral characteristic of collision-free molecules, which establishes S_n ($n \geqslant 2$) $\to S_1$ IC as an intramolecular process. This discussion is followed by a theoretical consideration of radiationless transitions, vis-à-vis the energy-gap law and the excess vibrational energy dependence of $(S_1 \to S_0$ IC$)/(S_1 \to T$ ISC$)$ branching ratio. An especially interesting case of radiationless transitions in nitrogen heterocyclic compounds with nearby $n\pi^*$ and $\pi\pi^*$ states is next considered theoretically, with particular attention to the critical role that vibronically active out-of-plane vibrations play in large electronic-gap radiationless transitions. In the last two Sections, the predictions of these theoretical considerations are compared with the experimental excess energy dependence of $S_1 \to S_0$ IC and $S_1 \to T$ ISC. We conclude the chapter with a comparison of the photophysical properties of benzene with those of other aromatic compounds, which provides clues to the identity of the so-called channel three decay in benzene.

II. INTRAMOLECULAR CHARACTER OF $S_n (n \geqslant 2) \rightarrow S_1$ IC AND SPECTRAL CHARACTERISTICS OF DISPERSED FLUORESCENCE IN COLLISION-FREE MOLECULES

The fact that $S_n \rightarrow S_1$ IC is an intramolecular process was essentially demonstrated by the early experiments of Pringsheim [5], which showed that the fluorescence of dilute anthracene vapor appears to originate from S_1 even when the molecule is initially excited into a higher energy singlet state. This important finding, which demonstrates the intramolecular character of $S_n \rightarrow S_1$ IC, has been confirmed and extended to a number of other molecules by subsequent workers. Figure 1 shows the dispersed fluorscence spectra of dilute naphthalene vapor (~ 0.07 torr), as obtained by excitation of the molecule through three different absorption systems [6]. Although excitation of the higher energy excited singlet states (S_2 and S_3) leads to the broadening of the fluorescence, the spectral position of the emission is clearly that which follows excitation of S_1. The fluorescence is therefore $S_1 \rightarrow S_0$ in character, independent of the electronic state to which the molecule was initially excited. Based on the $S_2 \rightarrow S_0$ radiative decay rate ($\sim 10^8 \, \text{s}^{-1}$) of naphthalene, deduced from the integrated $S_2 \leftarrow S_0$ absorption coefficient, and the fact that $S_2 \rightarrow S_0$ fluorescence with a quantum yield as small as 10^{-3} would have been detected under the experimental conditions, one obtains a lower limit of about $10^{11} \, \text{s}^{-1}$ for the $S_2 \rightarrow S_1$ IC rate. This rate is about five orders of magnitude greater than the hard-sphere collision rate ($\sim 10^6 \, \text{s}^{-1}$ for a 0.07-torr sample, based on a collision frequency of $\sim 10^7 \, \text{s}^{-1}/\text{torr}$), and establishes the $S_n (n \geqslant 2) \rightarrow S_0$ IC as a truly intramolecular phenomenon. The same conclusion is also obtained from experiments carried out under the collision-free conditions of supersonic molecular beams.

Since energy is conserved in the internal conversion from a higher excited singlet state to S_1, the S_1 molecules produced by the IC process will be vibrationally highly excited under the collision-free conditions. The spectral diffuseness of the $S_1 \rightarrow S_0$ fluorescence following excitation of higher energy excited singlet states is consistent with this expectation. Because the vibrational wave function high up in the S_1 potential energy surface peaks in the vicinity of its wall, the radiative transitions from the vibrationally highly excited levels of S_1 will terminate on the very steeply rising, anharmonic portions of the S_0 potential energy surface, as required by the Franck–Condon (FC) principle, Figure 2. The superposition of many such emissions from a distribution of isoenergetic vibrational levels of S_1 would necessarily

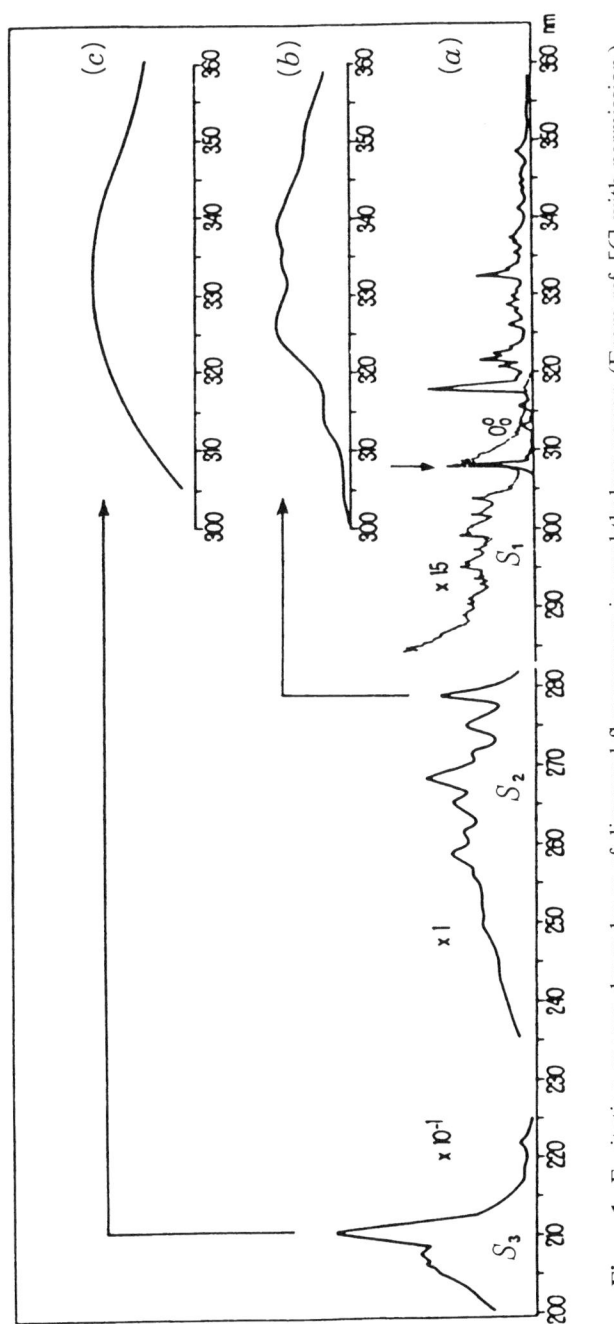

Figure 1. Excitation energy dependence of dispersed fluorescence in naphthalene vapor. (From ref. [6] with permission.)

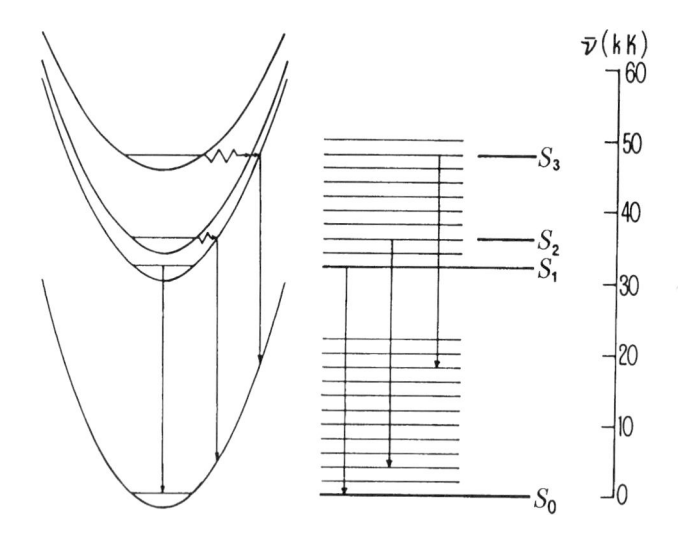

Figure 2. Schematic diagram illustrating the excitation energy dependence of dispersed fluorescence in collision-free naphthalene. (From ref. [6] with permission.)

be broad and congested. The degree of spectral congestion is expected to increase with an increasing degree of vibrational excitation in S_1, as has been observed experimentally.

Since the fluorescence originates from S_1, independent of the electronic state to which the molecule is initially excited, the excitation energy dependence of the quantum yield (Φ_F) and lifetime (τ_F) of the emission reflects the excess vibrational energy dependence of the radiative (k_r) and nonradiative (k_{nr}) decay rate constants. For molecules exhibiting exponential fluorescence decay, (see Section III) the magnitude of these decay rates can be determined from the lifetime [$\tau_F = 1/(k_r + k_{nr})$] and quantum yield [$\Phi_F = k_r/(k_r + k_{nr})$] of fluorescence through the expressions

$$k_r = \Phi_F/\tau_F \tag{1}$$

$$k_{nr} = (1 - \Phi_F)/\tau_F \tag{2}$$

Thus, by measuring the quantum yield and the lifetime as a function of excitation wavelength, it is possible to determine the manner in which the radiative and nonradiative decay rates vary with excess vibrational energy in S_1.

III. EXCESS ENERGY DEPENDENCE OF RADIATIONLESS TRANSITIONS: THEORETICAL CONSIDERATIONS

Historically, radiationless transitions from S_1 have been classified as small molecule, intermediate case, or statistical limit [1,7]. The classification is based on the density (ρ_ℓ) of the background (S_0 or T) manifold $\{|\ell\rangle\}$ relative to its decay width γ_ℓ [8]. In the small molecule limit, the density of the background manifold ρ_ℓ at the energy of the initial state $|s\rangle$ is very small compared to its decay width, that is, $\gamma_\ell \rho_\ell \ll 1$, whereas in the statistical limit $\gamma_\ell \rho_\ell \gg 1$. Nonradiative decay cannot occur in the small molecule limit, whereas an irreversible, exponential, nonradiative decay is expected in the statistical limit. In between these two limiting cases lies the intermediate case in which the background states, although moderately dense, are still well separated relative to their decay width, that is, $\gamma_\ell \rho_\ell < 1$. The theory of intermediate case radiationless decay [9] predicts that a coherent excitation of the wave packet leads to the decay of fluorescence that can be described as the sum of two exponential components. The shorter lived component represents the coherent decay of the initially prepared singlet state, whereas the longer lived component represents the incoherent decay of the molecular eigenstate, which is a superposition of the singlet and the background manifold (S_0 or T). If the spacing between interacting states is comparable to the bandwidth of the excitation, the theory also predicts that coherent excitation may lead to interference effects in the form of quantum beats. These predictions have been confirmed by experiment [10]. Because of the enormously large level density of the ground state associated with the large S_1-S_0 electronic gap, internal conversion from the S_1 state of aromatic molecules invariably belongs to the statistical limit. In contrast, the $S_1 \to T$ ISC of diazabenzenes with small S_1-T_1 electronic gaps can exhibit the intermediate case behavior when the excess energy of the S_1 molecule is small. As the level density in the triplet manifold increases by vibrational excitation in S_1, the $S_1 \to T$ ISC of the diazabenzenes (e.g., pyrazine) also displays the irreversible, exponential decay characteristic of the statistical limit [11]. Since the primary focus of this chapter is the $S_1 \to S_0$ IC and $S_1 \to T$ ISC of vibrationally excited molecules, we need only to consider the statistical limit of radiationless transitions.

In the large molecule statistical limit, the rate constant k_{nr} for the nonradiative (radiationless) transition from the initial state ψ_s of energy E_s to the final state ψ_ℓ of energy E_ℓ can be expressed by the Golden rule formula of time-dependent perturbation theory [1]

$$k_{nr} = \frac{2\pi}{\hbar}|\langle\psi_s|\hat{v}|\psi_\ell\rangle|^2\rho(E) \quad (3)$$

where $\hat{v}$ is the perturbation responsible for the nonradiative process and $\rho(E)$ is the density of states in the final state. To the first order, $\hat{v}$ is nuclear kinetic energy operator for internal conversion and spin–orbit coupling operator for intersystem crossing. By using the Born–Oppenheimer approximation, ψ can be factored into an electronic wave function ϕ, which depends on both the electronic coordinate q and nuclear coordinate Q, and a nuclear wave function χ

$$\psi(q,Q) = \phi(q,Q)\chi(Q) \quad (4)$$

The rate expression then takes the form

$$k_{nr} = \frac{2\pi}{\hbar}|\langle s|\hat{v}|\ell\rangle|^2|\langle\chi_s|\chi_\ell\rangle|^2\rho \quad (5)$$

where $\langle s|\hat{v}|\ell\rangle$ is the electronic matrix elements $v_{s\ell}$, and $\langle\chi_s|\chi_\ell\rangle$ is the vibrational overlap integral, whose square is the FC factor. It is conventional to represent the vibrational wave function of the final state χ_ℓ as a product of one quantum of *promoting* mode and as many quanta of various *accepting* modes as are necessary to meet the energy conservation requirement of the radiationless transition [12]. The promoting mode renders the electronic matrix elements finite via vibronic perturbation, whereas the accepting modes provide the FC factors. With this representation the rate expression becomes [13]

$$k_{nr} = \frac{2\pi}{\hbar}|v_{s\ell}^p|^2\prod_j F_j\rho(\{\bar{n}_j\}) \quad (6)$$

In Eq. 6, $v_{s\ell}^p$ is the electronic matrix element involving the promoting mode p; F_j is the FC factor for mode j; and $\rho(\{\bar{n}_j\})$ is the density of states weighted FC factor for the set $\{\bar{n}_j\}$, where $\{\bar{n}_j\}$ represents the most probable (i.e., optimal) distribution of the vibrational quanta in the final state. The function $\rho(\{\bar{n}_j\})$ accounts for the contributions to the FC factor of modes other than j. For the radiationless transition originating from the zero-point level of mode j, the FC factor is given by

$$F_j = |\langle\chi_j^s(0)|\chi_j^\ell(\bar{n}_j)\rangle| \quad (7)$$

Here $\bar{n}_j$ is related to the electronic energy gap and vibrational frequencies by the energy conservation condition

$$E_s - E_\ell - \hbar\omega_p = \sum_j \bar{n}_j \hbar\omega_j \tag{8}$$

For an electronic energy gap $E_s - E_\ell$ much larger than the vibrational spacing $\hbar\omega_p$, the FC factor in Eq. 7 decreases very rapidly with increasing values of $\bar{n}_j$. Thus, Eq. 6 accounts for the strong dependence of the radiationless decay rate on the electronic energy gap, that is, energy gap law. The inefficiency of $S_1 \to S_0$ IC relative to $S_n \to S_{n-1}$ ($n \geq 2$) IC or $S_1 \to T$ ISC can be qualitatively understood in terms of a large S_1–S_0 energy gap, and hence large $\bar{n}_j$, as can the fact that $S_1 \to S_0$ IC appears to be important only in molecules with a low-lying S_1 (e.g., pentacene).

To consider the effects of vibrational excitation on k_{nr} in the simplest possible terms, we assume that (1) the same normal mode description can be used for the vibrational wave functions of the initial and final electronic states, (2) the electronic matrix elements do not vary with vibrational excitation, and (3) the effective density of states $\rho(\{\bar{n}_j\})$ is a slowly varying function of the excess vibrational energy compared to the FC factor F_j. In addition, we will assume that the vibrational quanta m_j of the excess energy do not significantly affect the optimal distribution $\bar{n}_j$ for the vibrationless ($m_j = 0$) case, so that the excited quanta m_j can simply be added to $\bar{n}_j$ to give the new distributions $\bar{n}_j + m_j$ for nonradiative decay from the excited vibrational levels (model for retention of vibrational quanta [14]). With these assumptions, and neglecting the role of the promoting mode, the excess energy dependence of the radiationless decay rate from an initial electronic state with vibrational energy $\Sigma_j m_j \hbar\omega_j$ is given approximately by [14]:

$$\begin{aligned}
\frac{k(\{m_j\})}{k(0)} &\simeq \prod_j \frac{|\langle \chi_j^s(m_j)|\chi_j^\ell(\bar{n}_j + m_j)\rangle|^2}{|\langle \chi_j^s(0)|\chi_j^\ell(\bar{n}_j)\rangle|^2} \exp(m_j \Delta\omega_j \tau) \\
&= \prod_j \frac{\Gamma(m_j + \bar{n}_j + 1)}{\Gamma(m_j + 1)\Gamma(\bar{n}_j + 1)} \exp(m_j \Delta\omega_j \tau) \\
&= \prod_j \frac{(m_j + \bar{n}_j)!}{m_j! \, \bar{n}_j!} \exp(m_j \Delta\omega_j \tau)
\end{aligned} \tag{9}$$

where the quantity $\Delta\omega_j$ is the change in the frequency of the normal mode in going from the initial to the final states, namely, $\Delta\omega_j = \omega_j^\ell - \omega_j^s$, τ is a Lagrangian parameter, and Γ represents the gamma functions.

The expression containing the Γ functions or factorials is valid only in the so-called weak-coupling limit where the coupling constant $g_j^2 = m_j\omega_j\Delta R_j^2/2\hbar$, describing the changes in the equilibrium position ΔR_j, is smaller than unity. For displaced ($\Delta R \neq 0$) oscillators, the Lagrangian parameter τ, which controls the optimum distribution, must be determined from the equation [14]

$$E_s - E_\ell - \hbar\bar{\omega}_p = \sum_j \bar{n}_j \hbar\omega_j = \sum_j \hbar\omega_j g_j^2(s \to \ell)\exp(\omega_j\tau) \qquad (10)$$

Equations 9 and 10 provide a qualitative basis for understanding the dependence of the radiationless decay rate on the excess vibrational energies in S_1. For a very small electronic energy gap ($E_s - E_\ell \approx 0$), the occupation number $\bar{n}_j$ approaches zero and the ratio of the Γ functions (or factorials) approaches unity, so that Eq. 9 predicts little change in the nonradiative decay rate with increasing excess energy. Conversely, for large electronic gap radiationless transitions such as $S_1 \to S_0$ IC and $T_1 \to S_0$ ISC, Eqs. 9 and 10, predict strong increases in k_{nr} by vibrational excitation owing to the large $\bar{n}_j$ associated with the nonradiative processes. The increase in the nonradiative decay rate is expected to be especially large if the vibrational quanta m_j, which are due to the excess energy, correspond to modes that are good accepting modes for the radiationless transition. It should be noted that the presence of the frequency change term $\Delta\omega_j$ in Eq. 9 makes it possible for nonradiative decay rate to decrease with increasing excess energy when the frequencies of the accepting modes in the final electronic state are smaller than those in the initial state. However, this behavior is not likely to be observed for radiationless transitions involving large electronic energy gaps, since in these cases the increase in the factorials ratio is expected to more than compensate for the decrease in the exponential term.

Equation 9 also predicts an interesting deuterium isotope effect on the excess energy dependence of the nonradiative decay rate. It is well known that for large electronic energy gaps important contributions to the FC factor in radiationless transitions of aromatic hydrocarbons arise from high-frequency C—H stretching modes [1]. Replacement of H by D, leading to a decrease in the vibrational frequency, should result in the greater excess energy dependence of the decay rate since $\bar{n}_j$ increases considerably upon deuteration of the molecule. Thus, for sufficiently large excess energies (large m_j), the nonradiative decay rate of deuterated compounds may actually become greater than those of the corresponding protic compounds, despite the fact that the converse is generally true at zero excess energy [14]. However, such an isotope effect is expected to be small for radiationless transitions involving a small electronic energy gap.

An important conclusion to come out of the above discussions is that the $S_1 \rightarrow S_0$ IC rate increases with increasing vibrational excitation much more rapidly than the $S_1 \rightarrow T$ ISC rate, owing to the large S_1–S_0 electronic gap. Furthermore, since the frequency changes ($\Delta\omega_j > 0$) are generally greater between S_1 and S_0 than between S_1 and T, the excess energy dependence of the $S_1 \rightarrow S_0$ IC is expected to be greater than that of the $S_1 \rightarrow T$ ISC rate, even if the electronic energy gaps are comparable for the two transitions. We would therefore predict that the dominant photophysical decay channel of S_1 molecules, with large excess vibrational energies in S_1, is the internal conversion to the ground state [14,15]. A schematic diagram illustrating the disparity in the excess energy dependences of $S_1 \rightarrow S_0$ IC and $S_1 \rightarrow T$ ISC is given in Figure 3. It should be noted from the figure that the excess energy dependence of the total nonradiative decay rate ($k_{\text{ISC}} + k_{\text{IC}}$) is expected to increase rather sharply at higher energies due to the onset of the dominant internal conversion [14]. The threshold vibrational energy above which the $S_1 \rightarrow S_0$ IC surpasses the $S_1 \rightarrow T$ ISC would depend on the difference in the excess energy dependence of the two competing nonradiative decay channels as well as the disparity in their decay rates at zero excess energy. However, because of the extreme sensitivity of the IC rate to the S_1–S_0 electronic energy gap, one would expect the smaller aromatic molecules (with larger S_1–S_0 electronic gap, and hence larger excess energy dependence of the IC rate) to exhibit lower threshold energy as compared to larger aromatic molecules.

One important aspect of the excess energy dependence of radiationless transition, not accounted for by the simple theoretical treatment, is the "saturation" of the nonradiative decay rate that may occur at higher energies [16]. Whereas Eq. 9 predicts a simple exponential dependence of k_{nr} on excess energy [14] in the limit of complete vibrational energy redistribution (communicating states model [17]), deviations from the exponential energy dependence of k_{nr} may be expected due to the diabatic crossing between the potentials of the initial and final electronic states at higher energies, Figure 4. Below such a crossing, the molecule tunnels from the upper to the lower electronic states to make the radiationless transition, whereas above the crossing point the molecule hops from one surface to another without abrupt changes in nuclear position or momentum. As shown in the semiclassical treatment of radiationless transition by Heller and Brown [18], this change of mechanism from tunneling to hopping leads to a marked change in the excess energy dependence of k_{nr}, with a larger slope being in the tunneling region. The same conclusion also obtains from the quantum description of radiationless processes, as demonstrated by Sobolewski [19]. The ln k_{nr} versus excess energy plots may therefore tend to flatten out at high excess energies. This deviation from the exponential

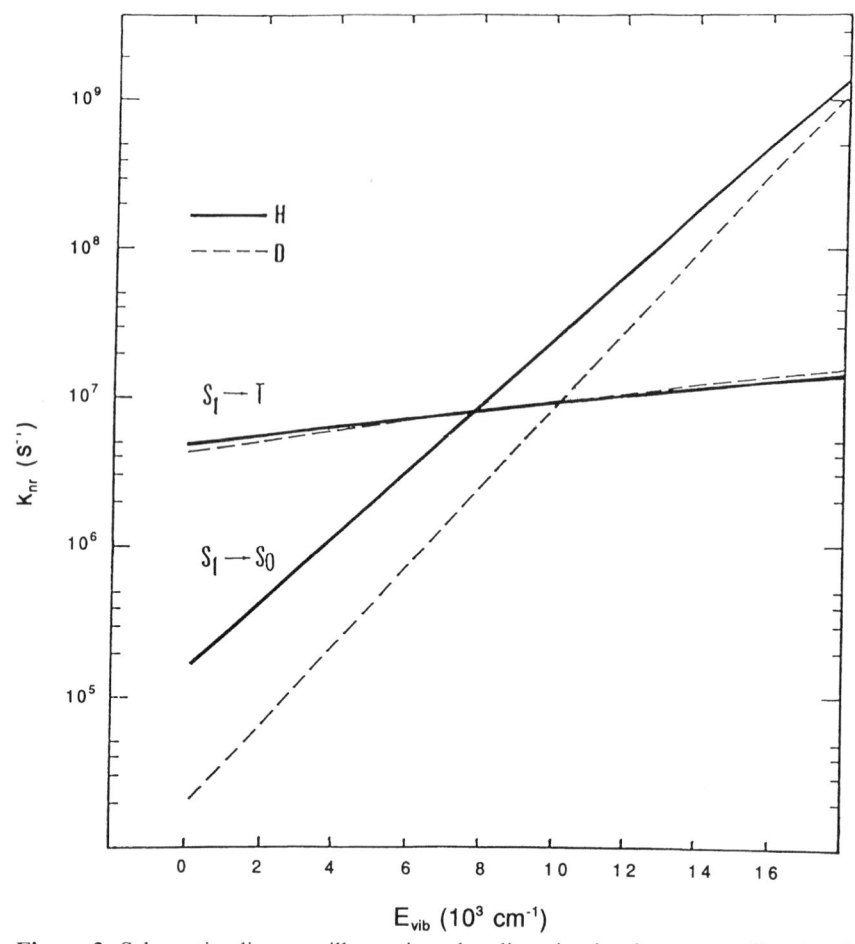

Figure 3. Schematic diagram illustrating the disparity in the excess vibrational energy dependence of $S_1 \to S_0$ internal conversion and $S_1 \to T$ intersystem crossing in naphthalene-d_0 and naphthalene-d_8. (After ref. [4] with permission.)

energy dependence of k_{nr} is expected to be especially serious in molecules with a shallow upper state potential and a steep lower state potential, since the diabatic crossing occurs at relatively low excess energies in such a molecule. Conversely, if the crossing occurs at very high excess energies, as would be the case when the two potential curves are nearly parallel, the exponential excess energy dependence of k_{nr} may be observed over a very

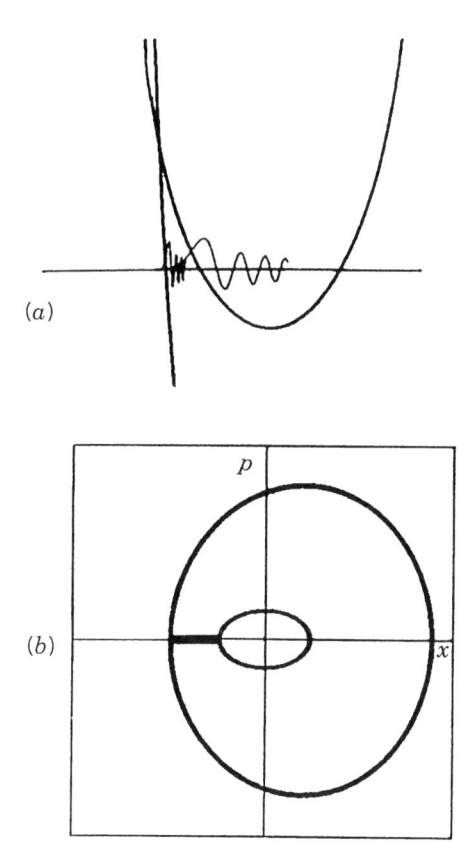

Figure 4. (a) A steep potential intersecting another, bound potential of higher electronic energy. Note that, at energies below the crossing point, the main contribution to the vibrational overlap integral (and hence the Franck–Condon factor for radiationless transition) comes from the turning point region of the steep potential. (b) A phase space picture corresponding to (a). The inner ellipse, which represents the path of the trajectory bound on the upper potential, tunnels to the outer ellipse (which describes a trajectory of the same energy on the lower potential) through the path shown. (From ref. [18] with permission.)

wide energy range. The former situation is exemplified by benzene and azabenzenes, and the latter by polycyclic aromatic hydrocarbons (see below).

We conclude this section by presenting the theoretical treatment of the excess energy dependence of k_{nr} due to Sobolewski [19], which is frequently used to simulate the experimental data (see below). The treatment starts with the "state-averaged" rate expression of Smalley and co-workers [20]

$$k_{nr}(E) = \frac{2\pi}{h} \langle \sum_{\ell} |v_{s\ell}|^2 \delta(E - E_\ell) \rangle \qquad (11)$$

which is valid when rapid IVR leads to distribution of excess energy over all the isoenergetic vibrational states of the initial electronic state on a time scale faster than the electronic radiationless transition. To investigate the dependence of k_{nr} on excess vibrational energy (E), Sobolewski recasts Eq. 11 in the form

$$k_{nr}(E) = [\rho_\Delta(E)]^{-1} \sum_s k_s(E_s) \delta_\Delta(E - E_s) \qquad (12)$$

where the sum runs over all the zero-order vibrational levels of the initial electronic state that are isoenergetic with E within the average coupling strength (Δ) for IVR, that is,

$$E - \frac{\Delta}{2} \leqslant E_s \leqslant E + \frac{\Delta}{2}$$

as indicated by the rectangular function $\delta_\Delta(E)$. In Eq. 12 $\rho_\Delta(E)$ is the vibrational density-of-state function in the initial electronic state calculated in the space of all $3N$-6 vibrational degrees of freedom

$$\rho_\Delta(E) = \sum_s \delta_\Delta(E - E_s) \qquad (13)$$

and $k_s(E_s)$ represents the single-vibronic-level (SVL) decay rate given by

$$k_s(E_s) = \frac{2\pi}{h} \sum_\ell |v_{s\ell}|^2 \delta(E_s - E_\ell) \qquad (14)$$

In order to simplify the calculation of the average decay rate in Eq. 12, Sobolewski [19] divides the vibrational degrees of freedom into three categories: promoting modes (p), accepting modes (a), and nonactive modes (n). The nonactive modes have vibrational wave functions that are orthogonal in the two electronic states (initial and final) and cannot promote the radiationless transition for symmetry reasons. Because the nonradiative decay rate depends on the vibrational quantum numbers of the accepting

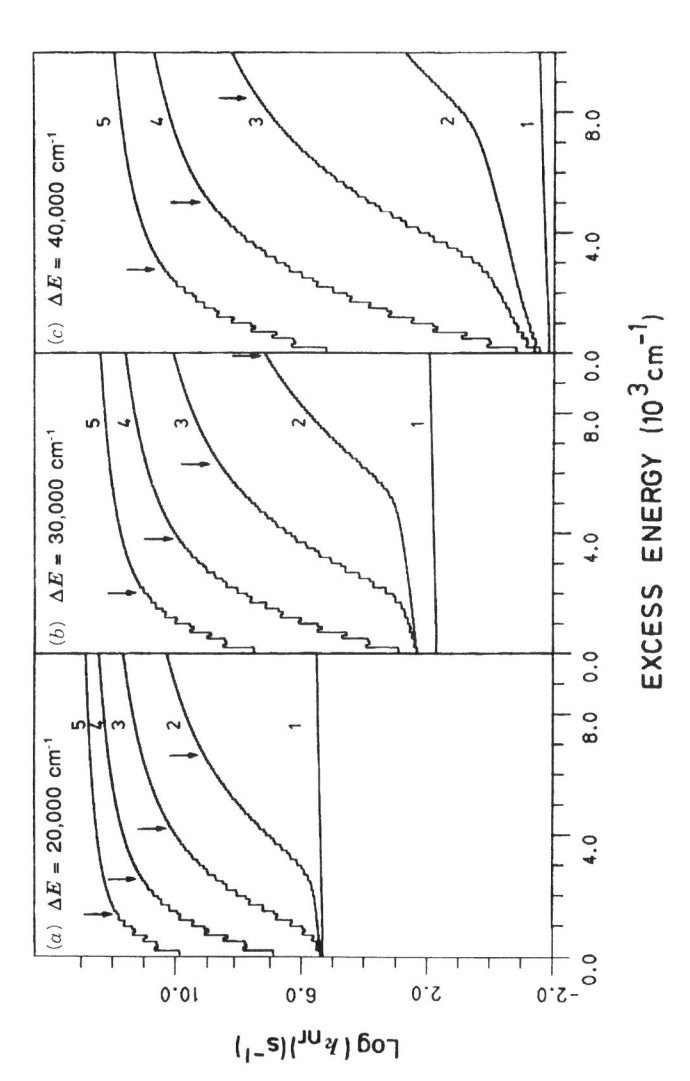

Figure 5. Dependence of the calculated nonradiative decay rates on the vibrational energies in the initial electronic state for three different electronic energy gaps ($\Delta E = E_s - E_l$). The frequency of the nontotally symmetric mode in the initial electronic state was fixed at 250 cm^{-1}, whereas that in the final electronic state was taken to be 250 cm^{-1} (1), 500 cm^{-1} (2), 600 cm^{-1} (3), 750 cm^{-1} (4), and 1000 cm^{-1} (5). Other vibrational parameters can be found in the original paper. Arrows indicate the energies of the diabatic crossing points between the potentials of the nontotally symmetric mode in the two electronic states. (From ref. [19] with permission.)

modes (n_a), but not on the quantum number of the nonactive modes, the number of the SVL rate constants that needs to be calculated is reduced dramatically by this partitioning of the vibrational degrees of freedom. The final rate expression obtained by Sobolewski [19] is

$$k_{nr}(E) = [\rho_\Delta(E)^{-1}] \sum_{n_p, n_a} (n_p + 1)k_s(n_a)\rho_\Delta^n(E - \varepsilon^p - \varepsilon^a) \quad (15)$$

where ε^p and ε^a denote, respectively, the vibrational energies of the promoting and accepting modes in the initial electronic state, and $\rho_\Delta^n(E)$ represents the density-of-states function calculated in the subspace of nonactive modes of the total number n. Equation 15 allows the calculation of the excess energy dependence of k_{nr} for molecules whose vibrational frequencies and displacements in the ground and excited states are either known or can be estimated. Results of model calculations [19], illustrating saturation of k_{nr} at higher vibrational energies, are shown in Figure 5.

IV. EXCESS ENERGY DEPENDENCE OF RADIATIONLESS TRANSITIONS IN MOLECULES WITH NEARBY $n\pi^*$ AND $\pi\pi^*$ STATES: THEORETICAL CONSIDERATIONS

The increase in the IC rate that accompanies the vibrational excitation is expected to be especially prominent for molecules in which one or more of the modes experiences a large frequency increase in going from the excited state to the ground state [21]. While the geometry and frequency changes accompanying electronic excitation are generally not large for aromatic hydrocarbons, they can be very significant for nitrogen heterocyclic and aromatic carbonyl compounds that possess a lowest energy $\pi\pi^*$ state close to a lowest energy $n\pi^*$ state. In such molecules, vibronic coupling between the close-lying $n\pi^*$ and $\pi\pi^*$ states can lead to large potential distortions and displacements along the vibronically active out-of-plane coordinates, as first shown by Hochstrasser and Marzzacco [22]. More specifically, the frequency of the out-of-plane mode decreases, and the molecule may become nonplanar, in the lower of the two vibronically coupled excited states [22, 23].

To examine the potential energy distortion and displacement arising from the $n\pi^*$–$\pi\pi^*$ vibronic coupling, we consider two electronic states $|m(q, Q_0)\rangle$ and $|n(q, Q_0)\rangle$, which are eigenfunctions of the electronic Hamiltonian (H_e):

$$H_e(q, Q_0) = T_e(q) + U(q, Q_0) \quad (16)$$

where T_e represents the kinetic energy operator for electrons (with coordinate q) and U represents the potential energy at the equilibrium nuclear configuration Q_0. The corresponding potential energy surfaces will be assumed to be simple harmonic potentials:

$$E_m^{(0)}(Q_p) = E_m^{(0)}(Q_0) + \tfrac{1}{2}k_{pm}^{(0)}Q_p^2 \tag{17}$$

$$E_n^{(0)}(Q_p) = E_n^{(0)}(Q_0) + \tfrac{1}{2}k_{pn}^{(0)}Q_p^2 \tag{18}$$

where $k_{pm}^{(0)}$ and $k_{pn}^{(0)}$ denote the force constants for the mode Q_p, which induces $|m(q, Q_0)\rangle - |n(q, Q_0)\rangle$ vibronic coupling. Expanding the potential energy of interaction in a power series of Q_p,

$$U(q, Q_p) = U(q, Q_0) + \left[\frac{\partial U(q, Q_p)}{\partial Q_p}\right]_{Q_0} Q_p + \frac{1}{2}\left[\frac{\partial^2 U(q, Q_p)}{\partial Q_p^2}\right]_{Q_0} Q_p^2 + \cdots \tag{19}$$

one obtains

$$U(q, Q_p) - U(q, Q_0) = \left[\frac{\partial U(q, Q_p)}{\partial Q_p}\right]_{Q_0} Q_p \tag{20}$$

as the dominant interaction term left out in the crude adiabatic electronic Hamiltonian of Eq. 16. Thus, the term $[\partial U(q, Q_p)/\partial Q_p]_{Q_0} Q_p$ can induce coupling of the state $|m(q, Q_0)\rangle$ with the state $|n(q, Q_0)\rangle$. In planar aromatic molecules, the mode Q_p that couples the $n\pi^*$ and $\pi\pi^*$ states is an out-of-plane bending mode, as n and π orbitals are symmetric and antisymmetric with respect to reflection through the molecular plane, respectively, and the operator $\partial U(q, Q_p)/\partial Q_p$ transforms as Q_p. Since vibronic coupling mixes $|m(q, Q_0)\rangle$ and $|n(q, Q_0)\rangle$, we write the adiabatic Born-Oppenheimer (ABO) electronic states as

$$|m(q, Q_p)\rangle = a_m(Q_p)|n(q, Q_0)\rangle + b_m(Q_p)|m(q, Q_0)\rangle \tag{21}$$

and

$$|n(q, Q_p)\rangle = a_n(Q_p)|n(q, Q_0)\rangle + b_n(Q_p)|m(q, Q_0)\rangle \tag{22}$$

Making use of Eqs. 17–22 and implementing the variation method, we solve for the ABO potential energy surfaces. The results are [22, 23]

$$E_m(Q_p) = E_m^{(0)}(Q_0) + \frac{1}{2}\left[k_{pm}^{(0)} + \frac{2|U_{mn}^{(p)}|^2}{\Delta E_{mn}}\right] Q_p^2 + \cdots \tag{23}$$

and

$$E_n(Q_p) = E_n^{(0)}(Q_0) + \frac{1}{2}\left[k_{pn}^{(0)} - \frac{2|U_{mn}^{(p)}|^2}{\Delta E_{mn}}\right]Q_p^2 + \cdots \quad (24)$$

where $U_{mn}^{(p)}$ is given by

$$U_{mn}^{(p)} = \left\langle m(q, Q_0) \left| \frac{\partial U(q, Q_p)}{\partial Q_p} \right| n(q, Q_0) \right\rangle \quad (25)$$

If $2[k_{pm}^{(0)} - k_{pn}^{(0)}] > |U_{mn}^{(p)}|^2/\Delta E_{mn}$ and the quantity in the square brackets of Eq. 24 is negative, the lower energy surface assumes a double minimum [22]. However, if the term in the brackets is positive, the quartic (Q^4) and higher order terms, not given explicitly in Eqs. 23 and 24, are simply anharmonic corrections to the harmonic surface. It should be noted from Eqs. 23 and 24 that, as a result of the vibronic interactions, the force constant (and therefore the frequency) of the vibronically active out-of-plane bending mode decreases in the lower electronic state, while that in the upper electronic state increases [22]. The potential energy distortion resulting from the vibronic interaction is schematically illustrated in Figure 6. The magnitude of the frequency decrease in the lower electronic state can be quite large as illustrated by the fact that the $v_{10a}(b_{1g})$ mode, which is active in the $^1B_{3u}(n\pi^*) - {}^1B_{2u}(\pi\pi^*)$ vibronic coupling of pyrazine (1,4-diazabenzene), undergoes a substantial reduction in its frequency (383 cm^{-1}) in the $^1B_{3u}(n\pi^*)$ state relative to that (919 cm^{-1}) in the ground electronic state [24]. Analyses of the fluorescence and absorption spectra of gaseous pyrazine by Ito and co-workers [25], Kanamaru and Lim [26], and Siebrand and co-workers [27] show that the greatly reduced frequency of the $v_{10a}(b_{1g})$ mode in S_1 is almost entirely due to vibronic coupling between $S_1({}^1B_{3u}, n\pi^*)$ and $S_2({}^1B_{2u}, \pi\pi^*)$, which are separated by about 7000 cm^{-1} [24]. This large reduction in frequency is accompanied by strong anharmonicity, consistent with the importance of the quartic and higher order terms in Eq. 24.

The potential energy distortion (frequency change) and displacement (geometry change) along the vibronically active out-of-plane bending modes are expected to have important effects on the nonradiative decay rate of the lowest excited state since the magnitude of the FC factor appearing in the rate expression of Eq. 6 depends sensitively on the frequency shift (cf. Eq. 9) as well as the displacement of the energy accepting modes [21]. For a displaced ($\Delta R_j \neq 0$) but undistorted ($\Delta \omega_j = 0$) oscillator, the FC factor takes the form [1(a)]

$$F_j = \exp(-\gamma)\gamma^n/n! \quad (26)$$

THEORETICAL CONSIDERATIONS 183

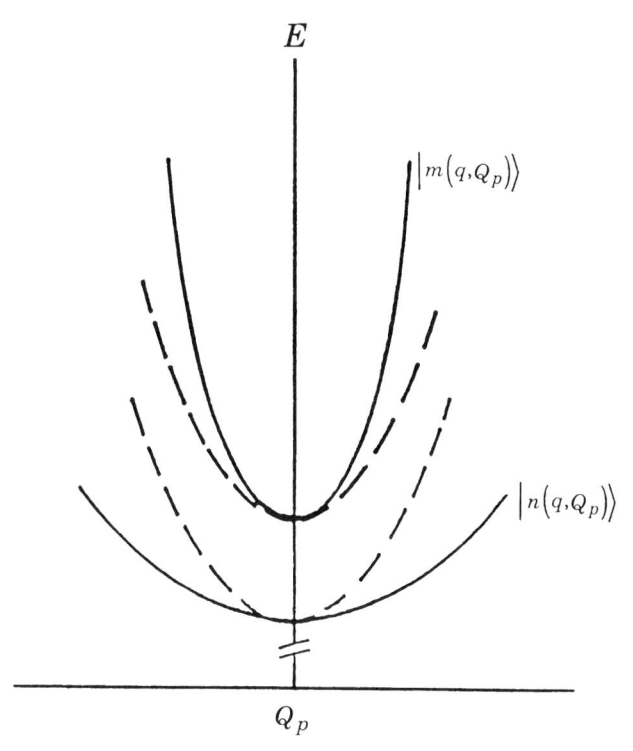

Figure 6. Schematic representation the potential energy distortion resulting from the vibronic interaction between two close-lying excited states. The dashed curves represent the potential energy curves in the absence of vibronic coupling. (From ref. [21] with permission.)

where

$$\gamma = \frac{1}{2} k_j (\Delta R_j)^2 / \hbar \omega_j \tag{27}$$

and k_j and ΔR_j are the force constant and the change in the equilibrium position, respectively. Note that the FC factor for the radiationless transition increases dramatically with the increasing magnitude of the potential displacement (ΔR). For a distorted ($\Delta \omega_j \neq 0$) but undisplaced ($\Delta R_j = 0$) harmonic oscillator, the FC factor for the radiationless transition is given by [1(a)]

$$F_j = (1 - \xi^2)^{1/2} \xi^n \frac{(1)(3)(5) \cdots (n-1)}{(2)(4)(6) \cdots (n)} \tag{28}$$

n even ($F = 0$ for n odd)

where

$$\xi^2 = [(\omega_\ell - \omega_s)/(\omega_\ell + \omega_s)]^2 \tag{29}$$

Thus, the modes that suffer large frequency increases in going from the initial electronic state $|s\rangle$ to the final electronic state $|\ell\rangle$ of the radiationless transition are expected to contribute most to the FC factors. Since the out-of-plane modes that are active in the $n\pi^*-\pi\pi^*$ vibronic coupling have lower frequencies in the excited electronic state than in the ground state (see above), these vibrations should be important accepting modes for the radiationless decay of the lowest excited state to the ground state.

To investigate the effect of $n\pi-\pi\pi^*$ vibronic coupling on radiationless decay rates more quantitatively, we have used the three-state model consisting of two excited electronic states, $|m\rangle$ and $|n\rangle$, coupled through a single vibronically active mode, and a third electronic state, $|0\rangle$, to which the lower of the two vibronically coupled states can decay nonradiatively. The results of the numerical investigations based on the known or estimated spectroscopic parameters show that, for a sufficiently large interaction strength and sufficiently small energy separation between the interacting states, the vibronically active out-of-plane bending mode could become a dominant accepting mode for the radiationless transition [21, 28–31]. The nonradiative decay rate is especially large if the lower excited state $|n\rangle$ adopts a double minimum potential. The radiationless decay rate was found to strongly increase with vibrational excitation of the out-of-plane bending modes in the lower of the two vibronically coupled states [29]. The vibrational enhancement of k_{nr} is much greater for the single minimum case than for the double minimum case [28]. Most significantly, the increase in the nonradiative decay rate by vibrational excitation is greater the larger the electronic energy gap for the radiationless transition [21, 28–31]. Since the S_1-S_0 electronic energy gap is much greater than the S_1-T gap, the $(S_1 \rightarrow S_0 \text{ IC})/(S_1 \rightarrow T_1 \text{ ISC})$ branching ratio is predicted to increase rather dramatically by vibrational excitation of the out-of-plane bending modes in S_1. We would therefore conclude that the $S_1 \rightarrow S_0$ IC would become the dominant S_1 decay channel for vibrationally highly excited nitrogen heterocyclic compounds [21, 28]. The threshold vibrational energy above which the IC process surpasses the $S_1 \rightarrow T_1$ ISC is expected to be rather small in these molecules as compared to the corresponding aromatic hydrocarbons.

The foregoing conclusions, based on the consideration of the FC factors, can be rationalized pictorially [32] using the so-called barrier width, which represents the horizontal distance (distance along the constant energy line) between the wall of the initial state potential energy (PE) surface and the wall of the final state surface. As shown by Ross and co-workers [33], the

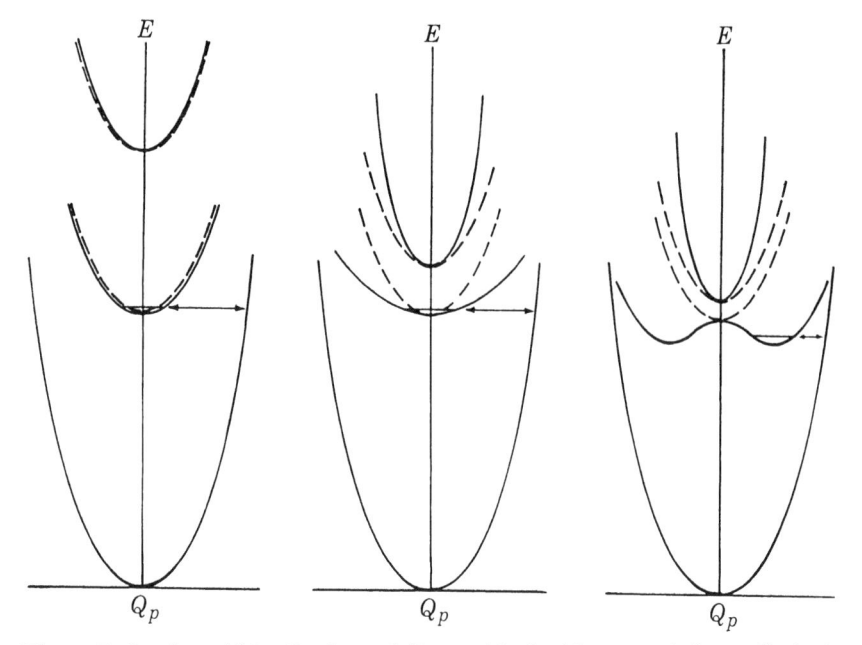

Figure 7. Barrier widths (horizontal lines with double arrows) for radiationless transition for cases in which the vibronic coupling is weak (left), strong (center), and very strong (right). (From ref. [32] with permission.) The dashed curves represent the potential energy curves in the absence of vibronic coupling.

barrier width is inversely related, albeit crudely, to the magnitude of the FC factors associated with the radiationless transition. In Figure 7, we compare the barrier widths for cases in which the vibronic coupling is weak (left), strong (center), and very strong (right). Note that the barrier width is smaller the stronger the vibronic interactions, consistent with the increasing FC factors. For the single minimum case (center), the barrier width decreases substantially with increasing vibrational excitation of the out-of-plane bending mode in the initial electronic state. The vibrational energy dependence of the barrier width is smaller for the double minimum case (right) than for the single minimum case (center), also in accord with the results of the mode calculations. It is also evident from Figure 7 that, if the out-of-plane mode has strongly reduced frequency in S_1 (or T_1) relative to S_0, the two PE surfaces can intersect at energies not far above the electronic origin of the excited state. In such a situation, the nonradiative decay rate to the ground state is expected to increase dramatically with vibrational excitation and saturate at rather low excess energies. Intersection of two PE surfaces is considered to be essential for an effective radiationless transition between two widely separated electronic states [21, 34].

It is important to point out that in order for the out-of-plane vibration to be an effective accepting mode for $S_1 \to S_0$ IC, its S_0 frequency cannot be too small relative to other good accepting modes, such as C—H stretching vibration. Thus, even if the frequency shift $\Delta\omega = \omega_{S_0} - \omega_{S_1}$ is large and positive, the barrier width for IC will remain relatively large if the S_0 frequency (and hence its force constant) is small for the out-of-plane mode. Model calculation of Hornburger [35] indicates that when the frequency of an out-of-plane mode is only one-tenth of the C—H stretching frequency, the out-of-plane vibration has a negligible influence on the IC even when the frequency distortion is unusually large.

V. EXCESS ENERGY DEPENDENCE OF RADIATIONLESS TRANSITIONS IN AROMATIC HYDROCARBONS: EXPERIMENTAL RESULTS

One of the most striking features of the excitation energy dependence of fluorescence lifetimes in gaseous aromatic hydrocarbons is that the lifetime decreases first slowly, then rapidly in an exponential fashion, as the excess energy in S_1 is increased [4, 36–39]. This finding is illustrated in Figure 8 for naphthalene and two derivatives. Since the quantum yield of fluorescence shows a very similar dependence on excitation energy, the variation of fluorescence lifetime must reflect the dependence of the nonradiative decay rate on excess vibrational energy, Eqs. 1 and 2. The slow initial increase in the radiationless decay rate with increasing excess energy implies that radiationless transitions with relatively small electronic energy gaps dominate for low excess energies. This result is in line with the prediction of Section III that $S_1 \to T_1$ ISC is the predominant channel of nonradiative decay for low excess energies owing to the more favorable FC factors.

The strong, exponential increase in the nonradiative decay rate at higher excitation energies suggests that there is a new relaxation channel, not important in the low excess-energy regime, which becomes dominant for molecules with large excess energies. This channel is most likely IC to the ground state, which is predicted to become increasingly important with increasing S_1 excess vibrational energies [14, 15]. The large excess energy dependence of the nonradiative decay rate is consistent with the large electronic energy gap between the lowest excited state and the ground state (Section III). This interpretation is supported by the deuterium isotope effect

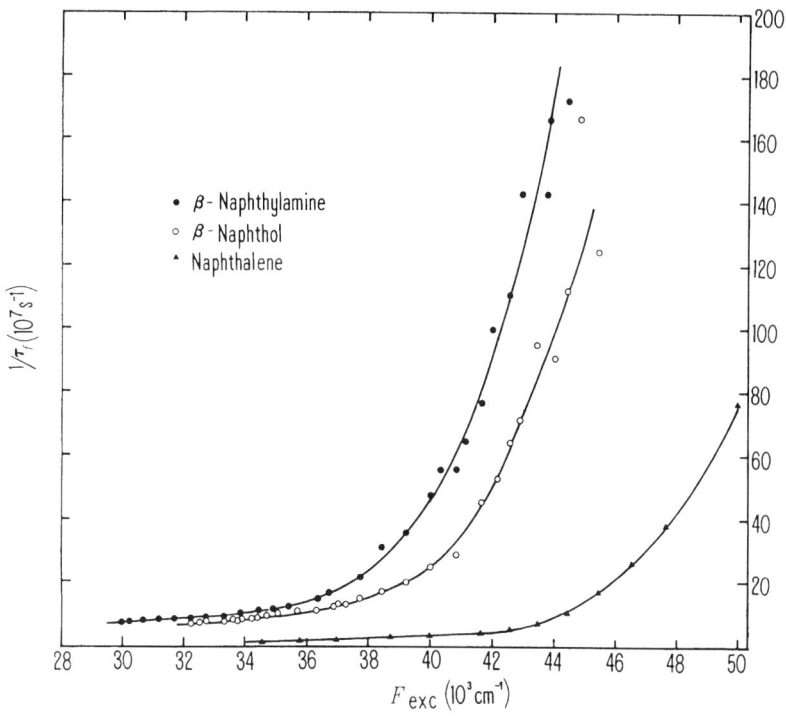

Figure 8. Excitation energy dependence of the S_1 decay rate (reciprocal of the measured fluorescence lifetime) in vapor-phase naphthalenes. Excitation source was a deuterium flash lamp. (From ref. [4] with permission.)

on radiationless decay rate (Fig. 9), which shows that the differences in the decay rates of naphthalene-d_0 and naphthalene-d_8 are small for small excess energy and large for large excess energy [4]. Interestingly, in the example shown in Figure 8, the onset of the large isotope effect practically coincides with the onset of the strong excess energy dependence of the nonradiative decay rate. The small isotope effect on the radiationless decay rates when the excess energy is small is not consistent with $S_1 \to S_0$ IC: the isotope effect on $S_1 \to S_0$ IC rate is expected to be normal and large for small excess energies. Likewise a large isotope effect on the decay rate when the excess energy is large cannot be rationalized in terms of $S_1 \to T_1$ ISC: The ISC with small electronic gap is expected to exhibit only a small normal isotope effect, or even an inverse isotope effect, at large excess energies. The prediction of Section III that the IC rate of deuterated compound should show a larger excess dependence than the corresponding protonated compound is also confirmed in the data of Figure 9. For naphthylamine, the disparity in the

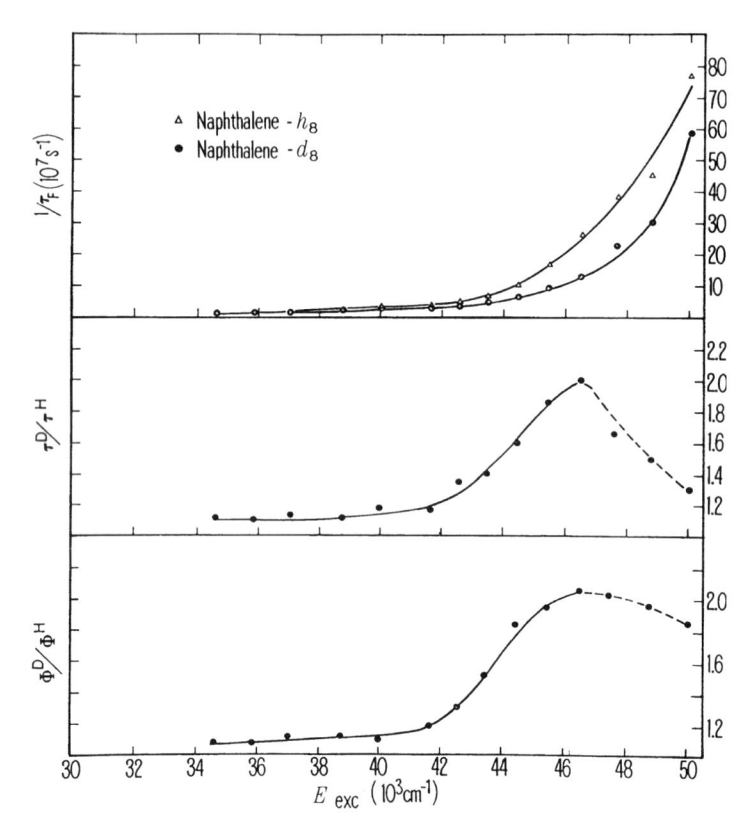

Figure 9. Excitation energy dependence of fluorescence (lifetime and quantum yield) in naphthalene-d_0 and naphthalene-d_8. (From ref. [4] with permission.)

excess energy dependence of the isotopomers is large enough to render the IC rate of the ring-deuterated species greater than that of the protic compound (inverse isotope effect) at high energies [37].

If the observation of a sharp change in the excess energy dependence of the nonradiative decay rate at high excitation energies is indeed due to the onset of the dominant $S_1 \to S_0$ IC, as we argue, then the excess energy dependence of the decay rate at higher energies should correlate with the S_1–S_0 electronic energy gap (see Section III). More specifically, the magnitude of the excess energy dependence of k_{nr} should decrease with decreasing size of the S_1–S_0 electronic energy gap. This expectation is indeed confirmed by the plot in Figure 10, which shows that the excess vibrational energy (E_{vib}) dependence of the radiationless decay rate, as measured by the quantity $\alpha \equiv d \ln k_{nr}/dE_{vib}$, is essentially a linear function of the S_1–S_0

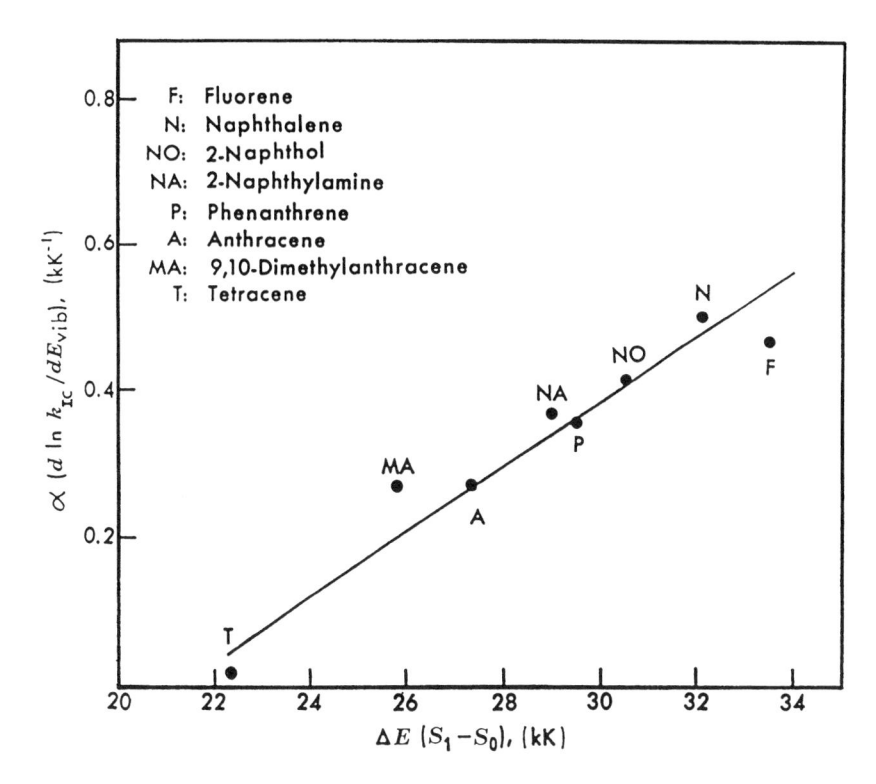

[Figure: plot with labels F: Fluorene, N: Naphthalene, NO: 2-Naphthol, NA: 2-Naphthylamine, P: Phenanthrene, A: Anthracene, MA: 9,10-Dimethylanthracene, T: Tetracene. Y-axis: α ($d\ln k_{IC}/dE_{vib}$), (kK^{-1}). X-axis: $\Delta E (S_1-S_0)$, (kK).]

Figure 10. A plot of $\alpha \equiv d\ln k_{IC}/dE_{vib}$ versus the S_1-S_0 electronic energy gap. (From ref. [38] with permission.)

energy gap [38]. This correlation, which applies to polycyclic aromatic hydrocarbons that are initially excited into S_2 or S_3, would not have been expected if a nonradiative channel other than S_1-S_0 IC were to dictate the excess energy dependence of the radiationless decay rate at high energies.

The dependence of α on the S_1-S_0 electronic gap is also evident in jet-cooled aromatic hydrocarbons, as demonstrated by Amirav [39]. Figure 11 displays the excess energy dependence of k_{nr}, as deducted from the quantum yield and lifetime of fluorescence measured in a supersonic free jet. Note that the slope of the log k_{nr} versus excess energy plot for higher energies decreases with increasing size of the aromatic hydrocarbon or the decreasing S_1-S_0 electronic energy gap. Furthermore, the zero-energy IC rate obtained from the extrapolation of the slope to zero excess energy increases with the increasing size of the molecule, consistent with the energy-gap law. Also in accord with the expectation, the threshold energy

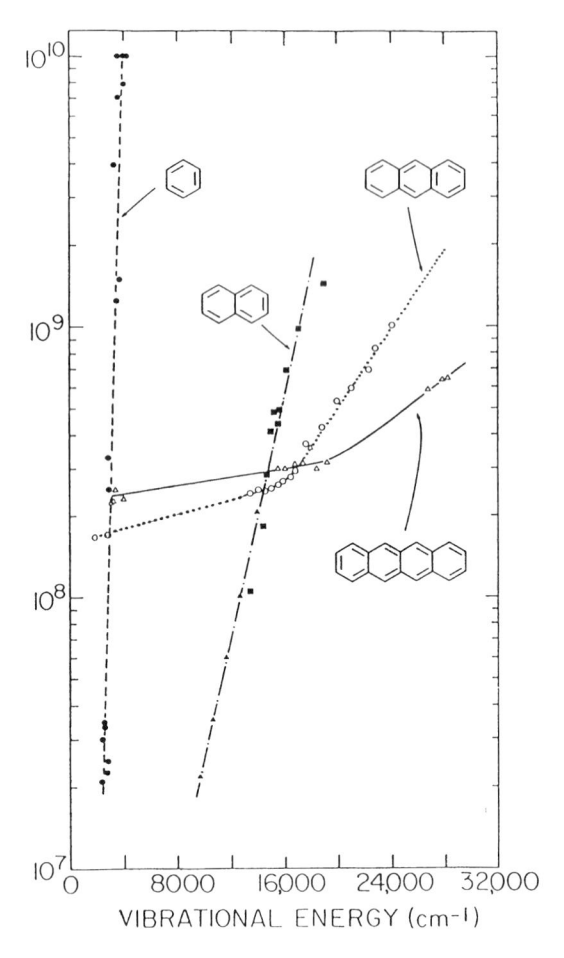

Figure 11. Excess vibrational energy dependence of nonradiative decay rates in jet-cooled aromatic hydrocarbons. (From ref. [39] with permission.)

above which the $S_1 \to S_0$ IC rate surpasses the $S_1 \to T$ ISC rate increases with increasing size of the molecule: Benzene ($\sim 3000\,\text{cm}^{-1}$) < naphthalene ($\sim 10{,}000\,\text{cm}^{-1}$) < anthracene ($\sim 16{,}000\,\text{cm}^{-1}$) < tetracene ($\sim 20{,}000\,\text{cm}^{-1}$). The excess energy dependence of radiationless transition in benzene is specifically discussed in Section VII.

The importance of internal conversion in molecules with large excess energies is substantiated further by the excitation energy dependence of triplet formation. Figure 12 shows the relative quantum yield of naphtha-

EXPERIMENTAL RESULTS 191

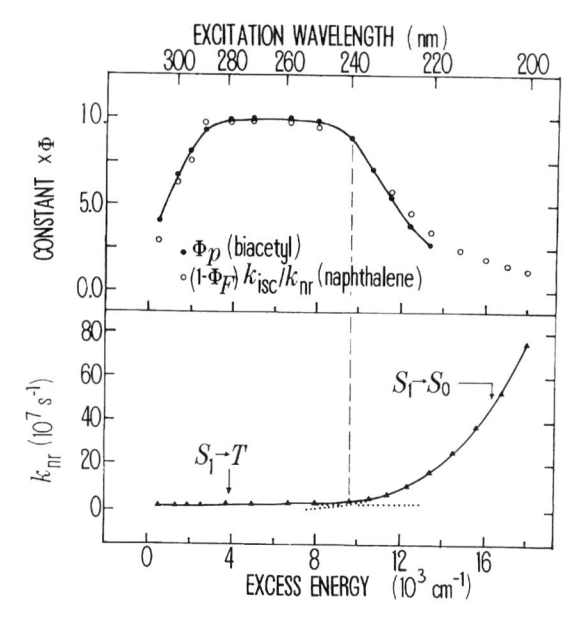

Figure 12. Relative quantum yield of sensitized phosphorescence of biacetyl (top), and nonradiative decay rate of naphthalene vapor (bottom), as a function of the excess vibrational energy in S_1 naphthalene. (From ref. [40] with permission.)

lene-sensitized biacetyl phosphorescence, as a function of excitation energy [40]. The experiment was carried out at room temperature using a mixture of 0.07 torr of naphthalene with approximately 0.10 torr of biacetyl. Under these conditions the mixture displays fluorescence, whose spectral shape and lifetimes are the same as those for naphthalene vapor (0.07 torr) without added biacetyl. This finding indicates that there is no singlet–singlet energy transfer between naphthalene and biacetyl under these experimental conditions, and that the sensitized phosphorescence of biacetyl comes entirely from energy transfer from the triplet state of naphthalene to the triplet state of biacetyl. The excitation energy dependence of the sensitized phosphorescence shows that the quantum yield of the emission, defined as the ratio of the number of quanta emitted by biacetyl to the number of quanta absorbed by naphthalene, decreases sharply when the excess vibrational energy in the naphthalene S_1 exceeds about $10,000\,\mathrm{cm}^{-1}$. This result demonstrates that either the triplet yield of naphthalene vapor becomes very small for large excess energies, or the biacetyl triplet states that are formed by energy transfer from naphthalene do not emit or emit with very low efficiencies, when their vibrational energy contents are large. However, the pressure effect on the excitation energy dependence of sensitized phosphorescence suggests that the latter of these two possibilities is unlikely to be the cause of the decreasing phosphorescence yield. Together with the photochemical stability of naphthalene and biacetyl, this result leads to the conclusion that the quantum yield of the $S_1 \to T$ ISC is very small for naphthalene vapor with large excess vibrational energies. Since the onset of the sharp drop in the triplet yield coincides with the onset of the sharp increases in the radiationless decay rate (Fig. 12), the assumption that the rapid increase in the decay rate at higher excess energies is due to the $S_1 \to S_0$ internal conversion is confirmed. Interestingly, if it is assumed that the excitation energy dependence of the nonradiative decay rate (k_{nr}) in the region of low excess energies represents the energy dependence of the intersystem crossing rate (k_{ISC}), then the excitation energy dependence of the intersystem crossing yield, as determined from $(1 - \Phi_F)k_{ISC}/k_{nr}$, parallels that deduced from the biacetyl sensitization [40] (Fig. 12).

In the foregoing discussions, we have shown how the nature of the dominant radiationless transition changes from $S_1 \to T$ ISC to $S_1 \to S_0$ IC as the excess vibrational energy within the S_1 state is increased. There is, however, one additional important piece of information that can be obtained from the study of the energy dependence of the nonradiative decay rate: the efficiency of IVR. The vibrational level distribution prepared by direct optical excitation of S_1 will, in general, be different from that generated by $S_2 \to S_1$ IC, since the former is determined by the FC factors for a vertical (isogeometric) transition while the latter is determined by the FC factor for a horizontal (isoenergetic) transition. Thus, if the vibrational relaxation in S_1 is incomplete during the S_1 lifetime, the change in the vibrational-level distribution following $S_2 \to S_1$ IC should lead to a change in the magnitude, and/or the excess energy dependence, of the S_1 decay rate compared to what it would have been for optically prepared S_1 states of the same energy. Experiments of this kind have shown that the nonradiative decay rate of S_1 as a function of excess energy, exhibits discernible (or even dramatic) changes when the S_2 excitation threshold is crossed [4, 36]. The results illustrated in Figure 13 suggest that $S_0 \to S_2$ excitation leads to changes in the vibrational distribution that are substantially maintained during the lifetime of the S_1 state. Considering the moderate vibrational densities of the S_1 manifold at the electronic origin of S_2 (S_1–S_2 electronic energy gap is relatively small), the lack of IVR leading to energy randomization is not unexpected at low excitation energies. As the excitation energy, and hence level density, increases it should be possible for extensive IVR to take place prior to electronic relaxation of S_1. Consistent with this expectation, naphthalene and related molecules do not exhibit discernible

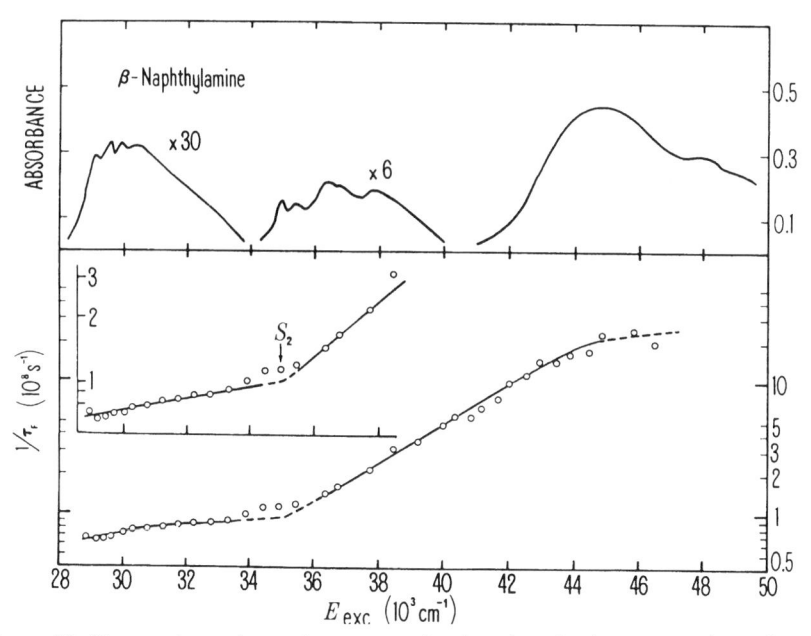

Figure 13. Vapor-phase absorption spectra (top) and excitation energy dependence of fluorescence decay rate (bottom) for β-naphthylamine. (From ref. [4] with permission.)

changes in the excess energy dependence of the S_1 nonradiative decay rate as the threshold for excitation of S_3 is crossed [4, 36]. This observation, illustrated in Figure 13, is consistent with the occurrence of very extensive IVR at high vibrational energies in the S_1 manifold.

From a theoretical point of view, observation of an exponential energy dependence of $S_1 \to S_0$ IC rate, as in Figures 8 and 11, is an indication of energy randomization. Fischer [41] and Lin [42] showed that the statistically averaged nonradiative decay rate resulting from a random distribution of vibrational levels is a simple exponential function of excess vibrational energy. Furthermore, the value of α ($\equiv d \ln k_{nr}/dE_{vib}$) has been predicted to depend on the electronic energy gap between the initial and the final electronic states of radiationless transition in a roughly linear fashion, consistent with the results in Figure 10. Thus, the observation of both the exponential excess energy dependence of the $S_1 \to S_0$ IC rate and the linear dependence of α on the S_1–S_0 electronic energy gap can be taken as strong evidence for vibrational energy randomization occurring in S_1 molecules with very large excess energies.

VI. EXCESS ENERGY DEPENDENCE OF RADIATIONLESS TRANSITIONS IN NITROGEN HETEROCYCLIC COMPOUNDS: EXPERIMENTAL RESULTS

The dramatic vibrational enhancement of $S_1 \to S_0$ IC expected for molecules with nearby $n\pi^*$ and $\pi\pi^*$ singlet states (Section IV) is clearly manifested in the photophysical behavior of the gas-phase nitrogen heterocyclic compounds.

Isoquinoline and quinoline offer good examples of azanaphthalenes that possess closely spaced, lowest energy $n\pi^*$ and $\pi\pi^*$ singlet states. In both of these molecules, the first excited singlet state is an $n\pi^*$ state that lies slightly below the second excited singlet state of $\pi\pi^*$ character. The S_1 $(n\pi^*)$–S_2 $(\pi\pi^*)$ electronic gap is only about 1100 cm^{-1} for isoquinoline and about 1800 cm^{-1} for quinoline in the vapor phase [43]. Because of the strong vibronic coupling between the close-lying S_1 $(n\pi^*)$ and S_2 $(\pi\pi^*)$ states, the first strong absorption system of isoquinoline vapor at about 3100 Å is best described as an electronic transition to a molecular eigenstate (ME), which is a mixture of the S_1 $(n\pi^*)$ and S_2 $(\pi\pi^*)$ states [44], that is, $\Psi = \alpha\psi_{n\pi^*} + \beta\psi_{\pi\pi^*}$. Electronic excitation of this S_2/S_1 ME state in isoquinoline vapor leads to the appearance of structured fluorescence that decays exponentially [45, 46]. The study of this fluorescence, as well as the sensitized phosphorescence of biacetyl, at room temperature has shown that the fluorescence yield and $S_1 \to T$ ISC yield of isoquinoline decrease rather dramatically with increasing energy of excitation above the origin [45, 47]. The results are illustrated in Figure 14, which displays the absorption spectrum and fluorescence excitation spectrum of isoquinoline, as well as the excitation spectrum of biacetyl phosphorescence sensitized by energy transfer from the T_1 isoquinoline [45]. Note that the isoquinoline absorption spectrum is characterized by diffuse spectral features that underlie the sharp, discrete bands. The three prominent sharp features, labeled as 0^+, 0^-, and $0^=$, arise from the near-resonance vibronic coupling of two out-of-plane vibrational levels of the S_1 $(n\pi^*)$ state with the vibrationless level of the S_2 $(\pi\pi^*)$ state [44]. There are two major differences between the absorption spectrum and the excitation spectra of Figure 14 that relate to radiationless transitions in collision-free isoquinoline. First, the diffuse spectral features that are very prominent in the absorption spectrum do not appear in the fluorescence excitation spectrum or in the excitation spectrum of the sensitized phosphorescence. The result indicates the S_2/S_1 vibronic levels prepared by the optical excitation of the diffuse absorption bands do not lead to appreciable fluorescence or $S_1 \to T$ ISC in isoquinoline. Since the

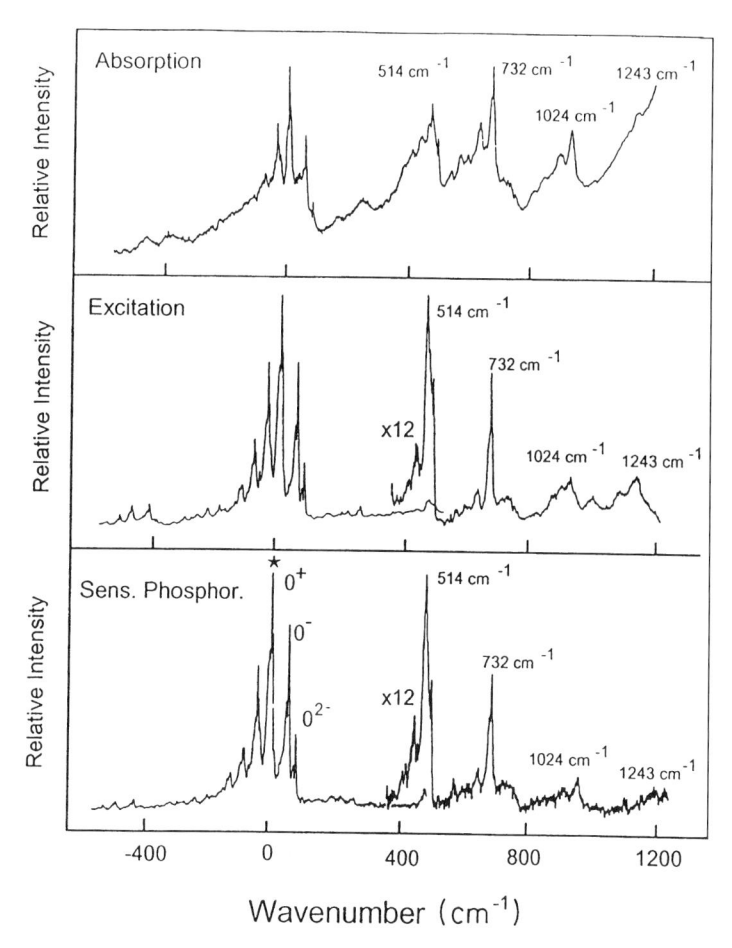

Figure 14. Absorption spectrum (top) and fluorescence excitation spectrum (middle) of isoquinoline vapor at room temperature, and the excitation spectrum of biacetyl phosphorescence (bottom), sensitized by energy transfer from the triplet isoquinoline. The band positions (in cm^{-1}) are relative to the starred 0^+ band at 31,925 cm^{-1}. (From ref. [45] with permission.)

diffuse bands do not appear with jet-cooled isoquinoline [48, 49], and since the only vibrations that can be thermally populated with high efficiency at room temperature are the low-frequency out-of-plane vibrations, it is reasonable to assume that the diffuse bands are composed of the sequence bands involving the out-of-plane vibrations. We may therefore conclude [45] that the $S_2/S_1 \to S_0$ IC yield increases dramatically by the optical excitation of the out-of-plane bending modes in S_1. This conclusion is consistent with the theoretical prediction of Section IV that the vibrations that adiabatically couple the S_1 ($n\pi^*$) and S_2 ($\pi\pi^*$) states act as efficient accepting modes for internal conversion. Second, the intensities of the sharp spectral features are much weaker in the excitation spectra than in the absorption spectrum for excess energies greater than about $500\,\text{cm}^{-1}$. This finding also indicates that the $S_2/S_1 \to S_0$ IC increases dramatically by vibrational excitation, and it becomes the dominant photophysical decay channel at higher excess energies. Within the context of the ME description of the optically prepared excited state, excitation at higher energies is expected to lead to population of the states with greater density of overtones and combinations of the out-of-plane modes that act as good accepting modes for the $S_2/S_1 \to S_0$ IC [45, 46]. Consistent with the large increase in the $S_1 \to S_0$ IC rate with excess vibrational energies, the fluorescence lifetime decreases very rapidly with increasing excitation energy in both static gas [45] and supersonic molecular beam [46].

A quantitative study of the excess energy dependence of the $(S_2/S_1 \to S_0)$ IC/$(S_2/S_1 \to T)$ ISC branching ratio for jet-cooled isoquinoline has been carried out by Cheshnovsky and co-workers [49], using the technique of surface electron ejection by laser excited metastables (SEELEM) [50] to detect the triplet isoquinoline. In this method, the laser excited metastables, that is, the T_1 isoquinoline produced by the $S_2/S_1 \to T$ ISC, travel downstream and strikes a metallic target coated with alkali metal vapor. This causes electron ejection from the low work function surface. The ejected electrons are detected by channel plates, and the signal intensity is recorded as a function of the laser wavelength to generate SEELEM. By using SEELEM, the fluorescence excitation spectrum and the absorption spectrum, together with the known fluorescence yield ($<10^{-2}$) and the radiative decay rate ($\sim 10^7\,\text{s}^{-1}$), Sneh et al. [49] were able to deduce the ISC and IC rates as a function of the excess vibrational energy in the S_2/S_1 ME state. Their results, shown in Fig. 15, clearly demonstrate that over the energy range of 0–800 cm^{-1} the $S_2/S_1 \to S_0$ IC rate increases more than two orders of magnitude, whereas the $S_2/S_1 \to T$ ISC rate remains essentially unchanged. The large discrepancy in the excess energy dependence renders the IC dominant decay channel for excess energies greater than about $500\,\text{cm}^{-1}$. These quantitative experimental data are in full conformity with the conclusions of Sections III and IV that the excess energy dependence of the nonradiative decay rate is small for the small energy-gap transitions (ISC) and large for the large energy-gap radiationless transitions (IC).

It should be noted that the excess energy dependence of the IC rate as measured by $\alpha \equiv d\ln k_{\text{nr}}/dE_{\text{vib}}$ is much larger for isoquinoline than for naphthalene [4, 38], and that the onset of the dominance of the IC channel

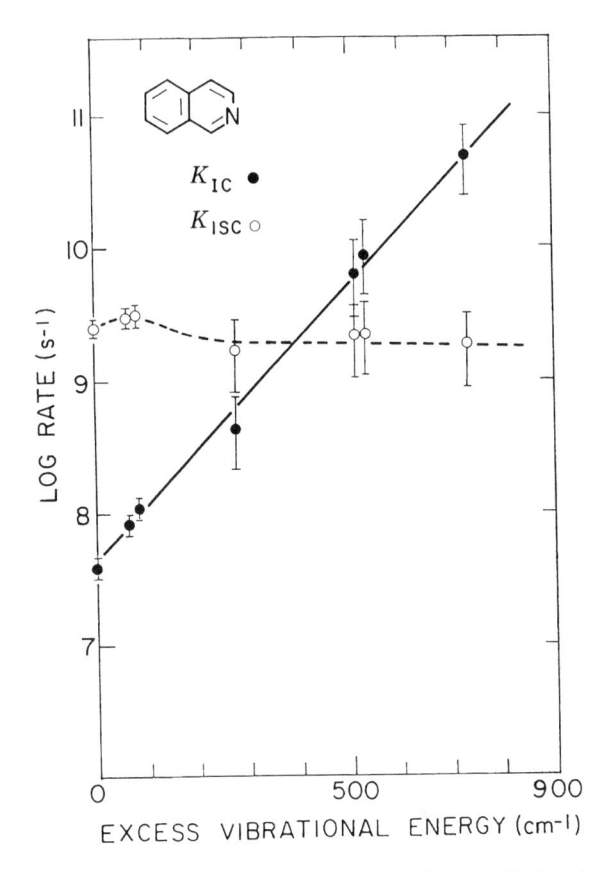

Figure 15. Excess vibrational energy dependence of nonradiative decay rates (k_{IC} and K_{ISC}) in jet-cooled isoquinoline. (From ref. [49] with permission.)

occurs at much smaller excess energy ($\sim 500\,\text{cm}^{-1}$) in isoquinoline than in naphthalene ($\sim 10{,}000\,\text{cm}^{-1}$) [4]. These differences can be traced to the vibronic coupling between the close-lying $n\pi^*$ and $\pi\pi^*$ states of isoquinoline, which makes the out-of-plane bending vibrations exceptionally good accepting modes for the IC process (Section IV). Since the energy accepting ability of the vibronically active out-of-plane modes is related to the frequency shifts of the modes in the excited state relative to the ground state (i.e., $\Delta\omega_j$) one would expect the excess energy dependence of IC rate as well as the onset of the dominance of the IC channel to correlate with the $S_1(n\pi^*)-S_2(\pi\pi^*)$ electronic energy gap for closely related molecules. In conformity with this expectation, quinoline, with a larger $S_1(n\pi^*)-S_2(\pi\pi^*)$ gap, exhibits smaller excess energy dependence of IC relative to isoquinoline

[47]. Moreover, the threshold vibrational energy at which the IC rate surpasses the ISC rate is substantially greater for quinoline than for isoquinoline [47].

The second major class of the nitrogen heterocycle compounds with nearby $n\pi^*$–$\pi\pi^*$ singlet states is the azabenzenes. As with the azanaphthalenes, the S_1 states of pyridine, pyrazine, and pyrimidine are $n\pi^*$ states. The $S_1(n\pi^*)$–$S_2(\pi\pi^*)$ electronic energy gaps are approximately 3600 cm^{-1}, 7000 cm^{-1}, and 9200 cm^{-1} for pyridine, pyrazine, and pyrimidine, respectively [24]. In the vapor phase at low pressures, the form of fluorescence decay in pyrimidine and pyrazine depends strongly on the excess vibrational energies in S_1. Thus, the fluorescence decay following picosecond laser excitation is typically biexponential in molecules with small excess energies [10], whereas it is a simple exponential at high excess energies [11]. The dependence of the form of fluorescence decay on excess energy can be attributed to the relatively small $S_1(n\pi^*)$–$T_1(n\pi^*)$ electronic energy gap in these molecules. Accordingly, the $S_1 \rightarrow T_1$ ISC behaves as an intermediate case or statistical limit (Section III), depending on the magnitude of the vibrational density of the T_1 states that are isoenergetic with the optically excited vibrational level of S_1. The $S_1(n\pi^*)$–$T_1(n\pi^*)$ electronic energy gaps are 2540 cm^{-1} in pyrimidine, 4060 cm^{-1} in pyrazine, and 5130 cm^{-1} in pyridine [24]. Consistent with expectations, the molecule with the largest S_1–T_1 gap (i.e., pyridine) exhibits the exponential fluorescence decay [11], characteristic of the statistical limit of intersystem crossing, even from the vibrationless level of S_1. A systematic study of the excess vibrational energy dependence of radiationless transitions for the gaseous azabenzenes has been made by Yoshihara and co-workers [11], by determining the excitation energy dependence of the quantum yield and lifetime of fluorescence, and of the quantum yield of $S_1 \rightarrow T_1$ ISC measured by the sensitized phosphorescence of biacetyl. From these data, the excess energy dependence of the $S_1 \rightarrow T_1$ ISC rate ($k_{ISC} = \Phi_{ISC}/\tau_F$) and of the rate of the "second nonradiative decay process" ($k_{FQ} = (1 - \Phi_{ISC} - \Phi_F)/\tau_F$) have been deduced, Fig. 16. Note that the excess energy dependence of the two nonradiative decay channels in the azabenzenes closely mimic those of the $S_1 \rightarrow T_1$ ISC and $S_1 \rightarrow S_0$ IC in isoquinoline. Thus, while ISC dominates the second nonradiative decay at low excess energies, the reverse is true at high excess energies. In each of these compounds, the threshold energy at which the rate of the second nonradiative decay channel surpasses the ISC rate occurs well below the electronic origin of the lowest energy $\pi\pi^*$ singlet state. Measurement of the single vibronic level $S_1 \rightarrow T$ ISC yield for jet-cooled pyridine, based on the photoionization of triplet pyridine, shows that the dramatic increase in the second nonradiative decay process occurs in between vibronic levels of energies 1543 and 1636 cm^{-1} [51]. Although Yamazaki et al. [11] attribute

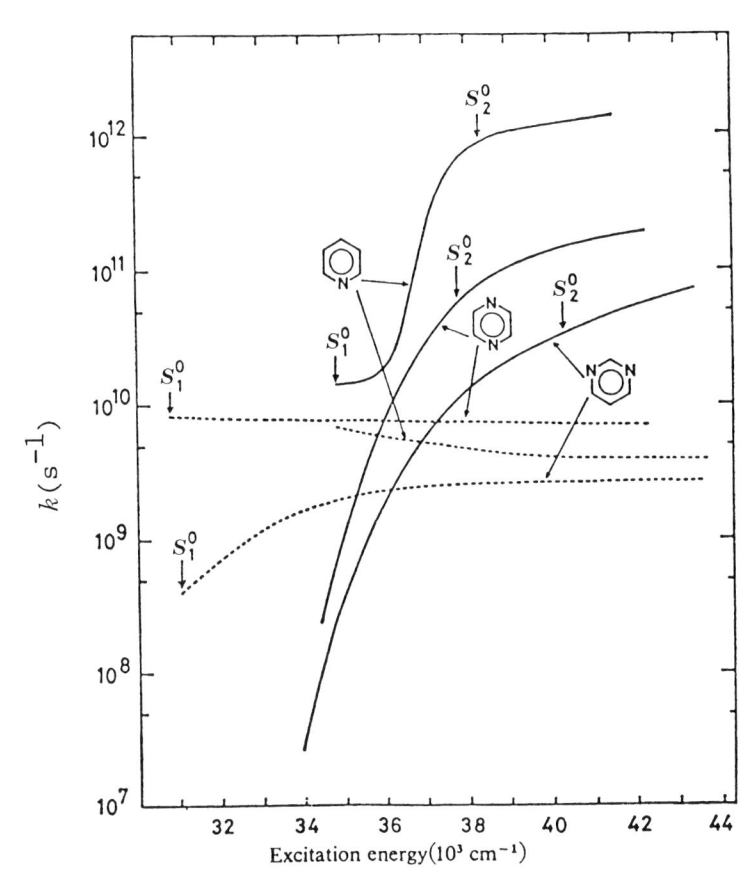

Figure 16. Excitation energy dependence of nonradiative decay rates in vapor-phase pyridine, pyrazine, and pyrimidine. The dashed lines represent $S_1 \to T$ intersystem crossing, whereas the solid lines represent the second nonradiative process attributed to photochemical reaction. (From ref. [11] with permission.)

the second nonradiative process to photoisomerization, we prefer to identify it with the direct $S_1 \to S_0$ IC for several reasons. First, unlike the polycyclic aromatic hydrocarbons [4, 39], the nonradiative decay rates of the azabenzenes exhibit nonexponential dependence on excess vibrational energy, and saturation at relatively low excess energies. These behaviors are consistent with a shallow S_1 potential intersecting a steep S_0 potential along the nuclear coordinates of good accepting modes for internal conversions (Section III and IV). Second, the magnitude of k_{nr} as well as its excess energy dependence vary systematically with the $S_1(n\pi^*)-S_2(\pi\pi^*)$ electronic gap.

Thus, both k_{nr} and its excess energy dependence are the largest for pyridine and the smallest for pyrimidine. Third, the threshold energy at which the second nonradiative decay process surpasses the $S_1 \to T$ ISC follows the sequence pyridine < pyrizine < pyrimidine. Finally, the nonradiative decay rate begins to saturate with excess energy most rapidly for pyridine and least rapidly for pyrimidine. Since the $S_1(n\pi^*)$–$S_2(\pi\pi^*)$ electronic gap follows the sequence pyridine < pyrazine < pyrimidine, these results strongly suggest that the nonradiative decay process that dominates at higher excess energies is $S_1 \to S_0$ IC, made efficient by vibronic interaction between close-lying $n\pi^*$ and $\pi\pi^*$ singlet states, the proximity effect [21, 32]. Consistent with this conclusion, the excess energy dependence of k_{nr} is greater in isoquinoline, which has a smaller $S_1(n\pi^*)$–$S_2(\pi\pi^*)$ electronic gap, than in the azabenzenes.

The identification of the second nonradiative decay process in the azabenzenes with the direct $S_1 \to S_0$ IC is further supported by the excess energy dependence of the rate of T_1 (lowest triplet state) nonradiative decay in jet-cooled pyrazines. Using the technique of time-of-flight SEELEM, Cheshnovsky and co-workers measured T_1 decay rates (reciprocal of T_1 lifetimes) of pyrazine-d_0 [52, 53], pyrazine-d_4 [53], and 2-methylpyrazine [54], as a function of $S_0 \to T_1$ excitation energy. Subtracting the contribution of the T_1 radiative decay rate to the measured rate, they obtained the dependence of the $T_1 \to S_0$ ISC rate on the excess vibrational energy in T_1. The results for pyrazine-d_0 and 2-methylpyrazine are given in Figure 17. It is evident from the data that the T_1 nonradiative decay is larger for 2-methylpyrazine than for pyrazine at lower excess vibrational energies. The $T_1 \to S_0$ ISC rate increases very sharply with increasing T_1 vibrational energy for both compounds, but the initial rate of increase is greater for 2-methylpyrazine than for pyrazine. Comparison of the data with the result in Figure 16 shows that the excess energy dependence of the $T_1 \to S_0$ ISC rate for the jet-cooled pyrazine is considerably greater than that of the $S_1 \to S_0$ IC rate for the pyrazine vapor.

The greater excess energy dependence of the $T_1 \to S_0$ ISC rate relative to the $S_1 \to S_0$ IC rate in pyrazine is consistent with the proximity effect in radiationless transitions [21, 32]. The $T_1(n\pi^*)$ state of B_{3u} symmetry lies just below ($\sim 1500\,\text{cm}^{-1}$) the $T_2(\pi\pi^*)$ triplet state of B_{1u} symmetry in pyrazine [24]. Because of the proximity, these two states are expected to strongly couple via the out-of-plane vibrations of b_{2g} symmetry. There are two vibrations of b_{2g} symmetry in pyrazine: v_5 mode with ground-state frequency of $983\,\text{cm}^{-1}$ and v_4 mode with frequency of $756\,\text{cm}^{-1}$ [24]. The former (v_5) is an out-of-plane CH bend, whereas the latter (v_4) is an out-of-plane ring mode. Calculations indicate that the matrix elements (U)

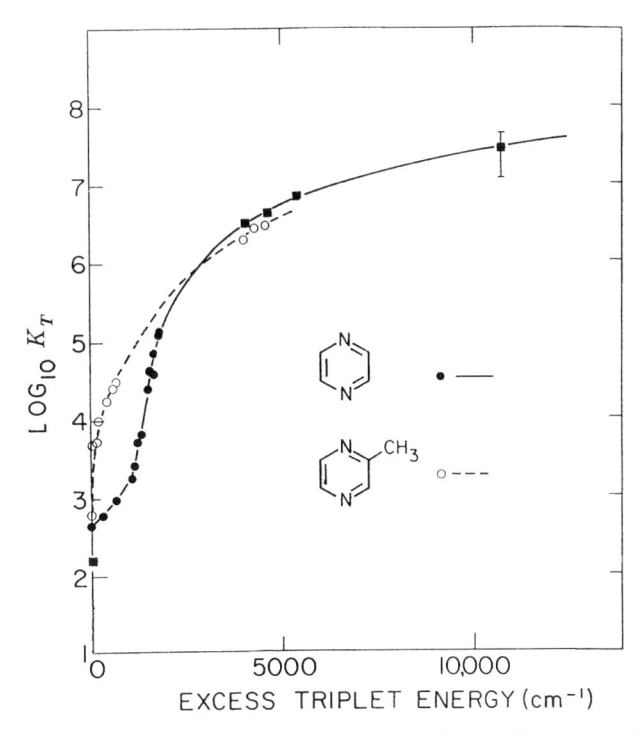

Figure 17. Measured (circles) excess energy dependence of T_1 nonradiative decay rates in jet-cooled pyrazine-d_0 and 2-methylpyrazine. (From ref. [54] with permission.)

of the T_1–T_2 vibronic coupling induced by v_4 and v_5 are comparable in magnitude as those of the $S_1(n\pi^*)$–$S_2(\pi\pi^*)$ vibronic coupling involving v_{10a} (b_{1g}) mode [55]. Since the decrease in the force constant of the vibronically active mode in the lowest excited state is determined by $U^2/\Delta E$, Eq. 24, the smaller electronic energy gap (ΔE) for the T_1–T_2 coupling ($\sim 1500\,\text{cm}^{-1}$) relative to that for the S_1–S_2 coupling ($\sim 7000\,\text{cm}^{-1}$) should lead to a greater frequency reductions of the out-of-plane modes in T_1 relative to v_{10a} in S_1. This in turn will lead to larger excess energy dependence of k_{nr} for the $T_1 \to S_0$ ISC as compared to the $S_1 \to S_0$ IC. The larger T_1 nonradiative decay rate of 2-methylpyrazine relative to pyrazine [54], which confirms the results from the condensed-phase experiments [56], is also consistent with the manifestation of the proximity effect [21, 32]. It is well known that the substitution of a methyl group for a hydrogen atom in nitrogen heterocyclic compounds leads to an increase in the energy of $n\pi^*$ states and a decrease in the energy of $n\pi^*$ states [57]. Thus, the $T_1(n\pi^*) - T_2(\pi\pi^*)$ electronic energy gap is expected to be smaller in 2-methylpyrazine than in pyrazine. The smaller $T_1(n\pi^*) - T_2(\pi\pi^*)$ energy gap in 2-methylpyrazine would lead to a more efficient $T_1 - T_2$ vibronic coupling, and hence to a greater ISC rate, relative to pyrazine [56]. The conclusion that the photophysical behavior of T_1 pyrazine is largely determined by the $n\pi^* - \pi\pi^*$ vibronic interactions is further supported by the simulation of the excess energy dependence of the $T_1 \to S_0$ ISC rate [52] based on the algorithm of Sobolewski [19] (Section III), which essentially reproduces the experimental data.

VII. CHANNEL THREE DECAY IN BENZENE

The excess energy dependence of $S_1 \to S_0$ IC and of $T_1 \to S_0$ ISC in nitrogen heterocyclic compounds, described in Section VI, are qualitatively very similar to that of $S_1 \to S_0$ IC in benzene, and it is appropriate here to consider the possible connection between them.

Much work, both experimental and theoretical, has been carried out on the radiationless transitions of collision-free S_1 benzene. At vibrational energies greater than about 3000 cm^{-1}, both the quantum yield [58] and lifetime [59] of fluorescence and the quantum yield of $S_1 \to T$ ISC [60–62] decrease dramatically relative to those at lower excess energies. Concomitantly, the $^1B_{2u}(S_1) \leftarrow {}^1A_{1g}(S_0)$ absorption system exhibits a sudden onset of line broadening [63]. Because it was thought that conventional nonradiative decay channels such as IC and IVR could not account for such step-wise increase in the nonradiative decay rate, Callomon et al. [63] introduced the term "channel three" to describe the "mysterious" nonradiative process. Although the channel three decay has been shown to lead to the highly excited vibrational manifold of S_0 by Nakashima and Yoshihara [64], and is therefore $S_1 \to S_0$ IC, it is not established whether the passage from S_1 to S_0 is a conventional two-state IC [65] or an IC process mediated by a photochemical intermediate or an intermediate electronic state [63, 66]. Recent experiments of Schlag and co-workers [67, 68], and Johnson and co-workers [69, 70] indicate that the channel three decay in benzene is a direct $S_1 \to S_0$ IC, made efficient by vibrational mixing of the optically excited vibronic states in S_1 with the background S_1 states that couple strongly to the highly excited vibronic states in S_0. These background S_1 vibronic states that have large IC rates have been identified as those involving an out-of-plane v_4 mode by Hornburger and Brand [65]. On the other hand, ab initio studies of the potential energy (PE) surfaces of

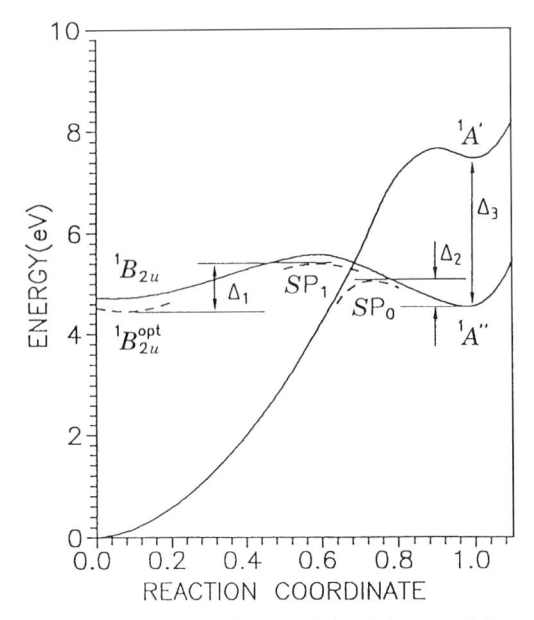

Figure 18. Schematic representation of potential minima, saddle points (SP) and barriers (Δ) on the S_0 and S_1 (B_{2u}) states of benzene along the reaction path to the ground state of prefulvene ($^1A''$). (From ref. [72] with permission.)

S_1 and S_0 benzene conducted by Kato [71], Sobolewski et al. [72], and Palmer et al. [73], strongly suggest that the channel three has its origin in the photochemically mediated IC. Thus, the calculations indicate that the $S_1(B_{2u})$ state crosses the $S_0(A_{1g})$ state along the benzene–prefulvene reaction coordinates. The planar S_1 minimum of benzene is separated by approximately 3000 cm^{-1} from the ground-state prefulvene minimum [72, 73], consistent with the energy threshold for the channel three decay. The proposed IC path, representing the channel three, starts from the $S_1(B_{2u})$ PE minimum with D_{6h} symmetry and proceeds over a transition state to the geometry of prefulvene, whereupon it enters the S_0 surface via S_1–S_0 conical intersection (Fig. 18). The lack of a PE barrier between the prefulvenic and aromatic forms of benzene provides an explanation for the low quantum yield of benzvalene formation through the intermediate prefulvene [71, 72]. In support of this conclusion, the dynamics calculations using semiempirical surface-hopping trajectories demonstrate that most trajectories lead back to S_0 benzene and do not reach the region of the prefulvene intermediate [74]. The calculated PE surfaces for the lowest energy $\pi\pi^*$ and ground states of pyrazine are qualitatively very similar to those for benzene, and the same photochemically mediated IC process involving the $S_2(\pi\pi^*)$ state has been proposed for the "channel-three-like" behavior of S_1 pyrazine [[72].

The proposal that the $S_1 \to S_0$ IC processes in the nitrogen heterocyclic compounds (azabenzenes and azanaphthalenes) are photochemically mediated poses serious problems. As noted in Section VI, the excess energy dependence of $S_1 \to S_0$ IC is much more dramatic in isoquinoline than in the closely related naphthalene. Thus, the onset of dominant IC occurs at much smaller excess energy in isoquinoline ($\sim 500 \mathrm{cm}^{-1}$) relative to naphthalene ($\sim 10,000 \mathrm{cm}^{-1}$), and the excess energy dependence of the IC rate, as measured by $dk_{\mathrm{IC}}/dE_{\mathrm{vib}}$, is almost two orders of magnitude greater for isoquinoline than for naphthalene. Similarly, while the $S_1 \to S_0$ IC in pyridine becomes the dominant decay channel at energies far below ($\sim 2000 \mathrm{cm}^{-1}$) the lowest energy $\pi\pi^*$ singlet state (Section VI), the onset of dominant IC in benzene occurs at about $3000 \mathrm{cm}^{-1}$ above the corresponding $\pi\pi^*$ singlet state. Since there is no compelling reason to expect large disparities in their ability to form prefulvene or prefulvene-like molecules, the dramatically different behavior of the nitrogen heterocyclic compounds relative to their hydrocarbon analogues cannot be easily attributed to the photochemically mediated, indirect IC process. Moreover, as shown in Section VI, both the magnitude of k_{nr} and its excess energy dependence for the azabenzenes correlate inversely with their $S_1(n\pi^*)-S_2(\pi\pi^*)$ electronic energy gap, consistent with the manifestations of the proximity effect [21, 32]. Barring coincidence, these correlations between the k_{nr} and $S_1(n\pi^*)-S_2(\pi\pi^*)$ electronic gap would be very difficult to rationalize in terms of the indirect $S_1 \to S_0$ IC involving a photochemical intermediate or intermediate electronic state. The more dramatic excess energy dependence of the $T_1(n\pi^*) \to S_0$ ISC in pyrazine relative to the corresponding $S_1(n\pi^*) \to S_0$ IC is also inconsistent with the radiationless processes involving photochemical conversion of excited molecules to isomeric species.

If the strong excess energy dependence of the nonradiative decay rates in the nitrogen heterocyclic compounds is due to the direct $S_1 \to S_0$ IC, made efficient by the large S_1-S_0 frequency distortions in vibronically active out-of-plane modes, it is natural to ask whether the same mechanism could also account for the channel three decay in benzene. Comparison of the excess energy dependence of k_{nr} in benzene with that of $S_1 \to S_0$ IC in other aromatic molecules, as well as with the model calculations, strongly suggest that the channel three decay is a conventional $S_1 \to S_0$ IC. Thus, the pump–probe photoionization experiments of Knee et al. [70] indicate that the excess energy dependence of the channel three decay is greater for C_6D_6 than for C_6H_6 under the cold environment of the supersonic expansion, consistent with the isotope effect on the $S_1 \to S_0$ IC of naphthalene [4, 37] and other polycyclic aromatic hydrocarbons [75]. Concomitantly, the onset of the channel three decay is lowered to about $2400\,\text{cm}^{-1}$ for C_6D_6 as compared to about $3300\,\text{cm}^{-1}$ for C_6H_6 [70]. The lower threshold energy for C_6D_6 has been interpreted as indicating that the dramatic increase in the IC rate is a result of an initial vibrational redistribution, which occurs at lower energies for C_6D_6 than for C_6H_6 due to the greater vibrational density of states in the deuterated compound. If confirmed [76], these results provide a strong support to the identification of the channel three as a direct $S_1 \to S_0$ IC, since such isotope effects are not expected for an indirect process involving a photochemical intermediate or an intermediate electronic state. The occurrence of the IVR in the channel three region is clearly evidenced in the rotationally resolved fluorescence and multiphoton ionization experiments of Schlag and co-workers [67, 68]. The results show that at about $3000\,\text{cm}^{-1}$ above the electronic origin of S_1, the optically excited rovibronic states couple with the background S_1 state via anharmonic and Coriolis (rotation–vibration) interactions. This IVR process, which leads to the population of out-of-plane modes that couple strongly to the S_0 vibronic manifold, was proposed [67, 68] to be responsible for the sudden decrease in the fluorescence [58, 59] and $S_1 \to T$ ISC [60–62]. The results [68] also demonstrate that the line widths deduced by Callomon et al. [63], which led to the postulation of the channel three, are more than two orders of magnitude greater than the homogeneous line widths observed in the rotationally resolved experiments. The rate calculations of Hornburger et al. [65, 77, 78], based on the communicating states model of Fischer and Schlag [17], indicate that the v_4 out-of-plane bending mode becomes the dominant accepting mode for the $S_1 \to S_0$ IC at excess vibrational energies greater than about $2500\,\text{cm}^{-1}$. The computed $S_1 \to S_0$ IC rate increases strongly with the increasing excess energy, in qualitative agreement with the vibrational energy dependence of the channel three decay. The excess energy dependence of the $S_1 \to S_0$ IC in pyridine [11, 51], described in Section VI, is remarkably similar to that in benzene [66] and the decay mechanism involving participation of IVR in electronic relaxation has also been proposed for the molecule [51]. The only significant difference between pyridine and benzene is that the threshold energy above which the IC rate surpasses the $S_1 \to T$ ISC rate occurs at much lower energy in pyridine ($\sim 1600\,\text{cm}^{-1}$) than in benzene ($\sim 3000\,\text{cm}^{-1}$). This difference can be attributed to the smaller frequencies of out-of-plane modes in S_1 pyridine [51] relative to S_1 benzene, which lowers the IVR threshold for pyridine, and to a greater zero-energy IC rate of pyridine as compared to benzene.

If there is any anomaly associated with the $S_1 \to S_0$ IC of collision-free benzene, it is only with respect to that of polycyclic aromatic hydrocarbons.

Specifically, the excess energy dependence of its rate is unusually large in that benzene exhibits a significant deviation from the $\alpha (\equiv d \ln k_{nr}/dE_{vib})$ versus $\Delta E(S_1-S_0)$ plot in Figure 10. The "exceptional" excess energy dependence of the IC rate in benzene is likely to be related to the fact that benzene has anomalously large S_1-S_0 frequency distortion (v_4, in particular) [65] and the largest S_1-S_0 electronic energy gap among aromatic hydrocarbons. The out-of-plane $v_4(b_{2g})$ mode, which has proper symmetry to vibronically couple $^1B_{2u}(S_1)$ with a $^1A_{1u}(\pi\sigma^*)$ state, exhibits a large frequency increase in going from $S_1(365\,\text{cm}^{-1})$ to $S_0(707\,\text{cm}^{-1})$ in benzene [79]. The large frequency shift in v_4, together with the large S_1-S_0 electronic energy gap, would lead to a strong excess energy dependence of direct $S_1 \to S_0$ IC in benzene, as the model calculations of Hornburger et al. [65, 77, 78] demonstrate. Although the location of the $\pi\sigma^*$ singlet state(s) is not known, it is reasonable to assume that the $^1\pi\sigma^*-S_1(\pi\pi^*)$ electronic energy gap is the smallest in benzene among aromatic hydrocarbons, owing to the high energy of the $S_1(B_{2u})$ state of benzene. Thus, it is quite likely that the large frequency reduction of v_4 in S_1 relative to S_0 has its origin in the vibronic coupling of the $S_1(\pi\pi^*)$ state with a higher lying $^1A_{1u}(\pi\sigma^*)$ state. The role of v_4 in the $S_1 \to S_0$ IC of benzene may therefore be very similar to that of the vibronically active out-of-plane bending modes in the $S_1 \to S_0$ IC of nitrogen heterocyclic compounds with nearby $n\pi^*$ and $\pi\pi^*$ singlet states. Based on these considerations, and the unmistakable correlation between the excess energy dependence of the IC rate and the $S_1(n\pi^*)-S_2(\pi\pi^*)$ electronic energy gap in aza-aromatic compounds, we would conclude that the channel three decay in benzene is a direct $S_1 \to S_0$ IC, which has been made efficient by vibronic coupling of S_1 with a close-lying excited state (proximity effect [21, 32]). The nonexponential excess energy dependence of the IC rate and the saturation of the rate that occurs at relatively low excess vibrational energies are consistent with a shallow S_1 potential intersecting a steep S_0 potential along a vibronically active out-of-plane coordinate. The intersection of two potential energy surfaces is considered essential for an effective radiationless transition between two widely separated electronic states [21, 34] (Section IV).

VIII. CONCLUDING REMARKS

In this chapter, we have shown that electronic relaxation of vibrationally highly excited aromatic molecules is distinctly different from that of molecules with low levels of vibrational excitation. The most notable difference is in the efficiency of $S_1 \to S_0$ IC relative to $S_1 \to T$ ISC. Thus, while $S_1 \to S_0$

IC is usually of minor or negligible importance in the condensed phase or in the gas phase under high pressures, the IC was shown to be the dominant decay channel for gaseous molecules with large excess energies in S_1. The importance of $S_1 \to S_0$ IC in vibrationally highly excited molecules has been shown to be due to the much greater vibrational enhancement of $S_1 \to S_0$ IC relative to $S_1 \to T$ ISC, as expected on theoretical grounds. The enhancement of the $S_1 \to S_0$ IC rate by vibrational excitation was shown to be especially dramatic for nitrogen heterocyclic compounds, which display out-of-plane vibrational frequencies that are small in S_1 relative to S_0. This observation is in accord with the theoretical conclusion that the frequency distortion, resulting from the vibronic coupling of S_1 with a nearby excited state, leads to a FC factor for $S_1 \to S_0$ IC, which increases dramatically with increasing excitation of the vibronically active out-of-plane modes. The similarity between the photophysical properties of benzene and other aromatic molecules suggests that the so-called channel three decay in benzene has its origin in a conventional $S_1 \to S_0$ IC. Although $S_1 \to S_0$ IC could also occur through a photochemical intermediate, as indicated by high-quality ab initio calculations [71–73], this is not likely to be of importance relative to the direct process. Finally, we suggest that the dramatic increase in the nonradiative decay rate with vibrational excitation and the concomitant saturation of the rate at relatively low excess energies, as observed for S_1 benzene, S_1 pyridine, and T_1 pyrazine, are symptomatic of a shallow excited-state (S_1 or T_1) potential intersecting the steep S_0 potential at energies not far above the electronic origin of the excited state.

The fact that $S_1 \to S_0$ IC becomes the dominant decay channel for vibrationally highly excited S_1 molecules has obvious photochemical implications. The vibrationally hot S_0 molecules produced by the IC process can decay by energy loss through infrared emission and collisional deactivation, or by dissociation and isomerization. Unfortunately, information presently available on the gas-phase photochemistry of aromatic molecules [80] is insufficient to allow any conclusion to be drawn concerning the deendence of photochemistry on photophysics. We can only make some obvious, general remarks concerning the photochemical relevance of the photophysics discussed in this chapter. As the excitation energy increases, the quantum yield of the unimolecular reactions involving vibrationally hot S_0 molecules, such as hydrogen atom elimination in benzene [81], would tend to increase owing to the increasing efficiency of $S_1 \to S_0$ IC. The reactions originating in S_1, for example, the valence isomerization of benzene to benzvalene through the intermediate prefulvene [71–73, 82], would also be affected by the competing $S_1 \to S_0$ IC. Moreover, mode selectivity on photochemistry may occur as the results of the vibronic level dependence of the ($S_1 \to S_0$ IC)/($S_1 \to T$ ISC) branching ratio [51]. Clearly, a systematic study of the gas-phase photochemistry as a function of the excess vibrational energy and the nature of optically prepared vibronic levels in S_1 is very important for gaining a detailed understanding of photochemistry in terms of photophysics. If this chapter has done no more than pointing out the need for such a study, it will have served its purpose.

ACKNOWLEDGMENTS

I am indebted to my collaborators Yohji Achiba, Aviv Amirav, Sighart Fischer, Brad Forch, Atsunari Hiraya, John Hsieh, Cheng-Schen Huang, Katsumi Kimura, Shigeo Okajima, Eliel Villa, and William Wassam, Jr. for their contributions to the work described in this chapter. I also thank Aviv Amirav for providing the unpublished data of Figure 11, Ori Cheshnovsky for providing the original figures of his work, Helmut Hornburger and Andrzej Sobolewski for very helpful correspondence and Helen Richter for going over the draft of this chapter. Work performed in the author's laboratory was supported by the National Science Foundation and the Department of Energy.

REFERENCES

1. See, for reviews: (a) B. R. Henry and W. Siebrand, in *Organic Molecular Photophysics*, J. B. Birks, Ed., Wiley-Interscience, London, 1973, Vol. 1, p. 153; (b) K. F. Freed, in *Topics in Applied Physics*, F. K. Fong, Ed., Springer-Verlag, Berlin Heidelberg, 1976, Vol. 15, p. 23; (c) P. Avouris, W. M. Gelbart, and M. A. El-Sayed, *Chem. Rev.* **77**, 793 (1977).
2. M. Kasha, *Faraday Discuss. Chem. Soc.*, **9**, 14 (1950).
3. See, for example, J. B. Birks, *Photophysics of Aromatic Molecules*, Wiley, London, New York, 1970.
4. J. C. Hsieh, C.-S. Huang, and E. C. Lim, *J. Chem. Phys.*, **60**, 4345 (1974).
5. P. Pringsheim, *Ann. Acad. Sci. Tech. Varsovie*, **5**, 29 (1938).
6. E. C. Lim, in *Photophysics and Photochemistry in the Vacuum Ultraviolet*, S. P. McGlynn, G. L. Findley, and R. H. Huebner, Eds., Reidel, Dordrecht, 1985, p. 855.
7. See, for a review, A. Tramer and R. Voltz, in *Excited States*, E. C. Lim, Ed., Academic, New York, 1979, Vol. 4, p. 281.
8. A. Nitzan, J. Jortner, and P. M. Rentzepis, *Proc. Roy. Soc.* (*London*), **A327**, 367 (1972).
9. F. Lahmani, A. Tramer, and C. Tric, *J. Chem. Phys.*, **60**, 4431 (1974).

REFERENCES

10. See, for a review, J. Kommandeur, W. A. Majewski, W. L. Meerts, and D. W. Pratt, *Annu. Rev. Phys. Chem.*, **38**, 433 (1987).
11. I. Yamazaki, T. Murao, T. Yamanaka, and K. Yoshihara, *Faraday Discuss. Chem. Soc.*, **75**, 395 (1983).
12. S. H. Lin and R. Bersohn, *J. Chem. Phys.*, **48**, 2732 (1968).
13. S. H. Lin, *J. Chem. Phys.*, **44**, 3759 (1966).
14. S. F. Fischer, A. L. Stanford, and E. C. Lim, *J. Chem. Phys.*, **61**, 582 (1974); S. F. Fischer and E. C. Lim, *Chem. Phys. Lett.*, **26**, 312 (1974).
15. G. S. Beddard, G. R. Fleming, O. L. J. Gijzeman, and G. Porter, *Chem. Phys. Lett.*, **18**, 481 (1973).
16. T. G. Dietz, M. A. Duncan, A. C. Puiu, and R. E. Smalley, *J. Phys. Chem.*, **86**, 4026 (1982); T. G. Dietz, M. A. Duncan, and R. E. Smalley, *J. Chem. Phys.*, **76**, 1227 (1982).
17. S. F. Fischer and E. W. Schlag, *Chem. Phys. Lett.*, **4**, 393 (1969).
18. E. J. Heller and R. C. Brown, *J. Chem. Phys.*, **79**, 3336 (1983).
19. A. L. Sobolewski, *Chem. Phys.*, **115**, 469 (1987).
20. M. D. Morse, A. C. Puiu, and R. E. Smalley, *J. Chem. Phys.*, **78**, 3435 (1983).
21. E. C. Lim, *J. Phys. Chem.*, **90**, 6770 (1986).
22. R. M. Hochstrasser and C. A. Marzzacco, in *Molecular Luminescence*, E. C. Lim, Ed., Benjamin, New York, 1969, p. 631.
23. A. J. Duben, L. Goodman, and M. Koyanagi, in *Excited States*, E. C. Lim, Ed., Academic, New York, 1974, Vol. 1, p. 295.
24. See, for a review, K. K. Innes, I. G. Ross, and W. R. Moomaw, *J. Mol. Spectrosc.*, **132**, 492 (1988).
25. I. Suzuka, N. Mikami, and M. Ito, *J. Mol. Spectrosc.*, **52**, 21 (1974); K. Kiamogawa and M. Ito, *J. Mol. Spectrosc.*, **60**, 277 (1976).
26. N. Kanamaru and E. C. Lim, *Chem. Phys. Lett.*, **35**, 303 (1975).
27. W. H. Hennecker, A. P. Penner, W. Siebrand, and M. Z. Zgierski, *J. Chem. Phys.*, **69**, 1884 (1978); Y. Udagawa, M. Ito, I. Suzuka, W. Siebrand, and M. Z. Zgierski, *Chem. Phys. Lett.*, **68**, 258 (1979).
28. W. A. Wassam, Jr., and E. C. Lim, *J. Chem. Phys.*, **68**, 433 (1978); *J. Mol. Struct.*, **47**, 1129 (1978).
29. W. A. Wassam, Jr., and E. C. Lim, *J. Chem. Phys.*, **69**, 2175 (1978).
30. W. Siebrand and M. Z. Zgierski, *J. Chem. Phys.*, **75**, 1230 (1981).
31. W. A. Wassam, Jr., and E. C. Lim, *Chem. Phys.*, **38**, 217 (1979).
32. E. C. Lim, in *Excited States*, E. C. Lim, Ed., Academic, New York, 1977, Vol. 3, p. 305.
33. J. P. Byrne, E. F. McCoy, and I. G. Ross, *Aust. J. Chem.*, **19**, 1589 (1965).
34. E. S. Medvedev, *Khim. Fiz.*, 1 (1982).
35. H. Hornburger, personal communication.
36. C.-S. Huang, J. C. Hsieh, and E. C. Lim, *Chem. Phys. Lett.*, **28**, 130 (1974).

37. C.-S. Huang, J. C. Hsieh, and E. C. Lim, *Chem. Phys. Lett.*, **37**, 349 (1976).
38. C.-S. Huang and E. C. Lim, *J. Chem. Phys.*, **62**, 3826 (1975).
39. A. Amirav, unpublished results.
40. J. C. Hsieh and E. C. Lim, *J. Chem. Phys.*, **61**, 736 (1974).
41. S. Fischer, *Chem. Phys. Lett.*, **4**, 33 (1969).
42. S. H. Lin, *J. Chem. Phys.*, **58**, 5760 (1973).
43. A. Hiraya, Y. Achiba, K. Kimura, and E. C. Lim, *J. Chem. Phys.*, **81**, 3345 (1984).
44. G. Fischer and R. Naaman, *Chem. Phys.*, **12**, 267 (1967).
45. B. E. Forch, S. Okajima, and E. C. Lim, *Chem. Phys. Lett.*, **108**, 311 (1984).
46. J. L. Knee, L. R. Khundkar, and A. H. Zewail, *J. Phys. Chem.*, **89**, 3201 (1985).
47. S. Okajima and E. C. Lim, *J. Chem. Phys.*, **69**, 1929 (1978).
48. A. Amirav and E. C. Lim, unpublished results.
49. O. Sneh, A. Amirav, and O. Cheshnovsky, *J. Chem. Phys.*, **91**, 3532 (1989).
50. O. Sneh and O. Cheshnovsky, *Chem. Phys. Lett.*, **130**, 53 (1986).
51. E. Villa, A. Amirav, and E. C. Lim, *J. Phys. Chem.*, **92**, 5393 (1988).
52. O. Sheh, D. Dünn-Kittenplon, and O. Cheshnovsky, *J. Chem. Phys.*, **91**, 7331 (1989).
53. I. Becker and O. Cheshnovsky, *J. Chem. Phys.*, **101**, 3649 (1994).
54. O. Sneh and O. Cheshnovsky, *J. Chem. Phys.*, **96**, 8095 (1992).
55. W. A. Wassam, Jr., and E. C. Lim, unpublished results.
56. S. L. Madej, G. D. Gillispie, and E. C. Lim, *Chem. Phys.*, **32**, 1 (1978).
57. L. Goodman and H. Shull, *J. Chem. Phys.*, **22**, 1138 (1954); *J. Chem. Phys.*, **27**, 1388 (1957).
58. C. S. Parmenter and M. W. Schuyler, *Chem. Phys. Lett.*, **6**, 339 (1970); C. S. Parmenter, *Adv. Chem. Phys.*, **22**, 365 (1972).
59. L. Wunsch, H. J. Neusser, and E. W. Schlag, *Z. Naturforsch.*, **36**, 1340 (1981).
60. R. B. Cundall and A. S. Davies, *Trans. Faraday Soc.*, **62**, 1151 (1966).
61. M. A. Duncan, T. G. Dietz, M. G. Liverman, and R. E. Smalley, *J. Phys. Chem.*, **85**, 7 (1981).
62. C. E. Otis, J. L. Knee, and P. M. Johnson, *J. Chem. Phys.*, **78**, 2091 (1983).
63. J. H. Callomon, J. E. Perkins, and R. Lopez-Delgado, *Chem. Phys. Lett.*, **13**, 125 (1972).
64. N. Nakashima and K. Yoshihara, *J. Chem. Phys.*, **77**, 6040 (1982).
65. H. Hornburger and J. Brand, *Chem. Phys. Lett.*, **88**, 153 (1982).
66. M. Sumitani, D. V. O'Conner, Y. Takagi, N. Nakashima, K. Kamogawa, Y. Udagawa, and K. Yoshihara, *Chem. Phys.*, **93**, 359 (1985).
67. V. Schubert, E. Riedle, and H. J. Neusser, *J. Chem. Phys.*, **84**, 6182 (1986).
68. E. Riedle, Th. Weber, U. Schuett, H. J. Neusser, and E. W. Schlag, *J. Chem. Phys.*, **93**, 967 (1990).
69. C. E. Otis, J. L. Knee, and P. M. Johnson, *J. Chem. Phys.*, **78**, 2091 (1983).

REFERENCES

70. J. L. Knee, C. E. Otis, and P. M. Johnson, *J. Chem. Phys.*, **81**, 4455 (1984).
71. S. Kato, *J. Chem. Phys.*, **88**, 3045 (1988).
72. A. L. Sobolewski, C. Woywod, and W. Domcke, *J. Chem. Phys.*, **98**, 5627 (1993).
73. I. J. Palmer, I. N. Ragazos, F. Bernadi, M. Olivucci, and M. A. Robb, *J. Am. Chem. Soc.*, **115**, 673 (1993).
74. B. R. Smith, M. J. Bearpark, M. A. Robb, F. Bernadi, and M. Olivucci, *Chem. Phys. Lett.*, **242**, 27 (1995).
75. C.-S. Huang and E. C. Lim, unpublished results.
76. The picosecond bulb experiments of Yoshihara and co-workers do not reveal a significant isotope effect on the threshold energy or the excess energy dependence of the channel three decay. See D. V. O'Conner, M. Sumitani, Y. Takagi, N. Nakashima, K. Kamogawa, Y. Udagawa, and K. Yoshihara, *Chem. Phys.*, **93**, 373 (1985).
77. H. Hornburger, H. Schröder, and J. Brand, *J. Chem. Phys.*, **80**, 3197 (1984).
78. H. Hornburger, C. M. Sharp, and S. Leach, *Chem. Phys.*, **101**, 67 (1986).
79. See, for example, L. D. Ziegler and B. S. Hudson, in *Excited States*, E. C. Lim, Ed., Academic, New York, 1982, Vol. 5, p. 42.
80. See ref. [73] for a summary of the photochemistry of benzene in the gas phase.
81. A. Yokoyama, X. Zhao, E. J. Hintsa, R. E. Continetti, and Y. T. Lee, *J. Chem. Phys.*, **92**, 4222 (1990).
82. D. Bryce-Smith and A. Gilbert, in *Rearrangements in Ground and Excited States*, P. de Mayo, Ed., Academic, New York, 1980, Vol. 3, p. 349.

LANTHANIDE COMPLEXES OF ENCAPSULATING LIGANDS AS LUMINESCENT DEVICES

Nanda Sabbatini, Massimo Guardigli, and Ilse Manet

Dipartimento di Chimica "G. Ciamician" dell'Università, 40126 Bologna, Italy

CONTENTS

I. Introduction, 214
II. Light conversion in Eu^{3+} and Tb^{3+} complexes, 215
III. Applications of luminescent Eu^{3+} and Tb^{3+} complexes, 219
IV. Overview on Eu^{3+} and Tb^{3+} complexes with encapsulating ligands, 225
    A. Cryptates, 227
    B. Complexes of macrocycles, 238
    C. Podates, 249
    D. Complexes of calixarenes, 252
V. Discussion of some photophysical properties of the complexes overviewed, 259
    A. Ligand absorption, 260
    B. Ligand–metal energy-transfer efficiency, 262

Advances in Photochemistry, Volume 23, Edited by Douglas C. Neckers, David H. Volman, and Günther von Bünau
ISBN 0-471-19289-9 © 1997 by John Wiley & Sons, Inc.

C. Metal luminescence efficiency, 263
D. Metal luminescence intensity, 267
VI. State of the art, 270
Acknowledgments, 273
References, 273

I. INTRODUCTION

Recent developments in the field of supramolecular chemistry consist of the design of new molecules characterized by a cavity capable of including a lanthanide ion if the cavity and the ion are complementary in size, shape, and binding sites [1, 2]. Encapsulation of these ions by these types of ligands represented an innovation in the coordination chemistry of lanthanide ions. In fact, encapsulating ligands have proven to be suitable for the formation of stable complexes with lanthanide ions. This finding is most interesting since these ions exhibit scarce complexation ability toward conventional ligands. If the encapsulating ligand incorporates chromophores, its complexes of luminescent lanthanide ions can act as luminescent molecular devices [3–5]. The light-induced function they perform consists of light emission by the metal ion. This function is realized upon conversion of light absorbed by the ligand moiety into light emitted by the metal ion via an intramolecular energy transfer. In these types of luminescent devices the ligand acts as an *antenna* by absorbing light and transferring the excitation energy to the metal ion [4, 5]. The design of these types of lanthanide complexes as luminescent devices allows us to overcome the drawback of the very low absorption coefficients of the lanthanide ions and to obtain intense metal luminescence.

Among the possible exploitations of lanthanide complexes with encapsulating ligands as luminescent devices, we focus on the use of Eu^{3+} and Tb^{3+} complexes as luminescent labels in bioaffinity assays. Bioaffinity assays utilizing luminescence signals for detection of the analyte have become very important in clinical diagnostics. In Section II, we report the processes involving electronically excited states, which occur in some Eu^{3+} and Tb^{3+} complexes after light absorption. Section III deals with the application of luminescent Eu^{3+} and Tb^{3+} complexes as labels in some bioaffinity assays. Section IV contains an overview of the Eu^{3+} and Tb^{3+} complexes with encapsulating ligands presented in different classes defined by the nature of the ligand. Furthermore, in Section V, we present a discussion of the complexes overviewed by analyzing the light absorption by the ligand in the complex, the efficiencies of the ligand–metal energy transfer and of the metal luminescence, and the intensity of the metal luminescence. In Section VI, we report the state of the art of Eu^{3+} and Tb^{3+} complexes with encapsulating ligands as luminescent devices.

II. LIGHT CONVERSION IN Eu^{3+} AND Tb^{3+} COMPLEXES

In Eu^{3+} and Tb^{3+} complexes of encapsulating ligands incorporating chromophores, light conversion occurs after a sequence of light absorption by the ligand, ligand–metal energy transfer, and metal luminescence (Fig. 1). The metal emissions we study are those originating from the Eu^{3+} $^5D_0$ and Tb^{3+} $^5D_4$ excited states because these emissions are the most intense and these states possess the longest lifetimes. The quantities that determine the intensity of the metal luminescence are (1) the efficiency of the light absorption by the ligand, (2) the efficiency of the ligand–metal energy transfer, and (3) the efficiency of the metal luminescence. The absorption efficiency is derived from the molar absorption coefficients of the ligand in the complex. The efficiency of the ligand–metal energy transfer, $\eta_{en.tr.}$, is obtained from the ratio between the metal luminescence quantum yield upon ligand excitation (hereafter indicated as quantum yield) Φ, and the

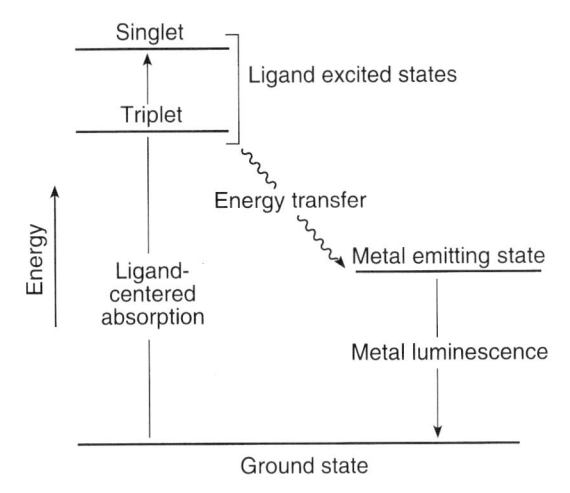

Figure 1. Schematic representation of the conversion of absorbed light into emitted light in Eu^{3+} and Tb^{3+} complexes.

metal luminescence quantum yield upon metal excitation, Φ_M (Eq. 1).

$$\eta_{en.tr.} = \Phi/\Phi_M \tag{1}$$

If Φ_M cannot be measured, it may be substituted, as in Eq. 2, by the efficiency of the metal luminescence (η_M), which is derivable from Eq. 3. Since in this study the radiative (k_r), and nonradiative (k_{nr}) rate constants are calculated from the experimentally obtained lifetimes, Eq. 3 may be replaced by Eq. 4 on the assumption that the decay of the metal emitting state in D_2O at 77 K is purely radiative.

$$\eta_{en.tr.} = \Phi/\eta_M \tag{2}$$

$$\eta_M = k_r/(k_r + k_{nr}) \tag{3}$$

$$\eta_M = \tau_{exp}/\tau_D^{(77K)} \tag{4}$$

In the presence of equilibria between the metal emitting state and other excited states deactivating to the ground state, Eq. 4 is not valid and substitution of $\tau_{exp}/\tau_D^{(77K)}$ for η_M in Eq. 2 gives only a lower limiting value of $\eta_{en.tr.}$.

Now, we will discuss the type of ligand excited state possibly involved in the ligand–metal energy transfer. Ligand absorption causes population of the ligand singlet excited states that undergo intersystem crossing to the ligand triplet excited states. In principle, both singlet and triplet ligand excited states may transfer the excitation energy to the metal ion. However, the lifetimes of the singlet excited states of aromatic compounds such as the chromophores of the ligands we reported in this study are of the order of nanoseconds. Thus, energy transfer from these states should be characterized by rate constants greater than 10^9 reciprocal seconds in order to be effective. The triplet excited states of aromatic compounds are generally much longer lived and, therefore, are expected, more than the singlet excited states, to transfer the energy to the metal. Only for the Tb^{3+} complex of ligand **2** has it been proved experimentally that a ligand triplet excited state is involved in the ligand–metal energy transfer and the efficiency of the triplet → metal energy transfer ($\eta_{triplet \to metal}$) was measured (Section V.B). The relationship between $\eta_{en.tr.}$ and $\eta_{triplet \to metal}$ is expressed by Eq. 5, where $\eta_{singlet \to triplet}$ is the efficiency of singlet → triplet intersystem crossing in the complex.

$$\eta_{en.tr.} = \eta_{singlet \to triplet} \eta_{triplet \to metal} \tag{5}$$

If the ligand singlet excited states do not decay to excited states different from the ligand triplet excited state, $\eta_{\text{singlet}\rightarrow\text{triplet}}$ is at least equal to the intersystem crossing efficiency of the free ligand (η_{isc}), or even higher, considering the heavy atom effect. Conversely, when the singlet excited states also undergo deactivation processes related to the presence of the metal, $\eta_{\text{singlet}\rightarrow\text{triplet}}$ may be lower than η_{isc} of the ligand, thus giving a lower limiting value of $\eta_{\text{en.tr.}}$ if $\eta_{\text{singlet}\rightarrow\text{triplet}}$ is substituted by η_{isc} in Eq. 5.

Finally, we deal with the metal luminescence efficiency in the complex via the discussion of the radiative and nonradiative decay processes of the $Eu^{3+}\ {}^5D_0$ and $Tb^{3+}\ {}^5D_4$ emitting states lying at 17,260 and 20,400 cm^{-1}, respectively. Nonradiative decay of excited states of lanthanide ions may occur via coupling of these states with suitable vibrational modes of the coordination environment [6-8]. The efficiency of the vibronic coupling process depends on the vibrational energy of the oscillators and on the energy gap between the electronic states involved [6, 7, 9, 10]. Nonradiative deactivation via vibronic coupling is very common for lanthanide ions coordinated by solvent molecules containing high-energy OH oscillators [11, 12]. Replacement of the OH oscillators by the low-energy OD ones renders this deactivation much less efficient. Through experiments in H$_2$O and D$_2$O solutions, Horrocks and Sudnick [13, 14] have shown that for the Eu^{3+} and Tb^{3+} complexes the number of water molecules coordinated to the metal ion n is given, with an estimated uncertainty of 0.5, by Eq. 6, where τ_H and τ_D are the experimental lifetimes (in milliseconds) of the $Eu^{3+}\ {}^5D_0$ and $Tb^{3+}\ {}^5D_4$ emitting states (hereafter called lifetimes) in H$_2$O and D$_2$O and q is 1.05 and 4.2 for the Eu^{3+} and Tb^{3+} compounds, respectively. The q values show that vibronic coupling is more efficient for Eu^{3+} than Tb^{3+}, as expected on the basis of the energy gap between the emitting state and the highest level of the ground-state multiplet (12,260 and 14,400 cm^{-1} for Eu^{3+} and Tb^{3+}, respectively).

$$n = q(1/\tau_H - 1/\tau_D) \tag{6}$$

On the assumption that methanol behaves like one-half of a water molecule, Holz et al. [15] have proposed the use of Eq. 7 to determine the number of coordinated methanol molecules, where τ_{CH_3OH} and τ_{CH_3OD} are the experimental lifetimes (in milliseconds) of the metal emitting states in CH$_3$OH and CH$_3$OD and r is 2.1 and 8.4 for the Eu^{3+} and Tb^{3+} compounds, respectively.

$$n = r(1/\tau_{CH_3OH} - 1/\tau_{CH_3OD}) \tag{7}$$

Nonradiative decay of the metal emitting state may also occur via thermally activated crossing to short-lived, upper lying excited states of other configurations, which deactivate nonradiatively to the ground state. For example, this type of deactivation pathway is very common for Eu^{3+} complexes in which thermally activated crossing from the emitting state leads to the population of ligand–metal charge-transfer (LMCT) excited states [16].

By taking into account the radiative and nonradiative decay processes described above, Eq. 8 gives the overall decay rate constant of the emitting state of the metal. In this equation, k_r is the radiative rate constant and k_{nr} and $k_{nr}(T)$ are the nonradiative temperature independent and temperature dependent decay rate constants, respectively.

$$k = 1/\tau = k_r + k_{nr} + k_{nr}(T) \qquad (8)$$

In OH-containing solvents, the most important contribution to k_{nr} comes from the vibronic coupling with the high-energy OH oscillators, $k_{nr}(OH)$. The overall decay rate constants can thus be rewritten as in Eq. 9.

$$k = 1/\tau = k_r + k_{nr}(T) + k_{nr}(OH) + k_{nr}(\text{other vibr}) \qquad (9)$$

The term $k_{nr}(\text{other vibr})$ accounts for the nonradiative decay via vibronic coupling with vibrational modes different from those of the OH oscillators. The decay rate constants can be obtained from the lifetimes in hydrogenated (τ_H) and deuterated (τ_D) solvents at different temperatures. Assuming that the term k_{nr} (other vibr) is negligible and the coupling with the OD oscillators is completely inefficient, and recalling the previous assumption on $\tau_D^{(77K)}$, the radiative rate constant k_r is given by Eq. 10 and the nonradiative rate constants $k_{nr}(T)$ and $k_{nr}(OH)$ at room temperature (hereafter indicated as 300 K) are given by Eqs. 11 and 12.

$$k_r = 1/\tau_D^{(77K)} \qquad (10)$$

$$k_{nr}(T) = 1/\tau_D^{(300K)} - 1/\tau_D^{(77K)} \qquad (11)$$

$$k_{nr}(OH) = 1/\tau_H^{(300K)} - 1/\tau_D^{(300K)} \qquad (12)$$

If there is an equilibrium between the metal emitting state and other excited states, a nonexponential decay of the metal emitting state is expected. Anyway, an exponential decay can be used in a first approximation and in this case Eq. 11 does not give $k_{nr}(T)$, but a number that reflects the importance of the equilibrium on the decay of the metal emitting state (Section V.C).

III. APPLICATIONS OF LUMINESCENT Eu^{3+} AND Tb^{3+} COMPLEXES

Among the lanthanide ions, Eu^{3+} and Tb^{3+} are the most interesting because they show intense emission in the visible and possess long-lived emitting states. These characteristics render Eu^{3+} and Tb^{3+} complexes important for bioaffinity assays since they allow us to perform time-resolved luminescence measurements that enhance the sensitivity of the assay via minimizing interference of the short-lived, background luminescence in the ultraviolet of the biological species and of scattered excitation light. Among the bioaffinity assays, immunoassays and DNA hybridization assays are the most common applications of luminescent Eu^{3+} and Tb^{3+} complexes [17].

Immunoassays rely on the immunoreaction occurring between an antibody used as a detection reagent and the antigen to be analyzed [18]. Immunoassays are most interesting as analytical tools thanks to the high specificity with which the antibody recognizes a certain antigen among several other biomolecules, even if these differ very little from the antigen at the molecular level. In order to detect the immunocomplex, a label capable of generating a signal is attached to the antibody. Adequate choice of the label renders immunoassays very sensitive. Nowadays, analytes present in concentrations up to femtomolar are quantified. Thanks to their high sensitivity and specificity, immunoassays are far superior to almost all other methods for in vitro determination of biomolecules. Immunoassays are classified according to several criteria. One criterion is based on either the presence or absence of a separation step, thus defining heterogeneous and homogeneous immunoassays, respectively [19]. Up to now, the heterogeneous immunoassays have been used more because of their high sensitivity, achieved via the physical separation of the fraction carrying the labeled immunocomplex from other interfering biological species in the sample. The heterogeneous immunoassays schematized in Figure 2, also called sandwich assays, make use of a liquid phase containing the antigen to be analyzed and a solid phase. A specific antibody immobilized on the solid phase binds the antigen. Washing is performed in order to eliminate other biological species present in the sample, which can interfere during the detection step. A solution of a labeled antibody is added in order to mark the antigen. In this way, the antigen is captured between the antibody immobilized on the solid phase and the labeled antibody. Then the signal of the label is detected. The field of immunoassays is huge, both in respect to variations in compounds to be analyzed and to concentration ranges, which may vary from millimolar to subpicomolar. The very optimal label should fulfill a number of requirements among which the high sensitivity is the most important.

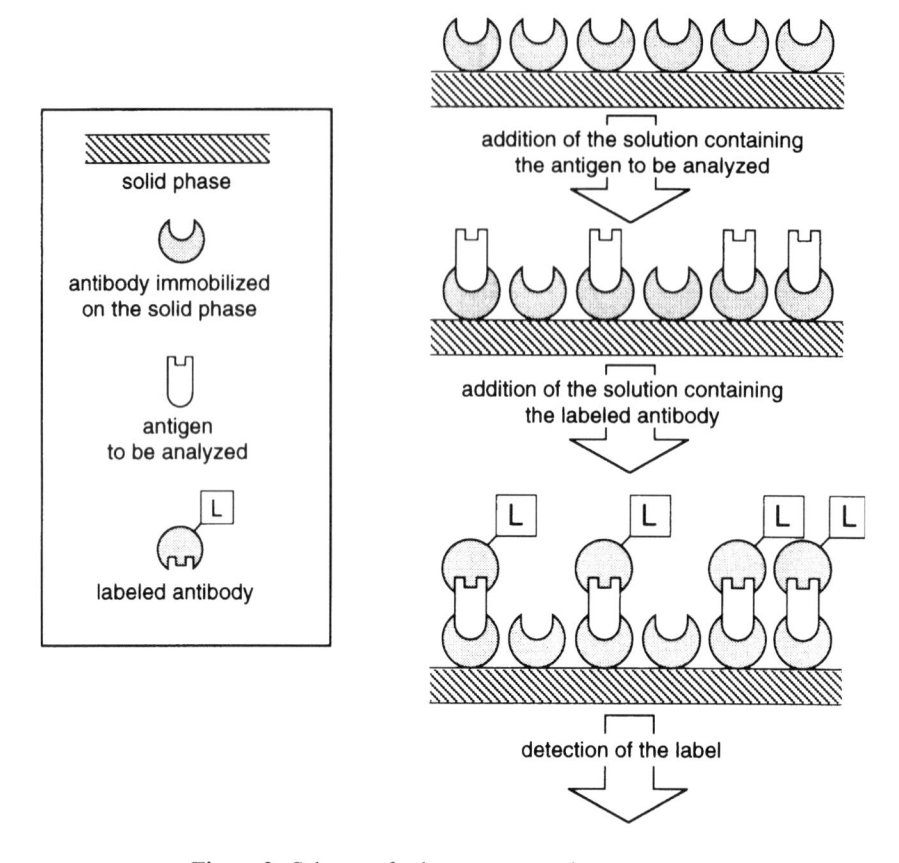

Figure 2. Scheme of a heterogeneous immunoassay.

Initially, immunoassays relying on labeling with radioisotopes gained wide acceptance as clinical analyses [20, 21]. The unquestionable disadvantages of the use of radioisotopes created a strong demand to obtain non-radioisotopic labels. In order to be competitive with radioisotopes, the alternative tracers should enable the detection of the labeled immunoreagent down to 10^{-15}–10^{-18} mol. Among the alternative tracers, enzymatic and luminogenic labels have been studied in particular.

In enzymatic immunoassays [22, 23], the enzyme used as the label does not generate any detection signal, but produces the signal-generating species upon the enzyme-catalyzed reaction. High turnover of the enzyme-catalyzed reaction gives rise to efficient amplification of the detection signal so that high sensitivities are achieved for this type of assay. However, these assays present some drawbacks. The detection signal depends very much on incubation time, temperature, and other physical and chemical conditions during substrate incubation. Enzymes can also be sensitive to interfering substances in the sample, such as endogenic enzymes or inhibitors. Finally, enzymes are molecules with a relatively large molecular weight and can cause steric hindrance during the coupling of the antibody to the antigen.

Immunological analyses based on the use of luminogenic labels are called fluoroimmunoassays (FIAs) [18, 24]. This kind of labeling is relatively inexpensive, not dangerous, and luminescence can be measured quickly and simply. Luminogenic labels are usually divided into three types: (1) photoluminescent species, which emit upon photonic excitation; (2) bioluminescent species, which emit upon an enzymatic reaction; and (3) chemiluminescent species, which emit upon a chemical reaction. We will treat photoluminescent labels in more detail because they are of interest for our research. The first photoluminescent labels used instead of radioisotopes were organic fluorophores such as rhodamine and fluorescein. The use of this type of label involved the disadvantage of decreased sensitivity due to light scattering and short-lived background fluorescence (lifetimes on the order of nanoseconds) of most biological species. A significant improvement has been obtained via FIAs based on time-resolved luminescence measurements, TR-FIAs, which make use of labels characterized by long-lived luminescent states. The compounds of Eu^{3+} and Tb^{3+} are most interesting for TR-FIAs because the Eu^{3+} $^5D_0$ and Tb^{3+} $^5D_4$ luminescent states are particularly long lived (lifetimes on the order of milliseconds). Moreover, their line-like emission bands lie in the visible, so that these bands can be distinguished from the background fluorescence usually characterized by broad bands in the ultraviolet. In order to minimize absorption of interfering biological species and to avoid the use in the instrumentation of deuterium lamps and expensive optic parts in fused silica, excitation of the Eu^{3+} and Tb^{3+} complexes used as labels should occur at wavelengths longer than 300 nm.

The DNA hybridization assays aim at the detection of a DNA sequence and generally rely on the hybridization of a target DNA single strand with a labeled probe DNA single strand. Rapid and precise registration of DNA sequences is of importance to characterize genetic material and, in particular, to recognize the presence of mutations [25, 26]. Up to now heterogeneous assays were most frequently used (Fig. 3). The target DNA single strand is immobilized on a solid matrix and then incubated with the labeled probe DNA single strand in order to obtain the hybridized duplex. Afterward washing is performed to eliminate the excess probe DNA and the

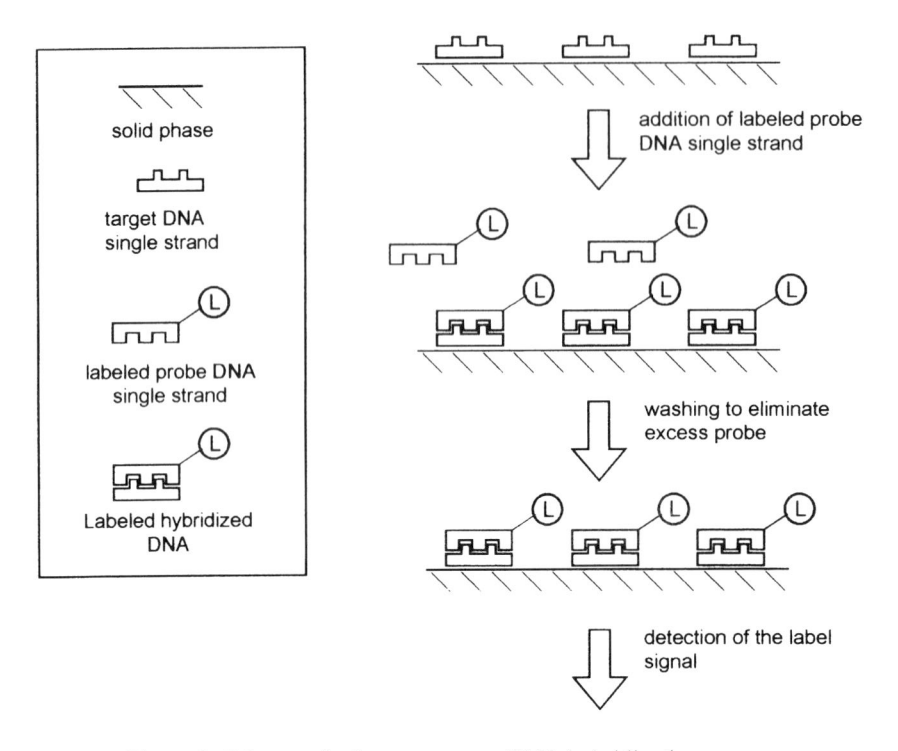

Figure 3. Scheme of a heterogeneous DNA hybridization assay.

detection is carried out. Luminescent Eu^{3+} and Tb^{3+} complexes have been used for labeling of the probe DNA allowing the detection of DNA sequences via time-resolved luminescence measurements. Nowadays, much research aims at the development of homogeneous assays because the hydribization kinetics is more favorable and simplified analytical procedures are attained. Homogeneous assays require special labeling of the DNA probes because the measurable signal is to be obtained only after hybridization occurred. Here we describe some different approaches to homogeneous assays for the detection of mutations in DNA sequences via time-resolved luminescence measurements of Eu^{3+} and Tb^{3+} complexes. Oser and Valet [26] described an assay involving a suitable pair of single-stranded DNA probes carrying the components for the formation of a Tb^{3+} complex after hybridization with the single-stranded target DNA. One probe carries a diethylenetriaminepentaacetate group complexing the Tb^{3+} ion and the second probe carries a salicylate group that can coordinate to the Tb^{3+} ion only after hybridization has taken place, and act as antenna in order to obtain metal luminescence. More recently, Coates et al. [27, 28] developed a new homogeneous assay involving a single-stranded DNA probe carrying a Eu^{3+} chelate of ethylendiaminetetraacetic acid (EDTA) and a single-stranded DNA target. Hybridization results in the formation of the duplex DNA but no metal luminescence is observed yet. Addition of a third component, incorporating a metal luminescence sensitizer attached to an intercalating group, can give rise to the formation of a luminescent Eu^{3+} complex after intercalation occurred. The approach capitalizes on the presence of mismatches of the probe with certain targets not having completely complementary nucleic acid sequences in the sites where intercalation preferably occurs. Then, the presence of mismatches in the duplex causes a reduced binding of the intercalator compared to the binding in a matching duplex. As a consequence the Eu^{3+} luminescence intensities are lower for duplexes with mismatches since a minor amount of the luminescent complex is achieved. The molecules used are the intercalating phenanthridinium linked to 2,9-dicarboxy-1,10-phenanthroline derivatives capable of coordinating the metal ion. The last approach we would like to mention consists of distance measurements in DNA sequences on the basis of the Förster theory on energy transfer. Selvin et al. [29] reported an assay in which energy transfer occurs between a Eu^{3+} complex exhibiting metal luminescence upon ligand excitation and an acceptor molecule. Depending on the position of these two species on the DNA double strand, energy transfer occurs and distances can be determined.

Chelates of lanthanide ions proved to be suitable for heterogeneous TR-FIAs and DNA hybridization assays. According to the way the lanthanide chelates are used, the assays can be divided into (1) assays relying on a separate fluorescence-enhancement step, (2) assays using stabilization techniques, and (3) assays using in situ fluorescent, stable chelates. Among these, the first practical assays using a lanthanide chelate for labeling were TR-FIAs based on a fluorescence-enhancement step. Two ways to perform the fluorescence enhancement, either by forming mixed-ligand chelates or by completely dissociating the labeling chelate prior to the formation of the fluorescent chelate, have been described. The latter approach proved to be more rapid and practical because this technique allows separate optimization of the immunoreagent labeling with the lanthanide chelate and of the conditions for the fluorimetric detection of the luminescent lanthanide chelate. This technique is available commercially under the trade name DELFIA (dissociation-enhanced lanthanide fluoroimmonoassay) and has been used for a great number of analytes [30, 31]. This analysis is performed as follows. The Eu^{3+} chelate of a polycarboxylic acid is linked to an immunoreagent at pH 7–9. After the immunoreaction and the separation of the labeled immunocomplex, the Eu^{3+} ion is released upon lowering the pH and the solution is treated with the fluorescence enhancement solution containing β-diketones, which form highly luminescent complexes with the Eu^{3+} ion. In order to make these complexes soluble in water solution the nonionic detergent Triton X-100 is added, which dissolves the Eu^{3+} complex in a micellar phase. Shielding of Eu^{3+} from water molecules, causing quenching of the luminescence, is obtained by the addition of trioctylphosphine oxide. This oxide occupies the coordination sites of the Eu^{3+} ion still available in the chelate. Finally, the Eu^{3+} luminescence is measured in a time-resolved mode. The CyberFluor system [32, 33] is an example of a heterogeneous assay using an in situ fluorescent, stable chelate. This system employs the 4,7-bis(chlorosulfophenyl)-1,10-phenanthroline-2,9-dicarboxylic acid chelating ligand capable of sensitizing the Eu^{3+} luminescence. The photosensitizer is attached to an antibody and, after saturation with the Eu^{3+} ion, metal luminescence, obtained upon ligand–metal energy transfer, is detected. Of course, neither DELFIA nor CyberFluor can be used for homogeneous assays. Much research has been dedicated to the development of simultaneously stable and luminescent Eu^{3+} and Tb^{3+} chelates in order to develop simplified heterogeneous assays and, possibly, to apply these chelates in homogeneous assays. Luminescent Eu^{3+} or Tb^{3+} chelates studied till now comprise mostly aromatic nitrogen heterocycles as energy absorbing and transferring moieties. The 4-(arylethynyl)pyridine [34], 2,2'-bipyridine [35], 1,10-phenanthroline [36, 37], and 2,2':6'2"-terpyridine [38, 39] moieties have been extensively used as energy absorbing and transferring units. Additional chelating groups used to obtain stable complexes are dicarboxylic acids, (methylenenitrilo)bis(acetic acids), and (methylenenitrilo)bis(methylphosphonic acids).

A different approach to obtain Eu^{3+}- and Tb^{3+}-based labels for TR-FIAs and DNA hybridization assays consists of the complexation of these ions with encapsulating ligands aiming at luminescent complexes stable in aqueous medium. These systems make the development of highly sensitive assays possible thanks to the intense metal luminescence that can be achieved via (1) the introduction of efficient chromophores in the ligand acting as *antenna*, (2) the efficient shielding of the metal ion by appropriate ligands from the solvent molecules quenching the metal luminescence, and (3) the substitution of solvent molecules in the first coordination sphere of the complexed metal ion by anions like fluorides and phosphates. In Section VI we will illustrate the state of the art on the use of Eu^{3+} and Tb^{3+} complexes with encapsulating ligands in TR-FIAs and DNA hybridization assays after a detailed discussion in Sections IV and V of the photophysical properties of the complexes studied to date.

IV. OVERVIEW ON Eu^{3+} AND Tb^{3+} COMPLEXES WITH ENCAPSULATING LIGANDS

Encapsulating ligands are revealed suitable for the formation of stable complexes with alkali, alkaline earth, and lanthanide ions [1,2]. What is most interesting since these ions exhibit scarce complexation ability toward conventional ligands. Complexation of these ions by encapsulating ligands occurs as a consequence of some fundamental principles of supramolecular chemistry. The encapsulating ligand plays the role of molecular receptor and the ion acts as substrate; the component whose binding is looked for. The formation of the complex depends on a process called *molecular recognition* that consists of selective binding of the substrate by the receptor on the basis of the information stored in the interacting components. Binding of the substrate by the receptor involves intermolecular, noncovalent bonds driven by electrostatic forces, van der Waals forces, hydrogen bonds, etc. and requires that the two species are complementary in size, shape, and binding sites. From this point of view, supramolecular chemistry may be considered generalized coordination chemistry, since it aims at extending to all kinds of substrates, that is, cationic, anionic, and neutral species of inorganic, organic, and biological nature. Among encapsulating ligands, mostly natural and synthetic macrocycles and polymacrocycles have been studied. Pedersen's macrocyclic polyethers (Fig. 4) combine the complexation ability of the ether functions toward cations of the hard type and the capacity of ion enclosure by the macrocycle [40]. Lehn's macrobicyclic ligands such as the cryptands shown in Figure 4 form even more stable complexes with several

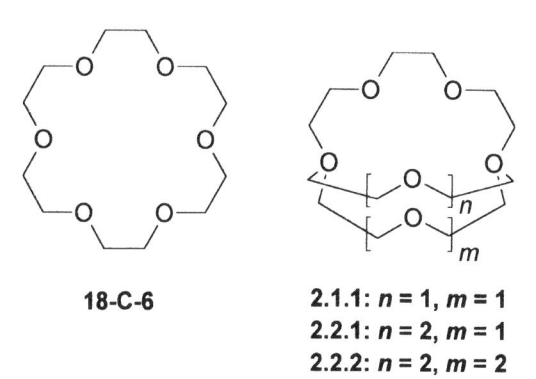

18-C-6

2.1.1: *n* = 1, *m* = 1
2.2.1: *n* = 2, *m* = 1
2.2.2: *n* = 2, *m* = 2

Figure 4. Schematic representation of the 18-C-6 crown ether and the 2.1.1, 2.2.1, and 2.2.2 cryptands.

cations [41]. These complexes have stability constants several orders of magnitude higher than those of the complexes of macrocyclic ligands. The strong complexation ability of the cryptands results from their macrobicyclic nature giving rise to a three-dimensional cavity well fitted for binding of cations. This type of complexation defines the *cryptate effect* characterized by high stability and selectivity, slow exchange rates, and efficient shielding of the ion from the environment. Furthermore, the cryptands show pronounced selectivity as a result of size complementarity between the cation and the intramolecular cavity of the ligand. For example, as the strands of the cryptands are lengthened from 2.1.1 to 2.2.2 (Fig. 4) the most strongly bound ions are Li^+, Na^+, and K^+, respectively. Cryptand 2.2.2 also displays a higher selectivity for Sr^{2+} and Ba^{2+} than for Ca^{2+}. Another important aspect of the encapsulation of cations is the amount of *preorganization* of the ligand [42]. Crystal structures of cryptands [43,44] and crown ethers [45–50] show that they contain neither cavities nor convergently arranged

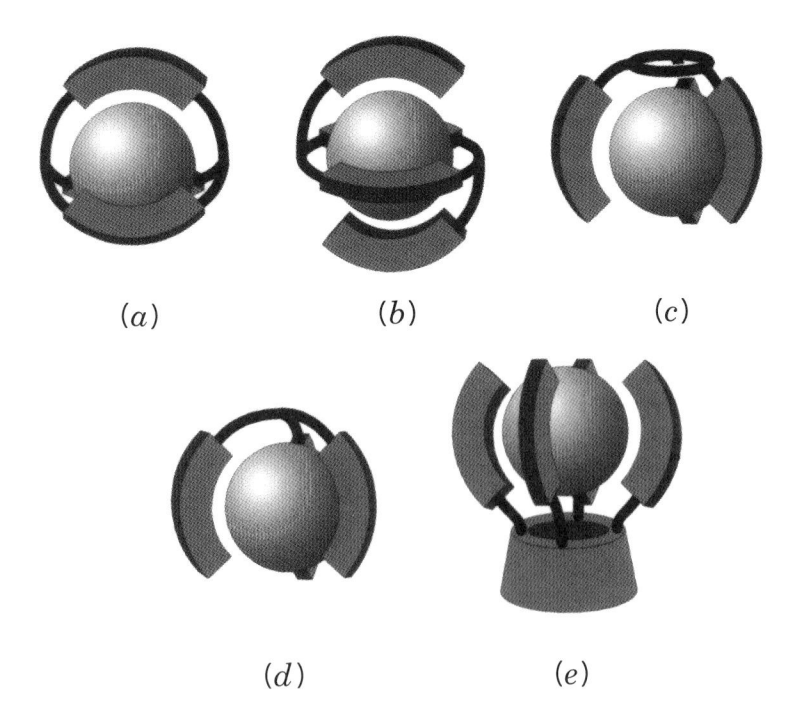

Figure 5. Schematic representation of different classes of Eu^{3+} and Tb^{3+} complexes with encapsulating ligands: (*a*) cryptates; (*b* and *c*) complexes of branched macrocyclic ligands; (*d*) podates; and (*e*) complexes of functionalized calixarenes.

binding sites prior to cation complexation. Other types of cage-type ligands (e.g., spherands) are after their synthesis completely preorganized for the cation complexation [51]. Comparison of the binding free energies for complexation of different cations by several types of cage-type ligands led to the conclusion that the more highly hosts and guests are organized for binding and the lower their solvation is prior to complexation, the more stable their complexes will be. In the following sections, we will report on Eu^{3+} and Tb^{3+} complexes obtained with encapsulating ligands of different types. The complexes are named by formulas consisting of the symbol of the metal ion followed by the number used for the ligand and are presented in different classes, schematized in Figure 5, which are defined by the nature of the ligand.

A. Cryptates

Figure 6 schematically shows the ligands of the Eu^{3+} and Tb^{3+} cryptates examined. Some photophysical data and the number of coordinated solvent molecules of these complexes are gathered in Table 1.

Cryptands containing polyoxyethylene chains as the 2.2.1 cryptand (**1**) can be considered three-dimensional analogues of crown ethers and were originally designed for complexation of alkali metal cations [1, 2]. Due to the similarities between the alkali cations and the trivalent lanthanide ions, it is not surprising that ligand **1** formed stable complexes with the latter ions. These complexes presented important characteristic like kinetic inertness in water and efficient shielding of the metal ion from solvent molecules. The first photophysical studies on luminescent Eu^{3+} and Tb^{3+} complexes of cage-like ligands have been performed on complexes of ligand **1** [52, 53, 57]. The absorption spectrum of Eu**1** presented two bands in the UV ($\varepsilon_{max} \sim 100 \, M^{-1} \, cm^{-1}$) attributed to LMCT transitions involving the ether oxygen atoms and the amine nitrogen atoms of the ligand. Analogous bands appeared in the metal luminescence excitation spectrum, together with the metal-centered bands of the Eu^{3+} ion. In this spectrum, the relatively high intensity of the metal-centered bands compared to that of the LMCT bands indicated that conversion of absorbed light into light emitted by the metal is more efficient when excitation is performed in the metal-centered bands. The absorption spectrum of Tb**1** did not differ appreciably from that of the Tb^{3+}_{aq} ion. Comparison of the lifetimes in H_2O and D_2O of both complexes indicated the presence of efficient nonradiative decay via vibronic coupling with the OH oscillators of the water molecules coordinated to the metal ion. This process is less efficient for Tb**1**, as expected considering that the energy gap between the emitting and ground state is higher for Tb^{3+} than Eu^{3+}

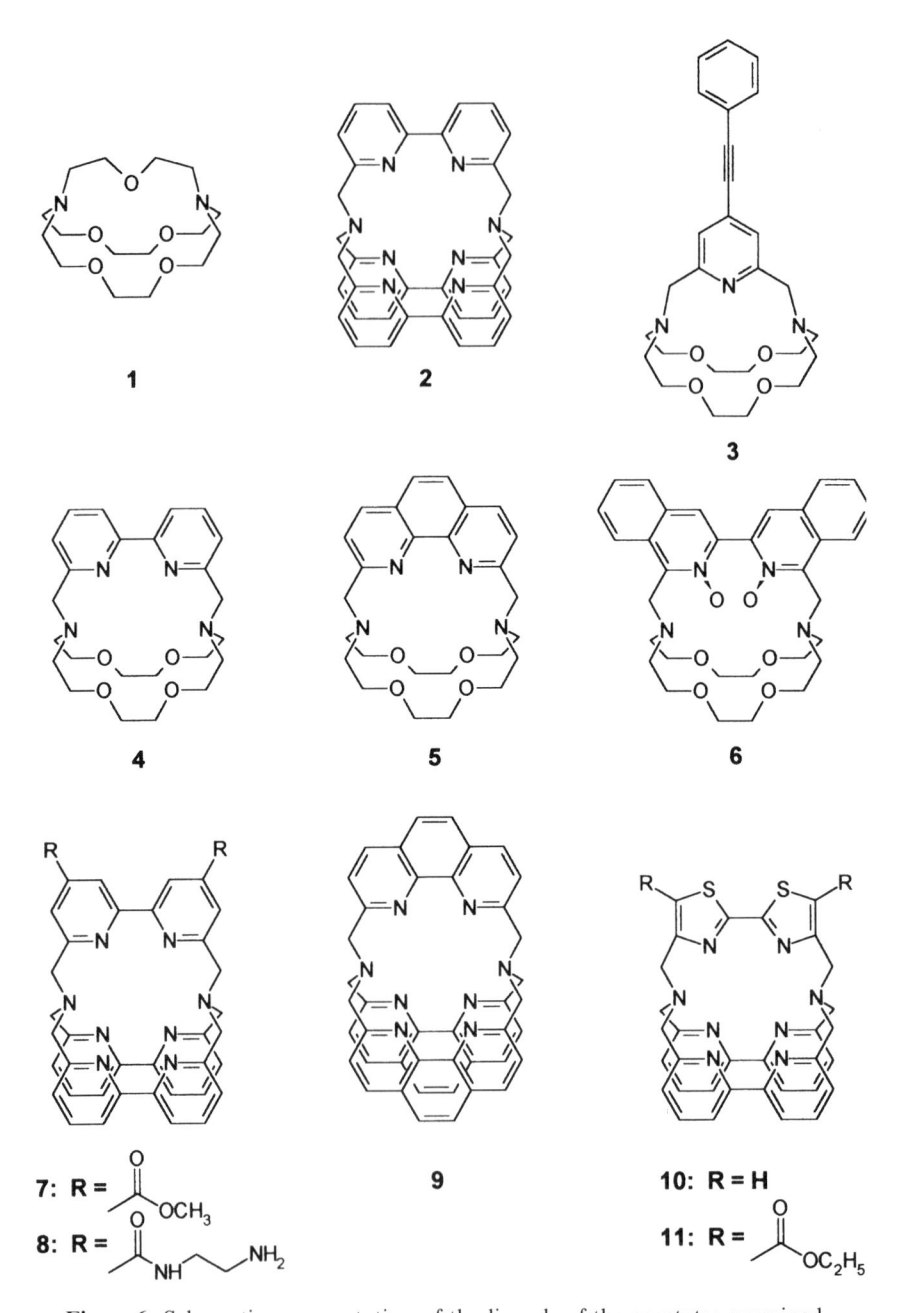

Figure 6. Schematic representation of the ligands of the cryptates examined.

12

13

14: R = H

15: R = ![acetate ester group]

16

17

18

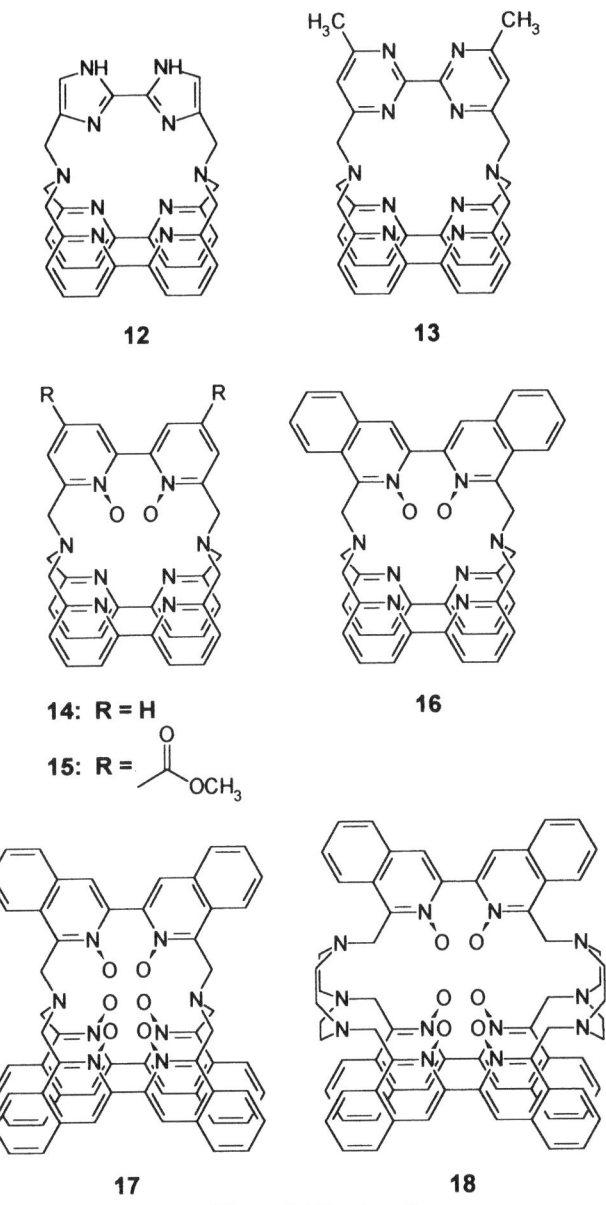

Figure 6 (*Continued*)

TABLE 1. Photophysical Data and Number of Solvent Molecules Coordinated to the Metal Ion for Some Eu^{3+} and Tb^{3+} Cryptates

Compound	Solvent	Absorption λ_{max}, ε_{max} (nm, M^{-1} cm^{-1})	$\tau_H^{(300 K)}$ (ms)	$\tau_D^{(300 K)}$ (ms)	$\tau_H^{(77 K)}$ (ms)	$\tau_D^{(77 K)}$ (ms)	$\Phi_H^{(300 K)}$	$\Phi_D^{(300 K)}$	n^b	References
Eu1	H_2O	298, 111^c	0.22	0.64	0.34	1.2		0.003	3.1	[52]
Tb1	H_2O	368, ~0.3^d	1.3	3.1	1.3	3.1	0.3		1.9	[53]
Eu2	H_2O	303, 28,000	0.34	1.7	0.81	1.7	0.02	0.10	2.5	[54]
Tb2	H_2O	304, 29,000	0.33	0.43	1.7	3.8	0.03	0.03	3.0	[54]
Eu4^e	H_2O		0.41				>0.01			[55]
Tb4^e	H_2O		0.72				>0.01			[55]
Eu5^f	H_2O		0.27				>0.01			[55]
Eu6^g	H_2O		0.70^h				0.17^h			[56]
	CH_3CN		0.62^i				0.057^i			
Eu7	H_2O	305, 26,000	0.35	1.5	0.73	1.5	0.02	0.08	2.3	[72]
Eu14	H_2O	304, 20,000	0.46	1.15	1.2	1.3	0.15	0.30	1.4	[78]
Eu15	H_2O	306, 17,000	0.40	0.70	0.80	1.1	0.09	0.17	1.1	[4]
Eu16	H_2O	304, 20,000	0.39	0.66	1.0	1.1	0.20	0.30	1.1	[78]
Eu18^g	CH_3CN		0.14				0.001			[56]

^{a}Excitation in the ligand at the λ_{max} values indicated in this table, unless otherwise noted. The lifetimes are measured in correspondence with the $^5D_0 \rightarrow {}^7F_2$ and $^5D_4 \rightarrow {}^7F_5$ emissions for Eu^{3+} and Tb^{3+}, respectively.
^{b}Calculated using Eq. (6).
^{c}LMCT absorption.
^{d}Metal-centered absorption.
^{e}Excitation at 310 nm.
^{f}Excitation at 280 nm.
^{g}Excitation at 337 nm.
^{h}Counterion $CF_3SO_3^-$.
^{i}Counterion Cl^-.

(Section II). By using Eq. 6, 3.1 and 1.9 water molecules were calculated to coordinate the metal ion for **Eu1** and **Tb1**, respectively, showing that the ligand partially shields the metal ion from the solvent (as is known, the Eu_{aq}^{3+} and Tb_{aq}^{3+} ions coordinate 9–10 water molecules). For **Eu1**, the temperature dependence of the lifetime of the metal emitting state indicated that a thermally activated nonradiative decay of this state occurs. This process was ascribed to the population of LMCT excited states from the Eu^{3+} $^5D_0$ emitting state followed by efficient nonradiative deactivation of these states to the ground state. It has been suggested that the LMCT excited states are in equilibrium with the emitting state. However, considering that the quantum yield upon LMCT excitation is one order of magnitude lower than that obtained upon excitation in the metal-centered $^5L_6$ state, the effect of the equilibrium on the decay kinetics is presumably small [4]. For **Tb1**, the lifetime values of the metal emitting state indicated that no thermally activated nonradiative decay of this state is present, as expected considering that this complex does not possess low-lying, excited-configuration states. Furthermore, 1:1 and 1:2 ion pairs between the **Eu1** and **Tb1** cryptates and fluoride anions were studied [58]. The fluoride anions were found to replace partially the water molecules coordinated to the metal ion and increase the lifetimes and quantum yields. The photophysical behavior of these ion pairs confirmed the conclusions drawn for the complexes on the role played by the nonradiative decays of the metal emitting states. Formation of ion pairs with phosphate anions, showing properties analogous to those of ion pairs with fluoride anions, has been reported [59]. In conclusion, the **Eu1** and **Tb1** complexes gave low metal luminescence intensity because of their weak absorption and, in the case of **Eu1**, of the low quantum yield.

In order to obtain more intense metal luminescence, further research pointed at the design of cryptates containing chromophores in the ligand, whose excitation would lead, possibly, to intense metal luminescence via an efficient ligand–metal energy transfer. Some lanthanide complexes of ligand **2** that contains three 2,2′-bipyridine (bpy) units as chromophores were synthesized from the Na**2** complex upon substitution of the Na$^+$ ion by lanthanide ions [60–62]. The crystal structure of the [Tb**2**]Cl$_3 \cdot 4H_2O$ complex (Fig. 7) showed that the cation is encapsulated by the ligand via binding by the eight nitrogen atoms [63]. The complex has wide open faces that are filled up by water and counterions, thus completing the first coordination sphere of the metal ion. The photophysical properties of the Eu**2** and Tb**2** complexes were extensively studied in water [54, 55, 62, 64, 65]. Among the complexes studied up to now, these have been most investigated and, therefore, they will be considered by us as prototypes of the Eu^{3+} and Tb^{3+} complexes with encapsulating ligands incorporating

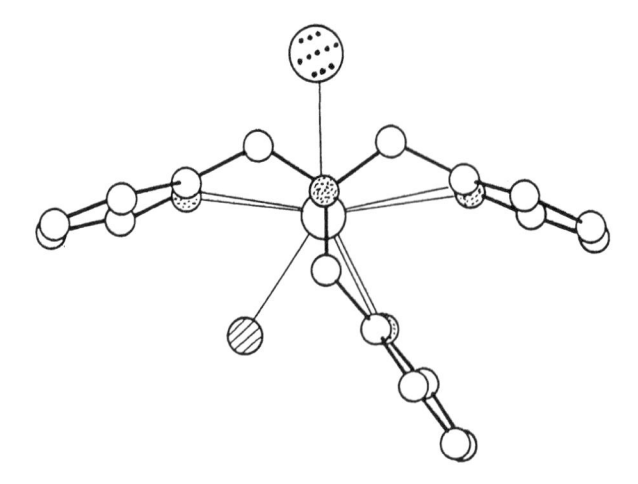

Figure 7. Crystal structure of [Tb2]Cl$_3$·4H$_2$O. The view is perpendicular to the mirror plane passing through the cation. The N atoms are small dotted circles; the O atoms are dashed circles; the Cl$^-$ anions are large dotted circles [63].

chromophoric units. Some photophysical data and the number of water molecules coordinated to the metal ion are shown in Table 1. The absorption spectra of Eu2 and Tb2 showed bands corresponding to $\pi\pi^*$ transitions in bpys. These bands were red shifted in comparison with the free bpy and the molar absorption coefficients were lower than the values corresponding to the sum of three free bpys (for free bpy $\varepsilon_{max} = 13{,}000\,M^{-1}\,cm^{-1}$ at $\lambda_{max} = 281$ nm [54]). Both effects have been attributed to the coordination of bpy to the metal ion. The large red shift has been considered indicative of a rather strong interaction between bpy and the metal ion [4]. Similarity of the absorption and the metal luminescence excitation spectra indicated that energy transfer from the ligand excited states to the metal emitting state takes place. On the basis of the lifetime values in H$_2$O and D$_2$O, it was found that an important decay process of the metal emitting states takes place via vibronic coupling with the OH oscillators of the coordinated water molecules. About three water molecules were estimated to coordinate the metal ion. Interestingly, the shielding of the metal ion in these cryptates is comparable with that of the cryptates of ligand **1**. No temperature dependence of the lifetime was observed for Eu2, indicating that excited states of other configurations are too high in energy to deactivate the Eu^{3+} $^5D_0$ emitting state at room temperature. The strong temperature dependence of the lifetime of Tb2 indicated that an important thermally activated nonradiative decay of the metal emitting state took place. It has been proved that the excited state thermally populated via this process is the ligand triplet excited state, which lies about 1200 cm^{-1} above the Tb^{3+} $^5D_4$ emitting state. The presence of an equilibrium between these states has been suggested, which could explain the quenching of the metal emitting state by oxygen (the lifetimes are 0.33 and 0.45 ms in aerated and deaerated water solution, respectively). In fact, the quenching should not take place by interaction of oxygen with the metal emitting state, as suggested by the absence of any oxygen effect on the luminescence of the Tb^{3+} ion, and could reflect the quenching of ligand triplet excited states by oxygen, which was observed in Gd2. Both Tb2 and Eu2 gave low quantum yields. The low value of Tb2 has been ascribed mainly to the efficient metal–ligand back energy transfer (Section V.B). In the case of Eu2, the low value has been explained by low efficiency of the ligand–metal energy transfer due to deactivation of the ligand excited states via LMCT excited states (Section V.B) and deactivation of the emitting state via vibronic coupling with the OH oscillators of the water molecules coordinated to the metal ion. The luminescence of the Eu2 and Tb2 complexes was investigated also in the solid state [66, 67]. High quantum yield, 0.5, was found at 300 K for Eu2, while for Tb2, high values were obtained only at low temperatures. In the latter case, the quantum yield decreased on raising the temperature, as expected considering the above mentioned deactivation of the metal emitting state via the thermally activated metal–ligand back energy transfer.

Results concerning other cryptates incorporating chromophoric groups are presented in the following. The synthesis of the Eu^{3+} complex of ligand 3 has been reported [68]. This complex was stable in aqueous solution and showed metal luminescence upon ligand excitation. The metal luminescence intensity was 56,000 times higher than that obtained for an equimolar solution of the Eu$^{3+}_{aq}$ ion upon excitation at the same wavelength. Ligands 4 and 5, which incorporated one bpy and one 1,10-phenanthroline (phen), respectively, as chromophores were obtained [60]. Preliminary photophysical results have been reported for Eu4, Tb4, and Eu5 [55]. These complexes proved to be stable in water, where they showed strong metal emission upon excitation in bpy and phen chromophores of the ligand. The lifetimes of Eu4 and Eu5 were longer than that of Eu1. This result has been attributed, at least in part, to a better shielding of the metal ion by the ligands. However, since lifetimes in heavy water were not reported, this hypothesis was not confirmed. The lifetime of Tb4 was shorter compared to that of Tb1. In this case, the effect due to a better shielding of the metal ion could be masked by efficient deactivation of the metal emitting state via thermally activated nonradiative decay involving the lowest triplet excited state of the phen unit. The quantum yields were of the order of 10^{-2}, that is, similar to those of Eu2 and Tb2. The 3,3'-biisoquinoline-N,N'-dioxide (biqO$_2$) chromophore coordinating the metal ion via the oxygen atoms was used to obtain ligand **6**. N-Oxides were found to be more powerful ligating groups for lanthanide ions than the corresponding amines and more efficient sensitizers of metal luminescence [69]. The photophysical properties of the Eu^{3+} complex of ligand **6** with several counterions were studied in acetonitrile [56, 70]. It was found that this ligand forms luminescent 1:1 complexes with the Eu^{3+} ion. The lifetimes and quantum yields depended on the nature of the counterion, the dependence being more pronounced for the quantum yields than for the lifetimes. This behavior may be due to different amounts of interaction between the counterion and the Eu^{3+} ion. For example, the highest quantum yield value was found in the case of $CF_3SO_3^-$, while in the presence of weakly coordination anions, like perchlorate and chloride, the quantum yields were lower. The Eu**6** complex was reported to be stable in water as well, but no luminescence data were given in this solvent. Ligand **7** [71], which was obtained upon introduction of carboxymethyl substituents in the 4 and 4′ positions of one bpy of ligand **2**, was synthesized. The Eu**7** complex was obtained [71] and its photophysical properties were studied [59, 72]. The absorption spectrum of Eu**7** in water was similar to that of the parent Eu**2**. Comparison of the metal emission spectra of Eu**7** and Eu**2** indicated that the carboxymethyl substituents do not affect the symmetry of the first coordination sphere of the Eu^{3+} ion. What is confirmed by the crystal structure of $[Eu7]Cl_3 \cdot 4H_2O$ shown in Figure 8 [63]. The lifetimes and quantum yields were similar to those of the parent Eu**2**, indicating that the presence of the carboxymethyl substituent slightly affects the photophysical

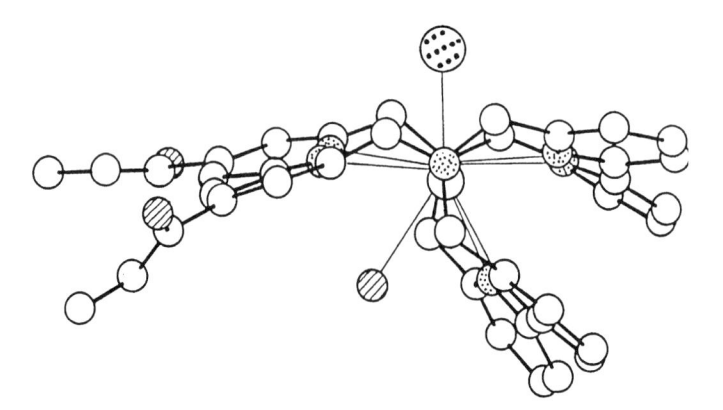

Figure 8. Crystal structure of $[Tb7]Cl_3 \cdot 4H_2O$. The view is perpendicular to the mirror plane passing through the cation. The N atoms are small dotted circles; the O atoms are dashed circles; the Cl^- anions are large dotted circles [63].

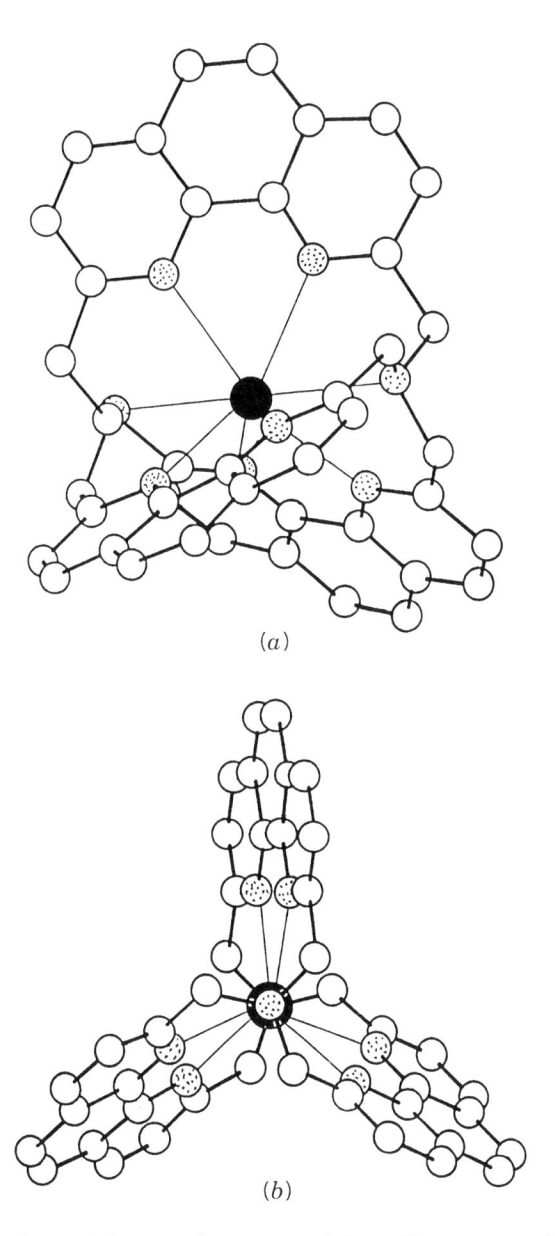

Figure 9. Two views of the crystal structure of the sodium cryptate Na**9**. (*a*) (Left) View into the cavity, showing the coordination of the Na$^+$ ion (solid block circle) to the eight N-sites (dotted circles). (*b*) (Right) Projection upon a plane perpendicular to the axis connecting the two aliphatic nitrogen atoms of the ligand [75].

properties. Also in the case of Eu8 [73], introduction of substituents did not cause relevant changes in the photophysical properties with respect to those of Eu2. The Eu^{3+} complex of ligand **9** containing three phen units was prepared [60]. It has been reported that its behavior in water is similar to that of Eu2, but quantitative data were not given [74]. Interestingly, the crystal structure of [Na9]Br·2CHCl$_3$ (Fig. 9) [75] showed that the symmetry is higher than that of Tb2. This result may be due to the rigidity of the phen unit. Ligands **10-13**, containing the heterocyclic chromophores 2,2'-bithiazole, 2,2'-biimidazole, and 2,2'-bipyrimidine, were prepared [76]. The Eu^{3+} complexes of these ligands were obtained and preliminary photophysical results in water have been reported for some of them [76]. For the Eu^{3+} complexes of ligands **10-12**, luminescence was much less intense than for Eu2, while Eu13 showed luminescence intensity similar to that of Eu2. Other cryptands incorporating N-oxides in one or more chromophores were obtained. First, ligand **14**, containing one 2,2'-bipyridine-*N*,*N*'-dioxide (bpyO$_2$) and two bpy units, was synthesized and its Eu^{3+} complex was obtained [77]. The photophysical properties of Eu14 were studied in water [77, 78]. Significant changes in the absorption spectrum with respect to that of Eu2 were observed. The number of water molecules coordinated to the metal ion was about 1.4 (that is smaller than that of Eu2), showing improvement of the shielding ability of the ligand upon introduction of N-oxides. The lifetimes showed that nonradiative decay of the metal emitting state takes place via both the vibronic coupling and the thermally activated processes. The thermally activated nonradiative decay absent in Eu2 was attributed to the population of LMCT excited states lying at lower energies in Eu14 than in Eu2. The lower energy was related to the weaker electron-donating properties of bpyO$_2$ compared to bpy, which leaves the Eu^{3+} ion more positive in Eu14 than in Eu2. Most interestingly, the quantum yield of Eu14 was much higher than that of Eu2. What has been attributed to better shielding of the metal ion mentioned above. Carboxymethyl groups were introduced in the bpyO$_2$ chromophore of ligand **14** thus obtaining ligand **15**. The Eu15 complex was synthesized [74] and its photophysical properties were studied in water [4]. Analogous to what was found for Eu7 and its parent Eu2, the symmetry of the first coordination sphere, obtained from high-resolution metal emission spectra, and the number of coordinated water molecules were similar for Eu15 and Eu14. The lifetime and quantum yield values were lower for Eu15 than for Eu14 and the decrease of the lifetime going from 77 to 300 K was higher for Eu15. What has been explained considering that the carboxymethyl substituents, due to their electron-withdrawing character, lower the energy of the LMCT excited states. Furthermore, ligand **16**, which contains one biqO$_2$ and two bpy units was synthesized and the Eu16 complex was obtained [77]. The crystal structure of the chloride salt of Eu16 (Fig. 10) [79] revealed that the

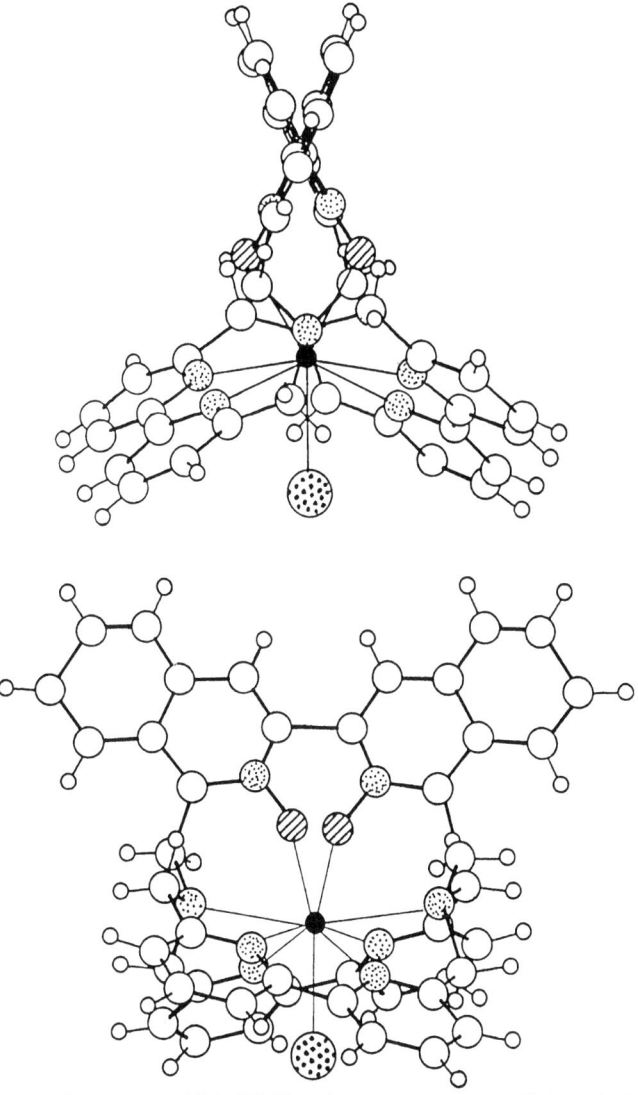

Figure 10. Crystal structure of [Eu16]Cl$_3$. The N atoms are small dotted circles; the O atoms are dashed circles; the Cl$^-$ anions are large dotted circles [79].

two isoquinoline moieties of biqO$_2$ are not coplanar and the two oxygen atoms block two faces of the macrobicyclic structure, preventing the approach of solvent molecules to the metal ion. The photophysical properties of **Eu16** were studied in water [78]. The absorption spectrum presented a band at 304 nm due to the bpy units and additional bands at about 270 and 360 nm, which have been attributed to absorption of the biqO$_2$ chromophore. The luminescence properties were similar to those of Eu14 and Eu15. The temperature dependence of the lifetimes indicated that the thermally activated nonradiative decay of the metal emitting state is more efficient than in Eu14 and Eu15. On the basis of the energy estimated for the lowest ligand triplet excited state, it has been proposed that this decay involves an equilibrium between this state and the metal emitting state. The quantum yield in H$_2$O was higher than those of Eu14 and Eu15, thanks to the more efficient ligand–metal energy transfer (Section V.B). Analogous to what was found for Eu14 and Eu15, the number of water molecules coordinated to the metal ion was low (1.1). Short lifetimes and low quantum yields (of the order of 10^{-3}) were obtained in acetonitrile [56] for the Eu^{3+} complexes of ligands **17** and **18**, which incorporated three biqO$_2$ units. These low values have been attributed to inefficient complexation, and therefore bad ligand–metal interaction, because the cavities of ligands **17** and **18** do not fit the size of the Eu^{3+} ion.

B. Complexes of Macrocycles

Complexes of the Eu^{3+} and Tb^{3+} ions with different types of macrocyclic ligands have been obtained and their properties, in particular luminescence, have been reported by different authors [4, 15, 80–82]. For this class of complexes, we discuss only the complexes of macrocyclic ligands incorporating chromophores, since such ligands can play the role of antennas. Figure 11 schematically shows the macrocyclic ligands of the Eu^{3+} and Tb^{3+} complexes examined. Some photophysical data and the number of coordinated solvent molecules of these complexes are gathered in Table 2.

One of the first approaches to the introduction of chromophores in macrocyclic ligands consisted of the synthesis of ligand **19** [86]. For its Eu^{3+} and Tb^{3+} complexes, metal emission following an energy transfer from the ligand excited states to the metal emitting states has been reported [87]. Ligands **20** and **21** incorporate the 3-aroylcoumarine chromophore in the 15-C-5 and 18-C-6 crown ethers, respectively. Metal luminescence was observed for the Eu^{3+} and Tb^{3+} complexes of these ligands upon ligand excitation [88]. A template synthesis gave the Eu^{3+} complex of ligand **22**, a hexaazamacrocycle containing two pyridines as chromophores [89]. A complete photophysical study has been reported for Eu22 in water [83]. The excitation spectrum of Eu22 showed the same pattern as the absorption spectrum, consisting of ligand-centered transitions. This comparison indicated that light absorption in the ligand was followed by an energy transfer to the metal emitting state. This state was quenched by vibronic

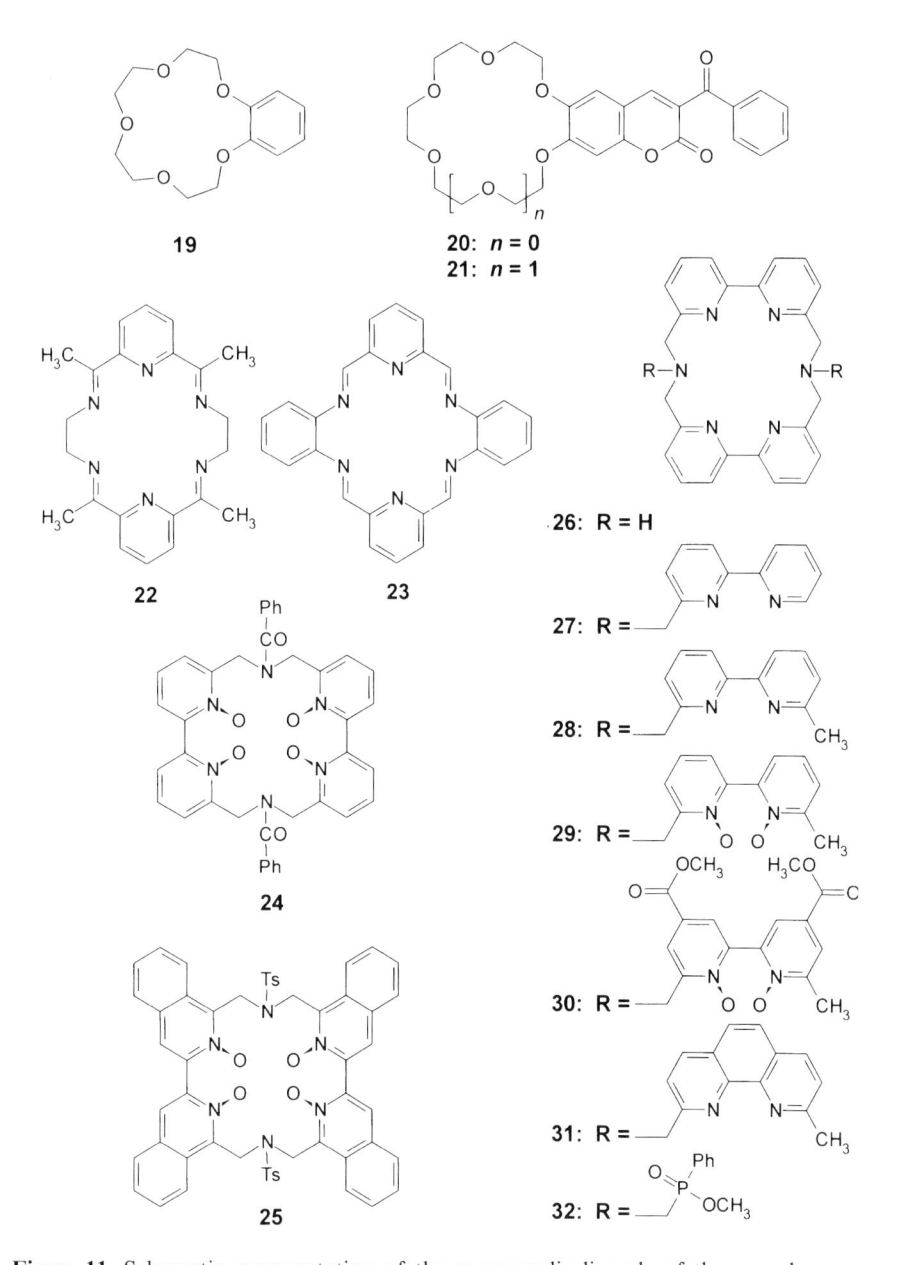

OVERVIEW ON Eu³⁺ AND Tb³⁺ COMPLEXES WITH ENCAPSULATING LIGANDS 239

Figure 11. Schematic representation of the macrocyclic ligands of the complexes examined.

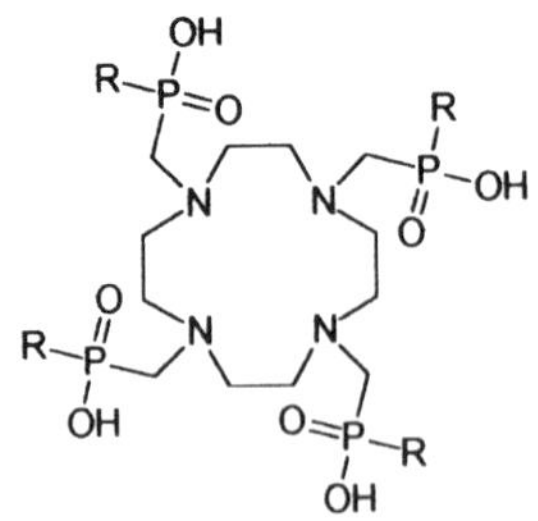

33: **R = CH_3**
34: **R = CH_2Ph**

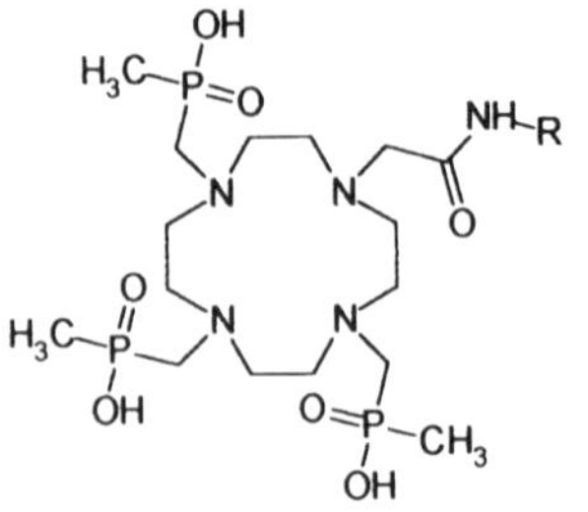

35: **R = CH_3**
36: **R =**

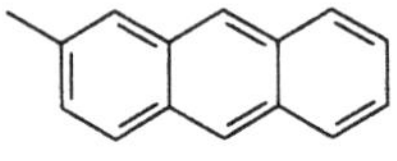

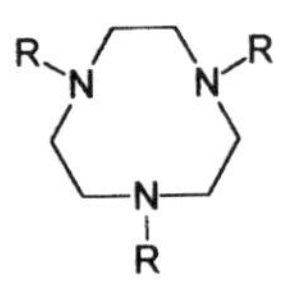

37: **R =**

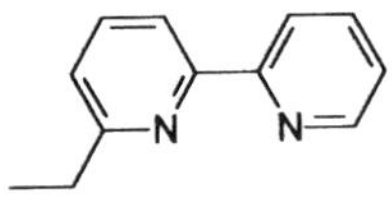

38: **R =**

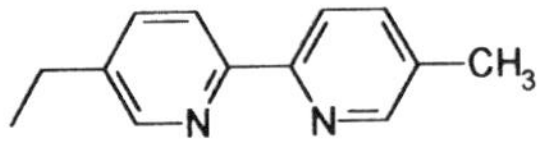

40: **R =**

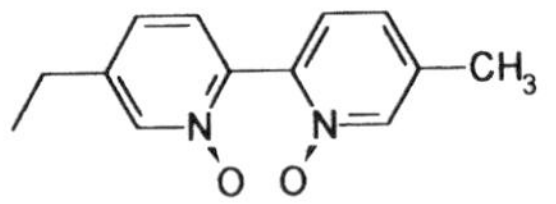

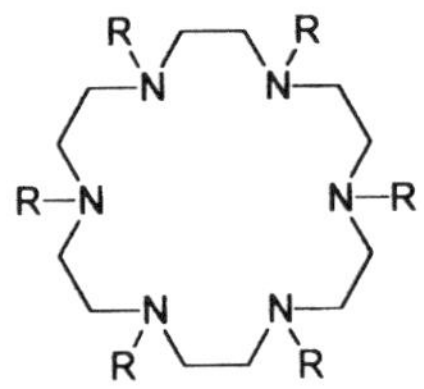

R =

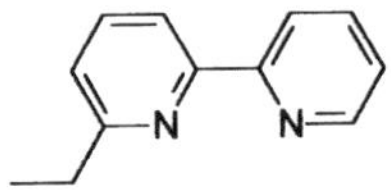

42

Figure 11. (*Continued*)

coupling with the OH oscillators. In fact, one water molecule was coordinated to the Eu^{3+} ion in the complex. Both the macrocyclic ligand and the coordinated acetate counterion account for the efficient shielding of the ion from the solvent molecules. Comparison between the quantum yields obtained upon ligand and metal excitation suggested that the energy transfer from the ligand excited states to the metal emitting state does not compete efficiently with the other deactivation processes of the ligand excited states. The Eu^{3+} complex of ligand **23** was obtained upon template synthesis and its metal luminescence was investigated [90]. The complex that was not luminescent itself developed Eu^{3+} luminescence upon addition of a β-diketonate in methanol. This luminescence has been attributed to replacement of counterions and/or solvent molecules by the sensitizer

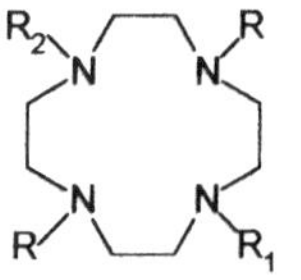

39: R = R₁ = R₂ = [5-ethyl-2,2'-bipyridine-5'-methyl group]

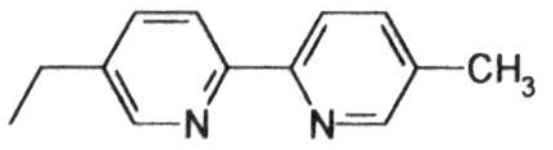

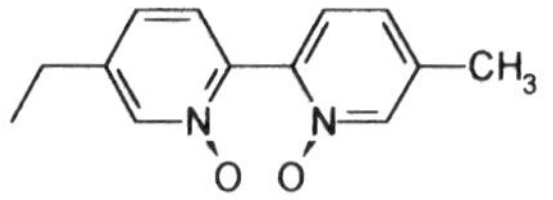

41: R = R₁ = R₂ = [5-ethyl-2,2'-bipyridine-5'-methyl N,N'-dioxide group]

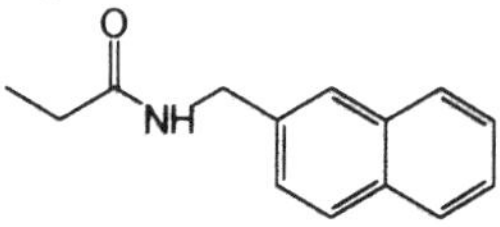

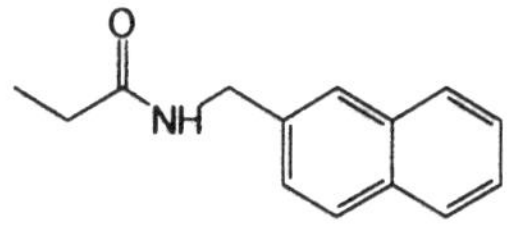

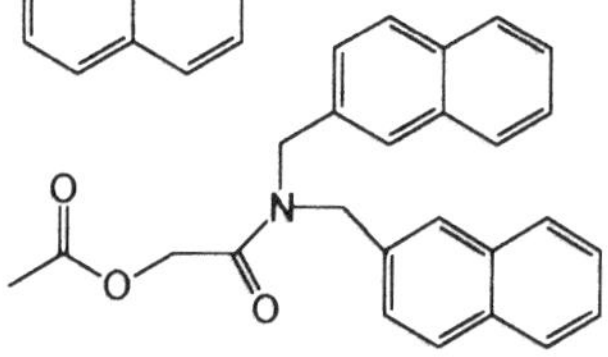

43: R = R₁ = CONHCH₃; R₂ = [CH₂C(O)NH-CH₂-naphthyl]

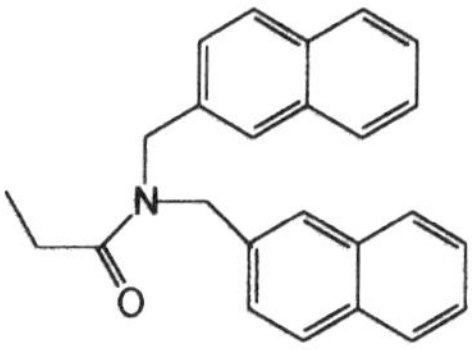

44: R = R₁ = R₂ = [CH₂C(O)NH-CH₂-naphthyl]

45: R = R₁ = CONHCH₃; R₂ = [CH₂C(O)O-CH₂C(O)N(CH₂-naphthyl)₂]

46: R₂ = R₁ = CONHCH₃; R = [CH₂CH₂C(O)N(CH₂-naphthyl)₂]

47: R = R₁ = CONHCH₃; R₂ = [carbostyril-type group with NHCH₃ amide]

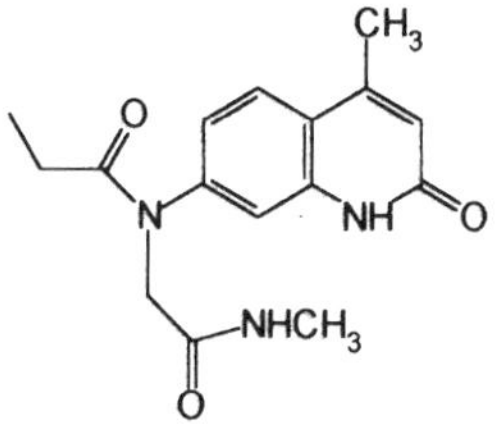

Figure 11. (*Continued*)

β-diketonate without disruption of the Eu³⁺ complex. The Eu³⁺ complexes of ligands **24** and **25** were prepared [56, 91]. The Eu24 complex, which was sparingly soluble in water, showed in this solvent metal luminescence upon ligand excitation. This complex had a quantum yield on the order of 10^{-3} [91]. The complex Eu25 was not stable in water and was studied in

TABLE 2. Photophysical Data and Number of Solvent Molecules Coordinated to the Metal Ion for Some Eu^{3+} and Tb^{3+} Compounds of Macrocyclic Ligands

Compound	Solvent	Absorption $\lambda_{max}, \varepsilon_{max}$ (nm, M^{-1} cm^{-1})	Emission[a] $\tau_H^{(300K)}$ (ms)	$\tau_D^{(300K)}$ (ms)	$\tau_H^{(77K)}$ (ms)	$\tau_D^{(77K)}$ (ms)	$\Phi_H^{(300K)}$	$\Phi_D^{(300K)}$	n[b]	References
Eu22	H_2O	300, 10,000	0.70	2.0	1.4	2.1		0.006	1.0	[83]
Eu27	H_2O	312, 36,000	1.5	1.9	1.8	1.9	0.10	0.20	<0.5	[84]
Tb27	CH_3OH	312, 44,000	1.1	1.2	2.0	2.9	0.14	0.35	0.6	[84]
Eu28	CH_3OH	299, 39,000	1.0	1.8	1.3	2.0	0.01	0.01	0.9	[84]
Eu29	H_2O	310, 15,500	0.5	1.2	1.2	1.45	0.015	0.035	1.2	[4]
Eu30	H_2O	286, 34,000	0.13	0.13	1.05	1.3	0.01	0.01	[c]	[72]
Eu31	H_2O	278, 46,000	0.27	1.85	0.3	1.9	0.013	0.09	3.3	[85]
							0.016[d]	0.10[d]		
Tb31	H_2O	278, 43,000	0.06	0.06	1.4	2.6	0.007[e]	0.007	2.7[f]	[85]
							0.007[d,e]	0.007[d]		
Eu32	H_2O	307, 15,100	0.6	2.1	1.2	2.0	0.08	0.30	1.3	[93]
Tb32	H_2O	306, 15,300	0.43	0.50	1.8	2.4	0.08	0.09	1.4	[93]
Eu33[g]	H_2O		1.25	1.85			0.0006	0.001	0.3	[94]
Tb33[g]	H_2O		2.96	3.71			0.01	0.009	0.3	[94]
Eu34[g]	H_2O		1.59	2.07			0.0009	0.0015	0.2	[94]
Tb34[g]	H_2O		4.13	4.44			0.44	0.49	0.1	[94]
Eu35[g]	H_2O		0.76	1.85			0.0005	0.0013	0.8	[94]

Compound	Solvent	λmax							Ref
Tb35[a]	H_2O		3.2			0.16	0.28	0.3	[94]
Eu36[b]	H_2O		0.07, 0.60[i]	0.17, 1.21[i]		0.0007	0.0032	0.9	[94]
Tb36[b]	H_2O		0.17, 1.63[i]	0.26, 0.80[i]		0.0006	0.004	1.4	[94]
Eu37	H_2O	311, 24,500	0.50	0.57	0.87	0.05	0.07	<0.5	[78]
Tb37	H_2O	311, 20,400	1.5	1.5	1.4	0.37	0.38	<0.5	[78]
Eu38	CH_3CN	304, 30,500	1.74			0.36			[97]
Tb38	CH_3CN	303, 31,200	0.42			0.08			[97]
Eu39	CH_3CN	302, 38,000	0.90			0.02			[97]
Eu40	CH_3OH	260, 38,000	0.52	0.76	0.58	0.05			[97]
Tb40	CH_3OH	260, 38,000			0.90			<0.7	[97]
Eu41	CH_3OH	261, 50,400	0.56	0.72	0.65	0.05	0.05		[97]
Tb41	CH_3OH	260, 51,000			0.82				[97]
Eu42	CH_3CN	306, 49,000	0.73		1.02	0.020			[98]
Tb42	CH_3CN	306, 60,400	1.36		1.74	0.015			[98]
Eu43	H_2O		0.58	2.40					[99]
Tb43	H_2O		0.80	2.00	1.00	0.11[j]			[100]
Eu44	CH_3OH		0.74	2.11	1.00	0.0012	0.0038		[99]
Eu47	H_2O		0.51	1.47	0.94	0.006	0.014	1.28	[101]
Tb47	H_2O		0.88	1.12	0.90	0.13	0.15	1.04	[101]

[a] Excitation in the ligand at the λ_{max} values indicated in this table, unless otherwise noted. The lifetimes are measured in correspondence with the $^5D_0 \to {^7F_2}$ and $^5D_4 \to {^7F_5}$ emissions for Eu^{3+} and Tb^{3+}, respectively.
[b] Calculated using Eqs. (6) and (7).
[c] The effects of the vibronic coupling with the OH oscillators are, most likely, masked by the thermally activated nonradiative decay.
[d] Excitation at 310 nm.
[e] In deaerated solution the values 0.040 and 0.028 were obtained for the quantum yields upon excitation at 310 nm and 278 nm, respectively.

acetonitrile. Metal luminescence was observed upon excitation of the ligand. The quantum yield was on the order of 10^{-2} [56].

The bpy chromophore has been used to synthesize several branched-macrocyclic ligands. Let us first discuss the complexes of the ligands incorporating the macrocycle **26**. Ligands **27** and **28** containing four bpy units were synthesized [92]. Complexes of these ligands with Eu^{3+} and Tb^{3+} were prepared and their photophysical properties were studied in water and methanol [84]. Similarity of absorption and metal luminescence excitation spectra indicated the presence of ligand–metal energy transfer. For Eu**27**, the lifetime and quantum yield values showed little deactivation via vibronic coupling with OH oscillators of coordinated water molecules and no temperature dependence. For Tb**27**, metal–ligand back energy transfer was found to be an important decay pathway of the metal emitting state, in analogy with Tb**2** (Section IV.A). Interestingly, the good quantum yield of Tb**27** indicated that this process plays a less important role than in Tb**2**. As for Eu**28**, lifetimes similar to those of the parent Eu**27** complex were obtained, but the quantum yields were much smaller. This behavior has been attributed to the methyl groups preventing the approach of the bpy units in the branches to the metal ion [4]. This interpretation is supported by the presence of a residual ligand phosphorescence [84]. Less than one solvent molecule was coordinated to the Eu^{3+} and Tb^{3+} ions complexed by ligands **27** and **28**. Interestingly, in these complexes the ion was more efficiently shielded than in the cryptates Tb**2** and Eu**2**, which still contained about three water molecules coordinated to the metal ion. The efficient shielding offered by ligands **27** and **28** toward solvent interaction suggested that, as envisaged by space-filling models, the metal ion is enclosed in the macrocyclic ring and encapsulated by the two branches of the ligand.

The N-oxides were introduced in the bpy units of the branches of ligand **28**, thus obtaining ligand **29** [74]. Remarkably, Eu**29** and Tb**29** were stable in water differently from Eu**28**. The stabilization introduced by the N-oxides has been explained considering that the oxygen atoms of these groups may approach the ion better than the bpy nitrogen atoms, thus reducing the steric hindrance by the methyl groups [4]. The absorption spectra of Eu**29** and Tb**29** showed the same bands of the parent complexes containing only bpy units and shoulders on the red side of the lowest energy bands due, most likely, to bpyO$_2$-localized ligand excited states. For Eu**29**, comparison of the lifetimes at 300 and 77 K in D$_2$O indicated that thermally activated nonradiative decay of the metal emitting state takes place. It has been proposed that this process involves LMCT excited states lying at lower energy than in Eu**27**, which contains only bpy units. As suggested for the cryptates, the lower energy of the LMCT excited states may be due to the weaker basicity of bpyO$_2$, which leaves the europium ion with a higher formal positive charge [4]. The Tb**29** complex showed metal emission only at 77 K [4]. The lack of metal luminescence at room temperature has been ascribed to nonradiative deactivation via the lowest ligand triplet excited state, which on the basis of the phosphorescence spectrum of the Gd**29** complex is localized at almost the same energy as the Tb^{3+} $^5D_4$ emitting state. Ligand **30** was obtained upon introduction of carboxymethyl substituents in the bpyO$_2$ branches of ligand **29** and the photophysics of its Eu^{3+} complex was studied in water [72]. Both lifetimes and quantum yield values of Eu**30** were lower than those of Eu**29** and the temperature dependence of the lifetime is stronger for Eu**30**. This behavior has been attributed to efficient thermally activated nonradiative decay via LMCT excited states lying at lower energy because of the presence of the electron-withdrawing carboxymethyl substituents.

In order to enhance the absorption efficiency, ligand **31**, containing two phen units in the branches, was synthesized and the photophysics of its Eu^{3+} and Tb^{3+} complexes was studied [85]. Interestingly, these complexes did not decompose in water differently from those of ligand **28** containing bpy instead of phen units. This finding has been related to the rigidity of the phen moiety. Similarity of the absorption and metal luminescence excitation spectra of the complexes indicated that ligand–metal energy transfer takes place from both the bpy and phen chromophores. Concerning the quantum yields upon excitation in the bpy or phen chromophores, lower values were obtained for the phen units indicating a lower efficiency of the ligand–metal energy transfer for this chromophore. The lifetimes and quantum yields of Eu**31** indicated that nonradiative deactivation via vibronic coupling with the OH oscillators of the coordinated water molecules is very efficient. Indeed, about 3.3 water molecules were found to coordinate the metal ion. The lack of any temperature dependence of the lifetime in heavy water suggested that thermally activated decay processes do not play an important role in the deactivation of the metal emitting state. As for the Tb**31** complex, the strong increase of the lifetime upon lowering the temperature indicated that thermally activated nonradiative decay of the Tb^{3+} emitting state via short-lived, upper lying excited states is very efficient. In analogy with the Tb**2** complex it has been suggested that the thermally activated decay may involve an equilibrium between the metal emitting state and the lowest ligand triplet excited state, which was found to lie at 20,800 cm^{-1} on the basis of the phosphorescence spectrum of the Gd^{3+} complex. In aerated solutions, the very efficient thermally activated nonradiative decay of the metal emitting state masked the quenching via vibronic coupling with OH oscillators, which could only be detected in deaerated solutions. By using Eq. 6, about 2.7 water molecules were calculated to coordinate the metal ion. The photophysics of these complexes was also studied in methanol [85].

The analysis of the photophysical data suggested that the complexation is worse in methanol than in water. This behavior has been ascribed to a weaker interaction in methanol between the phen branches and the metal ion.

An approach to optimize the stability of complexes of branched macrocyclic ligands in water consisted of the introduction of phosphinate esters in the branches of macrocycle **26** [93], since these groups have good ligating properties toward lanthanide ions. In fact, the Eu^{3+} and Tb^{3+} complexes of ligand **32** were found to be stable in water [93]. The absorption spectra of the complexes showed the typical bpy transitions and shoulders at 260 nm most likely due to the phenyl moiety. Comparison of the absorption spectra with the metal luminescence excitation spectra indicated that ligand–metal energy transfer takes place. This process clearly involved the bpy units while energy transfer from the phenyl moiety could not be unequivocally assessed. In any event, the involvement of phenyl or benzyl units attached to phosphonic or phosphinic groups in efficient energy transfer to Eu^{3+} and Tb^{3+} ions has been recently reported [94,95]. For both Eu**32** and Tb**32**, the lifetimes and quantum yields in H_2O and D_2O indicated that some vibronic coupling with the OH oscillators occurs. The number of water molecules coordinated to the metal ion was about 1.5. For Tb**32**, the strong temperature dependence of the lifetime indicated that a thermally activated nonradiative decay process plays an important role. In analogy with the other Tb^{3+} complexes containing the bpy chromophore, such a process most likely involved an equilibrium between the lowest bpy triplet excited state and the metal emitting state. Interestingly, the quantum yields were rather high compared to those of the prototypes Eu**2** and Tb**2** and were similar to those of the complexes of ligand **27**. As suggested for the latter, the branched macrocyclic structure provided better shielding from water molecules and, most likely, better complexation leading to more efficient ligand–metal energy transfer (Section V.B). Phosphinate esters were also attached to the tetraazacyclododecane macrocycle thus obtaining ligands **33–36** [96]. The Eu^{3+} and Tb^{3+} complexes of these ligands were obtained and their photophysics was studied in water [94]. The lifetimes of the Eu^{3+} and Tb^{3+} complexes of ligands **33** and **34** in water and heavy water were consistent with there being no water molecules bound to the metal ion, as also was suggested by the ^{17}O NMR and Gd relaxivity measurements. Low quantum yields were obtained for the Eu^{3+} and Tb^{3+} complexes of ligand **33** and stronger luminescence was observed for Tb**34** upon excitation in the phenyl moiety. For Tb**35**, it was reported that the absorption and metal luminescence excitation spectra match well and that the complex exhibits a relatively high quantum yield upon ligand excitation. The Eu^{3+} and Tb^{3+}

complexes of ligand **36**, which contain the anthryl chromophore, exhibited metal luminescence upon ligand excitation. It was claimed that the ligand–metal energy transfer involves the anthryl moiety [94]. However, considering that this chromophore has triplet excited states at energies lower than those of the Eu^{3+} and Tb^{3+} emitting states, they are expected to quench the metal luminescence. Indeed, the very low values of the quantum yields showed that the anthryl moiety is not a good sensitizer for Eu^{3+} and Tb^{3+} emission.

Several ligands were synthesized, which contained polyazamacrocycles carrying different types of chromophores. The triazacyclonane-based ligand **37**, containing three bpy units ligated to the ring through C6 [92], was used to encapsulate the Eu^{3+} and Tb^{3+} ions and the photophysical properties of the complexes were investigated in water [78]. The lifetimes of both complexes were almost the same in H_2O and D_2O, which showed that no water molecules are coordinated to the metal ions. For Eu**37**, the short lifetime, its temperature dependence, and the low quantum yield have been explained on the basis of low-lying LMCT excited states involving the aliphatic nitrogen atoms. The lifetime of Tb**37** did not depend on the temperature. This result was in agreement with the observation of the lowest ligand triplet excited state at $22,800\,cm^{-1}$, a value significantly higher than that of the Tb^{3+} emitting state. The lack of any thermally activated nonradiative decay and of coordinated water molecules led to a high quantum yield. Recently, three and four bpy units were ligated to the triazacyclononane and tetraazacyclododecane macrocycles through C5, thus obtaining ligands **38** and **39** [97]. Interestingly, a different behavior was observed for the complexes of ligand **38** [97] compared to the complexes of ligand **37**. The complexes of ligand **38** proved to be unstable in water conversely to those of ligand **37** and were found to be stable in anhydrous acetonitrile. Moreover, Eu**38** exhibited a much longer lifetime and a much higher quantum yield than Eu**37**. It is worthwhile noting that the quantum yield of Eu**38** (0.37) is the highest value found to date for Eu^{3+} complexes with cage-like ligands. Furthermore, the quantum yield of Tb**38** is much lower than that of Tb**37**. The following observations on the nonradiative deactivations may explain the different photophysical behavior of the complexes of ligands **37** and **38**. From the phosphorescence of the Gd^{3+} complexes of ligands **37** and **38**, the lowest ligand triplet excited states have been localized at 22,700 and $22,200\,cm^{-1}$, respectively, so that back energy transfer from the Tb^{3+} emitting state to the lowest ligand triplet excited state did not occur in Tb**37**, but may be present in Tb**38**. In the case of Eu**38**, deactivation via LMCT excited states involving the aliphatic nitrogen atoms may be less efficient than for Eu**37**. Indeed, such LMCT excited states are expected to lie at higher energies because of a longer distance between the Eu^{3+} ion and the aliphatic nitrogen atoms as appears also from the CPK models. The Eu^{3+} complex of ligand **39** was unstable in water and its photophysical properties were studied in acetonitrile [97]. The Eu**39** complex showed a shorter lifetime and a much lower quantum yield with respect to Eu**38**. The latter may be due to a worse metal–ligand interaction caused by steric hindrance that the four bpy units undergo when approaching the metal ion. The ligands **40** and **41** were obtained upon attachment of the bpyO$_2$ chromophore through C5 to the triazacyclononane and tetraazacyclododecane macrocycles, respectively, and their Eu^{3+} and Tb^{3+} complexes were prepared [97]. Their photophysics was studied in methanol where the complexes were stable [97]. For Eu**40** and Eu**41**, the lifetimes were rather low and the quantum yields were quite good. The values changed only little upon the deuteration of the solvent and the lowering of the temperature indicating that nonradiative deactivations of the Eu^{3+} emitting state via vibronic coupling and via thermally activated processes do not play a significant role. Complexes Tb**40** and Tb**41** were luminescent only at low temperature due, most likely, to efficient nonradiative deactivation at room temperature of the Tb^{3+} emitting state via ligand triplet excited states, which on the basis of the ligand phosphorescence spectrum of the Gd^{3+} complexes have been localized at almost the same energy (20,800 cm^{-1}) as the Tb^{3+} emitting state. Ligand **42** containing the hexaazacyclooctadecane ring carrying six pendant bpy units was synthesized [92] and the properties of its binuclear complexes were studied [98]. A 2:1 (metal/ligand) stoichiometry for the solid homonuclear Eu^{3+} and Tb^{3+} complexes of this ligand was found. These complexes decomposed in water and methanol and were stable in acetonitrile where they showed low quantum yields.

Lately, a series of new ligands was synthesized containing the tetraazacyclododecane carrying anthryl or carbostyril chromophores and amide esters as ligating groups [99–101]. The Eu^{3+} and Tb^{3+} complexes of ligands **43** and **44** were prepared and their photophysics was studied in different solvents [99, 100]. The lifetime values of Eu**43** and Eu**44** changed significantly upon deuteration of the solvent (water or methanol), indicating that vibronic coupling with OH oscillators plays an important role. Only the quantum yields of Eu**44** were reported and the values were low in both protonated and deuterated solvents. Parker and Wiliams [99] attributed these low values to efficient deactivation of the anthryl excited states via LMCT excited states, thus lowering the efficiency of the ligand–metal energy efficiency. For the Tb^{3+} complexes, Tb**43**, and Tb**44** exhibited only very weak metal luminescence at room temperature in acetonitrile, water, and ethanol. On the basis of lifetime measurements at different temperatures in aerated and deaerated ethanol, this behavior has been attributed to efficient back energy transfer from the Tb^{3+} emitting state to the lowest triplet excited state of the anthryl moiety. In degassed ethanol solution, a rather good quantum yield was obtained for **Tb43** [100]. The use of ligands **45** and **46** to prepare the complexes **Eu45** and **Eu46** has been reported [100]. Their luminescence intensity resulted to be similar to that of **Eu43** and **Eu44**. Finally, ligand **47** and its Eu^{3+} and Tb^{3+} complexes were synthesized and their photophysics was studied in water [101]. For both **Eu47** and **Tb47**, an increase of the lifetime values upon solvent deuteration showed the effect of vibronic coupling with OH oscillators. One water molecule was present in the first coordination sphere. Good quantum yield values were obtained for **Tb47**. Enhancement of both lifetimes and quantum yields of **Tb47** upon deaeration of the solution and the increase of the lifetime upon temperature lowering suggested that back energy transfer from the Tb^{3+} emitting state to the ligand excited states states occurs. The low quantum yields of **Eu47** have been attributed to deactivation of the carbostyril excited states via LMCT excited states.

C. Podates

The bpy unit was used as a chromophore in the podands **48–51**. Figure 12 schematically shows the podand ligands of the Eu^{3+} complexes examined. Some photophysical data and the number of coordinated solvent molecules of these complexes are gathered in Table 3.

The tripode and tetrapode ligands (**48** and **49**) were synthesized [103]. The Eu^{3+} complexes of these ligands were prepared and their photophysical properties were studied [102]. The complexes were stable in methanol and underwent decomposition in water, which was very slow for **Eu49**. The lifetimes showed that thermally activated nonradiative decay of the emitting state is negligible for both complexes and deactivation via vibronic coupling with the OH oscillators of the solvent is very efficient. The latter deactivation pathway played a more important role for **Eu49** than for **Eu48**, suggesting that the smaller ligand **48** leaves less room for solvent molecules in the first coordination sphere. It is interesting to notice that for **Eu49** in D_2O the lifetime is nearly the same as in CH_3OD while the quantum yield is one order of magnitude lower. This behavior has been attributed to a lower ligand–metal energy-transfer efficiency in water than in methanol due, most likely, to a longer distance in water between the Eu^{3+} ion and the bpy units (Section V.B). The tetrapode ligands **50** and **51** and their Eu^{3+} complexes were synthetized [97]. These complexes decomposed in water and methanol and were stable in anhydrous acetonitrile. The lifetimes of both complexes were rather long and temperature independent and the

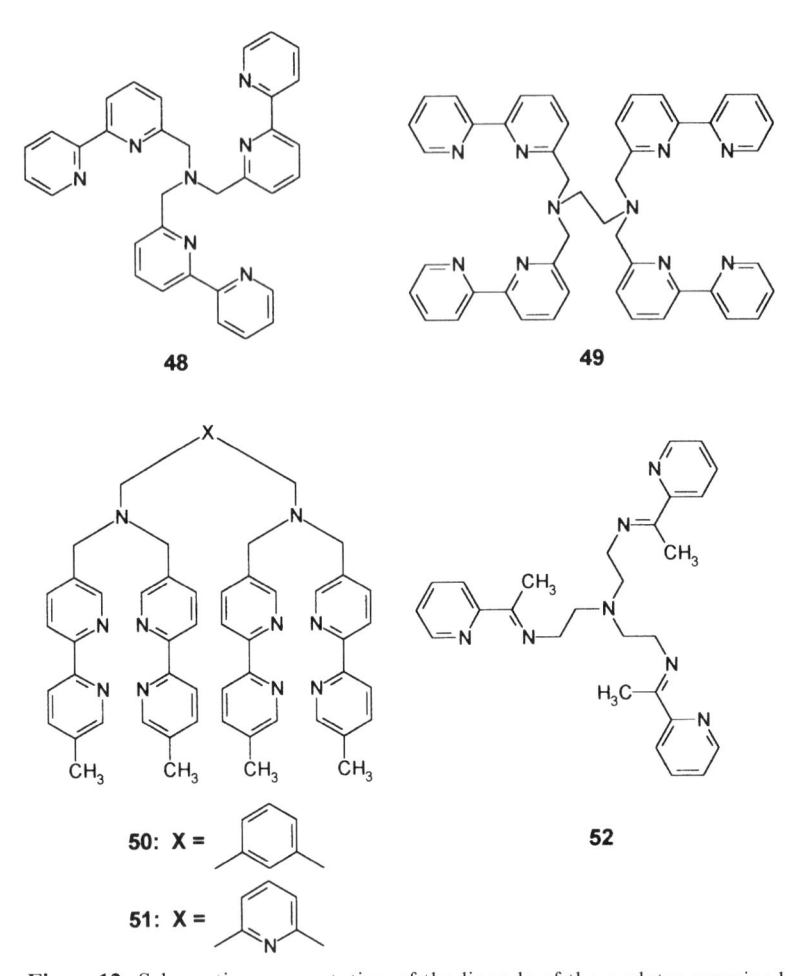

Figure 12. Schematic representation of the ligands of the podates examined.

quantum yields were good [5]. Some lanthanide complexes of ligand **52** were obtained while the free ligand could not be isolated [104]. These complexes were insoluble in the most common solvents, with the exception of dimethylsulfoxide (DMSO) and *N,N*-dimethylformamide (DMF). In DMF, weak Eu^{3+} and Tb^{3+} luminescence was observed upon ligand excitation and the lifetimes and quantum yields were given. However, since the complexes underwent partial decomposition, one has to be cautious with the values reported.

TABLE 3. Photophysical Data and Number of Solvent Molecules Coordinated to the Metal Ion for Some Eu^{3+} and Tb^{3+} Podates

Compound	Solvent	Absorption $\lambda_{max}, \varepsilon_{max}$ (nm, M^{-1} cm^{-1})	$\tau_H^{(300K)}$ (ms)	$\tau_D^{(300K)}$ (ms)	$\tau_H^{(77K)}$ (ms)	$\tau_D^{(77K)}$ (ms)	$\Phi_H^{(300K)}$	$\Phi_D^{(300K)}$	n^b	References
Eu48	CH$_3$OH	301, 32,000	0.95	1.80	0.94	2.0	0.06	0.10	1.0	[102]
Eu49	CH$_3$OH	301, 42,000	0.81	1.70	1.10	1.8	0.07	0.20	1.4	[102]
Eu50	CH$_3$CN	304, 26,500	1.2		1.12		0.04			[5]
Eu51	CH$_3$CN	308, 43,500	1.0		1.1		0.04			[5]

Emission^a

^{a}Excitation in the ligand at the λ_{max} values indicated in this table. The lifetimes are measured in correspondence with the $^5D_0 \to {}^7F_2$ and $^5D_4 \to {}^7F_5$ emissions for Eu^{3+} and Tb^{3+}, respectively.
^{b}Calculated using Eq. 7.

D. Complexes of Calixarenes

Encapsulation of the Eu^{3+} and Tb^{3+} ions by cage-like ligands was also achieved using functionalized calixarenes. Metal luminescence properties were studied mostly for complexes of calix[4]arenes, while few results are available for complexes of calix[6]arenes and calix[8]arenes, with the latter being dinuclear complexes. Figure 13 schematically shows the calixarene ligands of the Eu^{3+} and Tb^{3+} complexes examined. Some photophysical data and the number of coordinated solvent molecules of these complexes are gathered in Table 4.

First, complexes of the *p-tert*-butyl-calix[4]arene tetraacetamide ligand (**53**) were studied. Interestingly, Eu**53**, and Tb**53** were water soluble, and their photophysics was studied in water [105]. The absorption spectra of the complexes presented two bands due to ligand-centered transitions (λ_{max} = 273 and 282 nm, $\varepsilon_{max} \sim 1\,100\,M^{-1}\,cm^{-1}$). Comparison of the absorption and metal luminescence excitation spectra indicated that ligand–metal energy transfer from the calix[4]arene moiety to the metal ion takes place. About one water molecule was found to coordinate the metal ion, showing that ligand **53** efficiently shields the metal ion from solvent molecules. For Eu**53**, the lifetimes showed that the vibronic coupling and thermally activated nonradiative decay processes of the Eu^{3+} emitting state are not efficient. The very low quantum yield has been attributed to inefficient ligand–metal energy transfer due to deactivation of the ligand excited states via LMCT excited states (Section V.B). The lifetimes of Tb**53** showed no temperature dependence and a small effect of solvent deuteration. The former observation has been explained considering that the energy of the lowest ligand triplet excited state, which is about $4000\,cm^{-1}$ higher than that of the metal emitting state, renders metal–ligand back energy transfer inefficient. Interestingly, the quantum yields of Tb**53** were much higher than those of Eu**53** as a result of much more efficient ligand–metal energy transfer (Section V.B). The synthesis of ligands **54**–**57** and their complexation of Eu^{3+} and Tb^{3+} in methanol has been recently reported [111]. The Tb^{3+} complexes were strongly luminescent, while the Eu^{3+} complexes showed only weak emission. In all cases, addition of water caused quenching of the luminescence, the quenching being less efficient for complexes of ligands **56** and **57** based on calix[6]arenes. This effect has been explained considering that the larger, more flexible calix[6]arenes encapsulate the metal ion more efficiently. Some photophysical properties of the Tb^{3+} complexes of ligands **58**–**60** in water have been reported [106]. 1:1 and 2:1 (metal/ligand) stoichiometries were found for the complexes of ligands **59** and **60** and of ligand **58**, respectively. Comparison of the metal luminescence excitation spectra with the absorption spectra indicated that metal emission followed energy transfer from the phenol moiety of the ligand to the metal

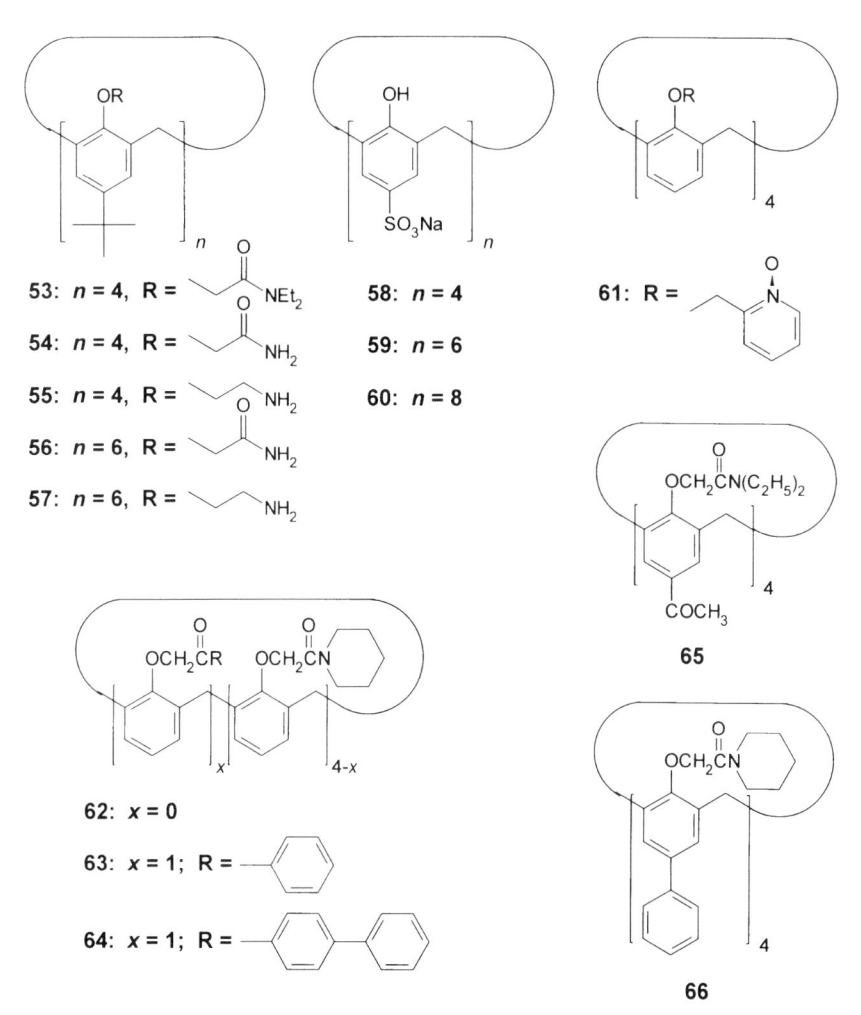

Figure 13. Schematic representation of the calixarene-based ligands of the complexes examined.

ion. The luminescence intensities were pH-dependent and reached the highest values at pH > 11. Interestingly, a high quantum yield was obtained for Tb**60**. For Tb**58** and Tb**59**, the quantum yields were determined in the presence of a ligand excess, which affects the efficiencies of light absorption by the complex and, therefore, the quantum yield values.

Because calixarenes have rather low molar absorption coefficients, more efficient chromophores were attached to calixarene moieties in order to obtain strongly absorbing ligands for Eu^{3+} and Tb^{3+} complexation. Complexes of ligand **61**, which contained four pyridine–N-oxide

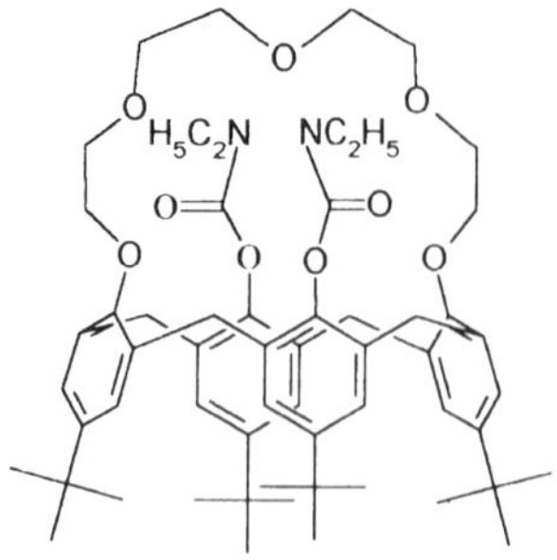

75

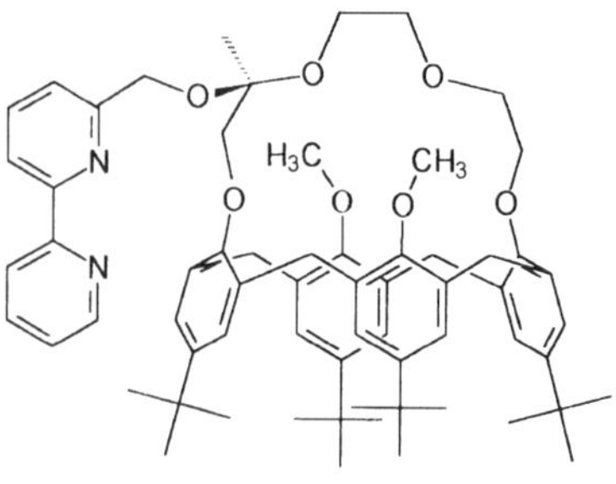

78

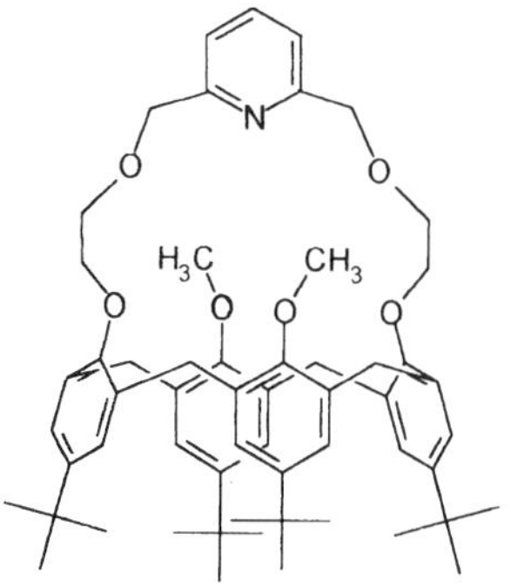

76

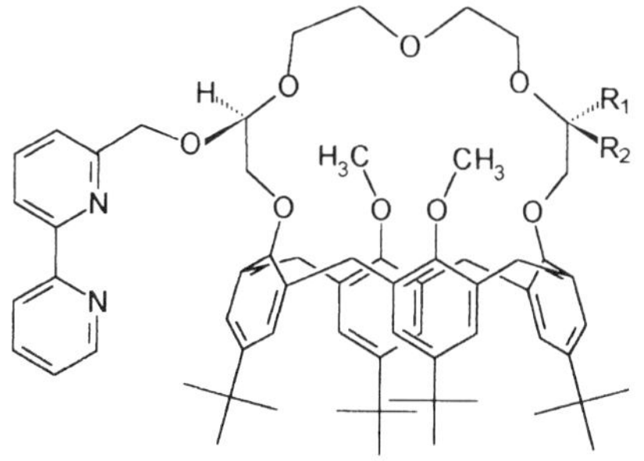

79: $R_1 = R_2 = H$

80: $R_1 = H$; $R_2 =$ 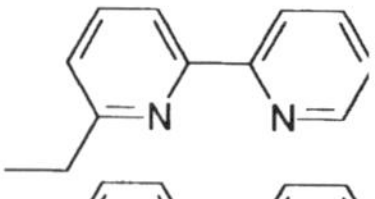

81: $R_2 = H$; $R_1 =$ 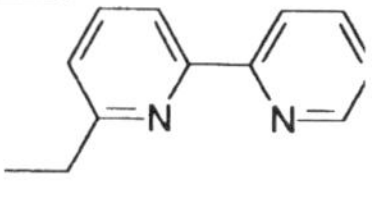

Figure 13. (*Continued*)

units as chromophores, showed metal luminescence upon ligand excitation in methanol [112]. The luminescence was completely quenched upon addition of water, suggesting the lability of the complexes in this solvent. The quantum yields in methanol were of the order of 10^{-3} for both Eu**61** and Tb**61**. The Eu^{3+} and Tb^{3+} complexes of ligands **62**–**66** containing

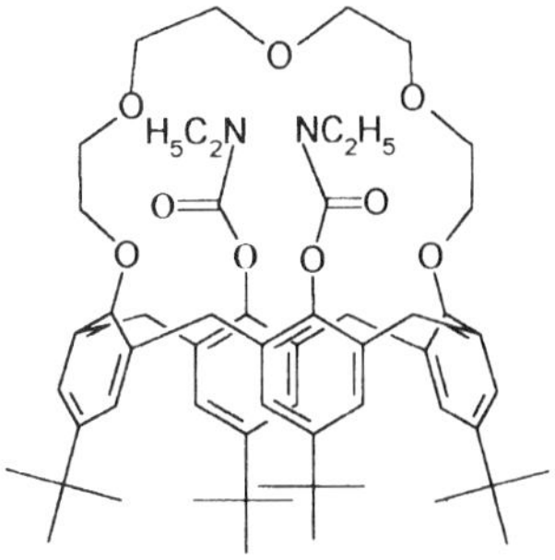

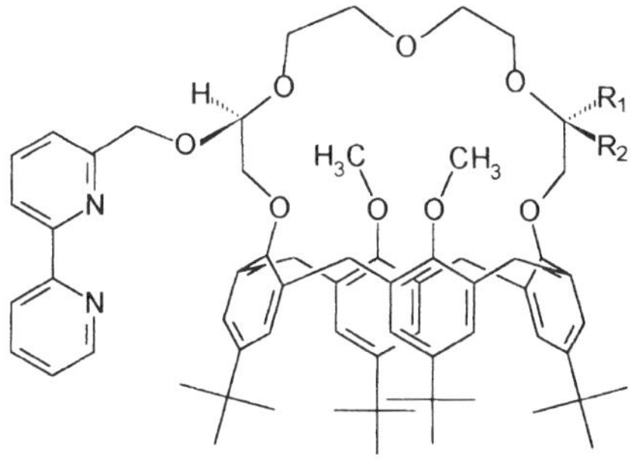

79: $R_1 = R_2 = H$

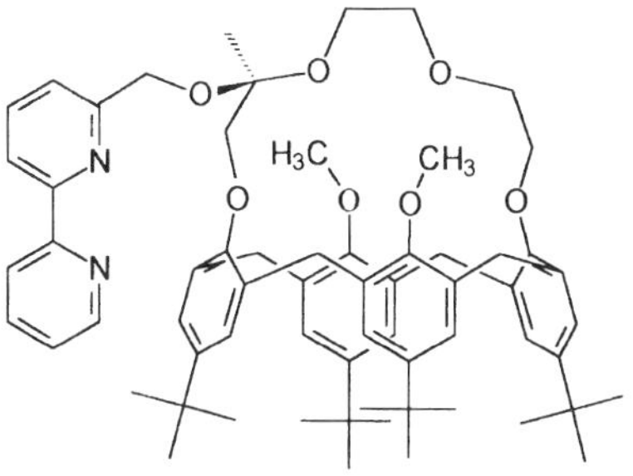

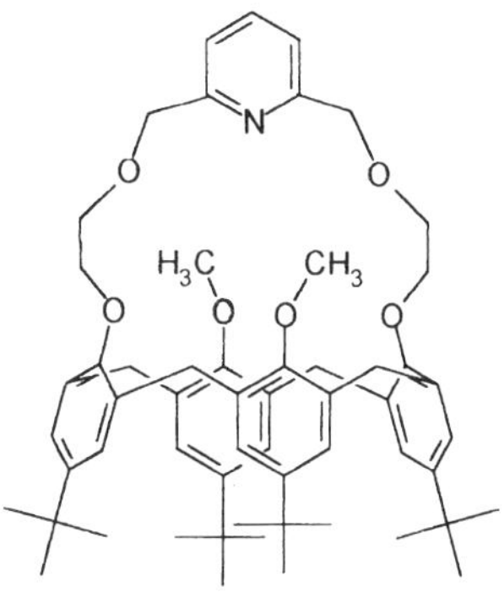

80: $R_1 = H$; $R_2 =$

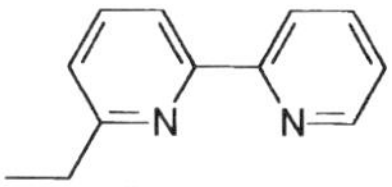

81: $R_2 = H$; $R_1 =$

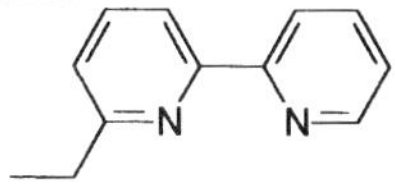

Figure 13. (*Continued*)

phenyl and diphenyl groups as chromophores were prepared in order to investigate the correlation between the metal luminescence intensity and the nature of the ligand and the chromophore [113, 114]. For all complexes, 1:1 stoichiometries were found in both acetonitrile and methanol. Metal luminescence upon ligand excitation was observed for almost all the complexes studied. High quantum yields (e.g., 0.27 for Tb**63**) and rather long lifetimes have been reported for some of the complexes in acetonitrile.

TABLE 4. Photophysical Data for Some Eu^{3+} and Tb^{3+} Complexes of Functionalized Calixarenes

Compound	Solvent	Absorption λ_{max}, ε_{max} (nm, M^{-1} cm^{-1})	Emission[a] $\tau_H^{(300K)}$ (ms)	$\tau_H^{(77K)}$ (ms)	$\Phi_H^{(300K)}$	References
Eu53	H$_2$O	273, 1,100	0.65[b]	1.8[b]	0.0002	[105]
Tb53	H$_2$O	273, 1,100	1.5[c]	1.6[c]	0.20	[105]
Tb60	H$_2$O	~290, ~25,000			0.20[d]	[106]
Eu67	CH$_3$CN	305, 28,000	0.65		0.04	[105]
Tb67	CH$_3$CN	306, 29,000	1.9		0.12	[107]
Eu68	CH$_3$CN	307, 50,000	0.65		<0.01	[107]
Tb68	CH$_3$CN	305, 46,700	1.4		0.02	[107]
Eu70	CH$_3$CN	309, 22,500	1.2		0.03	[108]
Tb70	CH$_3$CN	301, 23,400	0.52	1.03	0.001	[108]
Eu71	CH$_3$CN	306, 26,700	1.26		0.14	[108]
Tb71	CH$_3$CN	305, 30,000	0.81	0.98	0.0013	[108]
Eu72	CH$_3$CN	305, 39,600	1.6		0.16	[108]
Tb72	CH$_3$CN	302, 42,800	0.57	1.18	0.007	[108]
Eu73	CH$_3$OH		0.29			[109]
Tb73	CH$_3$OH		0.65			[109]

Eu74	CH$_3$OH		0.50	[109]
Eu75	CH$_3$CN	282,2200	e	[110]
Tb75	CH$_3$CN	282,2000	3,30 0.15	[110]
Eu76	CH$_3$CN	281,2100	0.65 0.0002	[110]
Tb76	CH$_3$CN	281,2100	1.85 0.01	[110]
Eu77	CH$_3$CN	276	0.50 0.0005	[110]
Tb77	CH$_3$CN	276	1.00 0.01	[110]
Eu78	CH$_3$CN	305,13200	0.95 0.18	[117]
Tb78	CH$_3$CN	305,13000	1.85 0.32	[117]
Eu79	CH$_3$CN	305,14200	1.38 0.32	[117]
Tb79	CH$_3$CN	305,13000	1.88 0.35	[117]
Eu80	CH$_3$CN	305,22700	1.24 0.23	[117]
Tb80	CH$_3$CN	304,23400	1.86 0.39	[117]
Eu81	CH$_3$CN	305,21000	1.29 0.28	[117]
Tb81	CH$_3$CN	305,22700	1.83 0.37	[117]

[a]Excitation in the ligand at the λ_{max} values indicated in this table, unless otherwise noted. The lifetimes are measured in correspondence with the $^5D_0 \rightarrow {}^7F_2$ and $^5D_4 \rightarrow {}^7F_5$ emissions for Eu^{3+} and Tb^{3+}, respectively.
[b]The values 1.9 ms and 2.0 ms are obtained in D$_2$O at 300 K and 77 K, respectively.
[c]The values 2.6 ms and 2.8 ms are obtained in D$_2$O at 300 K and 77 K, respectively.
[d]Excitation at 255 nm.
[e]No metal luminescence was observed at room temperature.

However, it must be noticed that the method [11] used to obtain the quantum yields foresees that the quantum yield of the standard is considered at the excitation wavelength also used for the sample. Since this is not the case for the quantum yields reported, one has to be cautious with these values. Several calix[4]arenes carrying the bpy chromophore were synthesized. Ligands **67** and **68**, which contained two bpy units and two amides, and four bpy units, respectively, in which the bpy units are ligated to the calixarene moiety through C6 were obtained [107, 108]. These ligands gave 1:1 complexes with Eu^{3+} and Tb^{3+} in acetonitrile with stability constants on the order of $10^5 M^{-1}$. The photophysical properties of these complexes were studied in solutions containing the ligand and an excess of the Eu^{3+} or Tb^{3+} salts [107, 108]. The absorption spectra of the ligands and the complexes were characterized by intense bands in the UV due to bpy absorption. Comparison of the absorption and the metal luminescence excitation spectra indicated that energy transfer from the bpy moiety to the metal ion takes place. Good quantum yields were obtained for the complexes of ligand **67**. The lower quantum yields of the complexes of ligand **68** were attributed to a less efficient ligand–metal energy transfer due to the steric hindrance that the four bpy units may undergo when approaching the metal ion (Section V.B). No metal luminescence was found upon ligand excitation in acetonitrile solutions containing the Eu^{3+} or Tb^{3+} ions and ligand **69** [107]. Ligands **70–72** in which the bpy units are ligated to the calixarene moiety through C5 were synthesized and their Eu^{3+} and Tb^{3+} complexes were isolated [115]. The photophysical properties were studied in anhydrous acetonitrile [108]. For all complexes, metal luminescence was observed upon bpy excitation. The Eu^{3+} complexes showed longer lifetimes and higher quantum yields than the Tb^{3+} complexes. Interestingly, Eu**71** and Eu**72** gave particularly high quantum yields.

A series of calix[4]arene ligands was synthesized upon attachment of one naphthalene, phenanthrene, or triphenylene chromophore to the calix[4]arene [109]. For the Eu^{3+} and Tb^{3+} complexes of these ligands, metal luminescence was observed upon excitation of the chromophore [109]. The Eu^{3+} and Tb^{3+} of ligands **73** and **74** complexes were characterized by low lifetimes in methanol. Deaeration of the solutions did not change the lifetime values of the complexes but led to an increase of the intensity of the metal luminescence. According to Steemers et al. [109], this behavior indicates that ligand–metal energy transfer is slow, so that the ligand excited states are quenched by oxygen instead of contributing to the population of the metal luminescent states [109].

Recently, a new type of calixarene ligands, the calixcrowns, were examined for complexation of lanthanide ions. The calixcrowns **75** and **76** were synthesized and ligand **77** was obtained as a secondary product in the synthesis of **76** [110]. Complexation of Eu^{3+} and Tb^{3+} was obtained in acetonitrile and the complexes were characterized by a 1:1 stoichiometry [110]. The Tb**75** complex showed a long lifetime and a quite high quantum yield. Considering the energy of the lowest ligand triplet excited state (26,700 cm^{-1}), it has been suggested that thermally activated nonradiative decay is not very efficient. Complex Eu**75** was not luminescent at 300 or at 77 K. This behavior was attributed to efficient nonradiative decay via LMCT excited states involving the ether oxygen atoms. The quantum yields of Tb**76** and Tb**77** were one order of magnitude lower than that of Tb**75**. Considering that the lowest ligand triplet excited state lies at 24,000 cm^{-1}, it has been suggested that thermally activated nonradiative decay is most likely inefficient and the lower quantum yields may be due to a little efficient ligand–metal energy transfer. Complex Eu**76** and Eu**77** gave very low quantum yields. This result has been attributed to nonradiative deactivations of the ligand and metal excited states via low-lying LMCT excited states involving the ether oxygen atoms. In order to increase the molar absorption coefficients of the calixcrowns, ligands **78**–**81** were synthesized incorporating the bpy chromophore [116]. The Eu^{3+} and Tb^{3+} complexes of these ligands formed in acetonitrile and a 1:1 stoichiometry was established [117]. The absorption spectra of the ligands and the complexes were characterized by intense bands due to bpy absorption. Metal luminescence was observed upon bpy excitation for all complexes. The Tb^{3+} complexes of the ligands **78**–**81** were characterized by rather long lifetimes and high quantum yields. These values suggested that thermally activated nonradiative decay via ligand triplet excited states is not very efficient. This hypothesis is confirmed by the energy of the lowest ligand triplet excited states ranging from 22,500 to 24,000 cm^{-1}. For the Eu^{3+} complexes of the ligands **78**–**81**, quite long lifetimes and high quantum yields were measured. The lower quantum yield of Eu**81** was ascribed to the smaller crown rendering the ion more exposed to water molecules present in the Eu^{3+} salt used for titration or in the solvent.

V. DISCUSSION OF SOME PHOTOPHYSICAL PROPERTIES OF THE COMPLEXES OVERVIEWED

In the following discussion, we analyze some factors responsible for the Eu^{3+} and Tb^{3+} luminescence intensity, that is the efficiencies of the ligand absorption, of the ligand–metal energy transfer, and of the metal luminescence. Furthermore, we compare the intensities of the metal luminescence estimated using the molar absorption coefficients and the quantum yields. Only the complexes for which the necessary experimental data are available will be treated.

A. Ligand Absorption

The absorption properties of the complexes (Table 5) and, for the sake of comparison, of the free chromophores incorporated in the ligands are discussed. We focus on the absorption bands at lowest energy because they are important for applications (Section III). In all cases, where the ligands contain only the bpy chromophore, complexation gives rise to a red shift of the bpy absorption (for free bpy, $\lambda_{max} = 281$ nm in water). The observed differences may be related to the amount of ligand–metal interaction in the complex. The effect is similar for complexes of the cryptands and podands and the branched macrocyclic ligands 28 and 38, while a more pronounced shift is observed for complexes of the branched-macrocyclic ligands 27 and 37. This effect may be explained considering that in the complexes of the last two ligands the approach between the chromophore and the metal ion may be closer than in the other complexes, as illustrated by the space-filling models. This hypothesis is supported by the kinetic behavior in solution (Section IV.B). A decrease of the molar absorption coefficients of bpy is usually observed upon complexation of the metal ion. The ε_{max} values obtained span the range of 7000–11,000 M^{-1} cm^{-1} for one bpy, while for the free bpy $\varepsilon_{max} = 13,000$ M^{-1} cm^{-1}. The ε_{max} decrease and the λ shift show the same trend, as expected considering that both factors reflect the amount of metal coordination by the ligand. In the complexes of ligand 31, only a small red shift of the phen absorption is observed (for free phen, $\lambda_{max} = 270$ nm in water). The small shift may be attributed to scarce metal coordination due to steric hindrance induced by the presence of the methyl substituents in the phen moiety. This hypothesis is supported by the luminescence data (Section IV.B). A decrease of the molar absorption coefficients upon metal complexation is observed in correspondence with the phen absorption. This decrease cannot be unequivocally attributed to complexation of the phen moiety because in this spectral region the bpy absorption is also present, which is presumably more red shifted than the phen absorption. The introduction of N-oxides in the bpy units gives rise to substantial changes in the absorption spectra of the complexes with respect to those of the analogous complexes containing only bpy units. For the complexes of ligands 14, 15, 29, and 30, the intensities of the absorption maxima at about 305 nm decrease. This result is due to the low absorption intensity of the 6,6'-dimethyl-bpy-N,N'-dioxide chromophore at 305 nm.

TABLE 5. Absorption Data and Ligand-to-Metal Energy-Transfer Efficiencies[a]

Compound	λ_{max} (nm)	ε_{max} ($M^{-1} cm^{-1}$)	$\eta_{en.tr.}$[b]
Eu2	303	28,000	0.10
Eu7	305	26,000	0.08
Eu14	304	20,000	0.40[c]
Eu15	306	17,000	0.25[c]
Eu16	304	20,000	0.55[c]
Eu27	312	36,000	0.20
Eu28[d]	299	39,000	0.02
Eu29	310	15,500	0.05[c]
Eu30	286	34,000	0.10[c]
Eu31	278	46,000	0.09
Eu32	307	15,100	0.27
Eu37	311	24,500	0.14[c,e]
Eu38[f]	304	30,500	[g]
Eu40[d]	260	38,000	0.08
Eu41	260	50,400	0.08
Eu48[d]	301	32,000	0.12
Eu49[d]	301	42,000	0.20
Eu53	273	1,100	0.0006
Tb2	304	29,000	0.34[h]
Tb27[d]	312	44,000	0.80[h]
Tb31	278	43,000	0.30[h]
Tb32	306	15,300	0.45[h]
Tb37	311	20,400	0.40[e]
Tb38[f]	303	31,200	[g]
Tb53	273	1,100	0.35

[a]In water solution, unless otherwise noted. λ_{max} and ε_{max} values are those reported in Tables 1–4.
[b]Obtained from Eqs. (2)–(4), using the lifetime and quantum yield values reported in Tables 1–4.
[c]Calculated on the assumption that the thermally activated decay process does not involve any equilibrium between the metal emitting state and other excited states (Section III).
[d]In methanol solution.
[e]A higher value is expected considering that the radiative lifetime may be longer than $\tau_D^{(77K)}$ introduced in Eq. (4) if vibronic coupling involving CH bonds occurs (Section V.B).
[f]In acetonitrile solution.
[g]The value cannot be calculated because lifetime values at 77 K are not reliable.
[h]The value reported is a lower limiting value because of the equilibrium between the emitting state of the Tb^{3+} ion and the triplet excited state of the ligand (Section III).

Note that the latter chromophore shows an absorption band at higher energy [$\lambda_{max} = 260$ nm [97]] compared to the bpy chromophore. The absorption spectra of the Eu^{3+} and Tb^{3+} complexes of ligands **40** and **41** have maxima at 260 nm, as the free chromophore, with very high molar absorption coefficients [97].

For the Eu^{3+} and Tb^{3+} complexes of the calix[4]arene-based ligands **53** and **75**, their absorption spectra are characterized by low molar absorption coefficients, $\varepsilon_{max} \sim 1100\, M^{-1}\, cm^{-1}$ at $\lambda_{max} = 270$ nm and $\varepsilon_{max} \sim 2000\, M^{-1}\, cm^{-1}$ at $\lambda_{max} = 282$ nm. Also, the complexes of ligand **76** containing a pyridine unit have low molar absorption coefficients. Importantly, complexes of ligands **67**, **68**, **70**–**72**, and **78**–**81** obtained upon attachment of chromophores to the calix[4]arene moiety have much more intense absorption with ε_{max} values in the range of $20{,}000$–$50{,}000\, M^{-1}\, cm^{-1}$.

B. Ligand–Metal Energy-Transfer Efficiency

The efficiencies of energy transfer from the singlet excited state of the ligand populated upon absorption to the emitting state of the metal ion ($\eta_{en.tr.}$) obtained from Eqs. 2–4 (Section II) are presented in Table 5.

The Tb^{3+} complexes of ligands **2**, **27**, **31**, and **32** are characterized by high values of $\eta_{en.tr.}$. It must be pointed out that these values are lower limiting values since the equilibrium between the lowest ligand triplet excited state and the Tb^{3+} $^5D_4$ emitting state plays an important role in the deactivation of the metal emitting state (Section II). Indeed, one has to be cautious when comparing the $\eta_{en.tr.}$ values for these complexes. For example, the value of Tb**27**, which is higher than the other ones, is most likely related to the lower value of $k_{nr}(T)$ and thus to an equilibrium playing a less important role. Therefore, the lower $\eta_{en.tr.}$ values corresponding to higher $k_{nr}(T)$ values are not necessarily connected with a lower energy-transfer efficiency but could reflect a more important role of the equilibrium in the deactivation of the metal emitting state. For the Tb**2** complex, the value of $\eta_{triplet \to metal}$ has been obtained experimentally [54]. This value is about 1, showing that in this complex the energy transfer to the metal overcomes the other deactivation processes of the lowest ligand triplet excited state. A value close to unity is also obtained for $\eta_{en.tr.}$ by using Eq. 5 if $\eta_{singlet \to triplet}$ is substituted by the η_{isc} value of the free bpy (~ 1). This substitution appears to be justified if one considers that the Tb^{3+} ion should not introduce low-lying excited states deactivating the bpy singlet excited state and the heavy metal ion could only increase the intersystem crossing efficiency. Noticeably, $\eta_{en.tr.}$ equal to unit is three times higher than the value calculated using Eqs. 2–4. This result shows how the latter is affected by the presence of the equilibrium. For the Tb**37** complex whose lifetime is not temperature dependent,

$\eta_{\text{en.tr.}}$ is lower than for the other Tb^{3+} complexes. In this case, however, the lifetime in D_2O at 77 K used in Eq. 4 may be shorter than the radiative lifetime because of a vibronic coupling contribution due to the CH oscillators [78]. If the value commonly obtained for the radiative lifetime of Tb^{3+} complexes is used, $\eta_{\text{en.tr.}}$ and $\eta_{\text{triplet}\rightarrow\text{metal}}$ increase by a factor of 2. For Tb53, $\eta_{\text{en.tr.}}$ is 0.35. Since η_{isc} of the free ligand is not known, by using Eq. 5 we can only estimate that $\eta_{\text{triplet}\rightarrow\text{metal}} \geqslant 0.35$.

Let us now consider the Eu^{3+} complexes. First of all, it is noticeable that the energy-transfer efficiency of Eu2 is lower than that of Tb2. Such a behavior has been explained by possible deactivation of the ligand singlet and triplet excited states via low-lying LMCT excited states in Eu2. Analogous considerations may be valid for the other Eu^{3+} complexes. Complexes containing only bpy and phen units show similar efficiencies with the exception of Eu28 for which the much lower value may be ascribed to scarce metal–ligand interaction due to steric hindrance of the methyl groups. For the cryptates incorporating N-oxides $\eta_{\text{en.tr.}}$ is considerably higher. Note that this quantity is never unity, suggesting that LMCT excited states still play a role. However, this may be less important because of a more efficient population of the lowest ligand triplet excited state due to the presence of $bpyO_2$- and $biqO_2$-localized levels or because a closer approach between the ligand and the metal ion could give rise to a higher value of $\eta_{\text{triplet}\rightarrow\text{metal}}$. Analogous to the cryptates, introduction of the $bpyO_2$ chromophore in Eu29 and Eu30 gives rise to an increase of $\eta_{\text{en.tr.}}$ compared to the parent Eu28. Nevertheless, for these three complexes, the $\eta_{\text{en.tr.}}$ values are rather low, most likely because of steric hindrance of the methyl groups in the branches. The exceedingly low value of $\eta_{\text{en.tr.}}$ for Eu53 suggests a very efficient deactivation of the ligand excited states via LMCT excited states considering that $\eta_{\text{en.tr.}}$ of the analogous complex Tb53 is about three orders of magnitude higher.

C. Metal Luminescence Efficiency

In order to discuss the efficiency of the metal luminescence in the complex, the radiative and nonradiative decay rate constants of the Eu^{3+} $^5D_0$ and Tb^{3+} $^5D_4$ emitting states (Table 6) have been calculated using Eqs. 10–12.

The radiative rate constants are higher for the Eu^{3+} than for the Tb^{3+} complexes. This observation is in line with what is usually observed, because of the mixing of the $^5D_0$ emitting state with LMCT excited states in Eu^{3+} complexes and the much less efficient mixing of the $^5D_4$ emitting state with f–d excited states in Tb^{3+} complexes [16]. The k_r values differ from one another more for Eu^{3+} than for Tb^{3+} complexes, with the exception of Tb37, which is in agreement with the larger effect of the ligand on the energy

TABLE 6. Decay Rate Constants of the Metal Luminescent States and Number of Solvent Molecules Coordinated to the Metal Ion[a]

Compound	k_r[b] (s^{-1})	$k_{nr}(T)$[c] (s^{-1})	$k_{nr}(OH)$[d] (s^{-1})	n[e]
Eu2	590	<50	2,400	2.5
Eu7	670	<50	2,200	2.3
Eu14	770	100[f]	1,300	1.4
Eu15	900	500[f]	1,100	1.1
Eu16	900	600[f]	1,000	1.1
Eu27	530	<50	140	<0.5
Eu28[g]	500	50	450	0.9
Eu29	700	150[f]	1,200	1.2
Eu30	770	6,900[f]	[h]	[h]
Eu31	530	<50	3,200	3.3
Eu32	500	<50	1,200	1.3
Eu37	900[i]	850[f]	250	<0.5
Eu40[g]	1,000	315	600	1.1
Eu41[g]	1,060	325	400	1.1
Eu48[g]	500	50	500	1.0
Eu49[g]	550	<50	650	1.4
Eu53	500	<50	1,000	1.1
Tb2	260	2,100[j]	700	3.0
Tb27[g]	350	500[f]	80	0.6
Tb31	390	16,000[f]	630[k]	2.7
Tb32	420	1,600[f]	330	1.4
Tb37	670[i]	<50	<50	<0.5
Tb53	360	<50	280	1.2

[a] In water solution, unless otherwise noted. The lifetime values used to calculate the k and n values are those reported in Tables 1–4.
[b] Obtained from Eq. (10).
[c] Obtained from Eq. (11), regardless of the presence of equilibria (Section III). Cases in which the equilibrium may be present are indicated.
[d] Obtained from Eq. (12).
[e] Obtained from Eqs. (6) and (7).
[f] It cannot be excluded that the excited state involved in the thermally activated nonradiative decay is in equilibrium with the metal emitting state.
[g] In methanol solution.
[h] The lifetimes in H_2O and D_2O are equal (Table 2). However, the effects of the vibronic coupling with the OH oscillators are, most likely, masked by the very fast thermally activated nonradiative decay.
[i] A lower value is expected considering that vibronic coupling involving CH oscillators may occur (Section V.C).
[j] From kinetic data in deaerated solution a value of $1.6 \times 10^5 \, s^{-1}$ was obtained for the rate constant of the thermally activated back energy transfer from the $^5D_4$ Tb^{3+} emitting state to the lowest triplet excited state of the ligand (Section IV.A).
[k] In deaerated solution. No deuteration effect on the lifetimes is observed in aerated solution.

of the LMCT excited states in Eu^{3+} complexes than on the energy of the $f-d$ states in Tb^{3+} complexes. In particular, the Eu^{3+} complexes containing bpyO$_2$, biqO$_2$, and electron-withdrawing substituents in the ligand show higher k_r values than the analogous complexes containing only bpy units as was expected on the basis of the lower energies of their LMCT excited states (Section IV.A). The values of k_r for Eu**37** and **Tb37** may be overestimated because the lifetimes at 77 K could be shortened by the contribution of vibronic coupling involving the CH oscillators in the ligand [78].

As to the rate constant of the thermally activated nonradiative decay, it is worthwhile to recall that in the presence of equilibria between the emitting states and other excited states its value cannot be obtained from Eq. 11. Nevertheless, we think that for a homogeneous series of complexes the values obtained from this equation may be used to discuss the relative importance of the thermally activated nonradiative decay. Following this approach, the values obtained from Eq. 11 are presented in Table 6 regardless of whether the equilibrium is present or not. The cases in which the equilibrium may be present are indicated. Among the Eu^{3+} complexes, the $k_{nr}(T)$ term is negligible for complexes containing only bpy or phen units with the exception of Eu**37** and becomes important in complexes containing N-oxides. An explanation proposed for such a behavior relies on the deactivation of the Eu^{3+} emitting state via low-lying LMCT excited states, which might involve an equilibrium. For Eu**37**, low-lying LMCT excited states most likely involve the aliphatic nitrogen atoms of the macrocycle. For the complexes of ligands containing N-oxides, the LMCT excited states are expected to lie at lower energies than in complexes containing only bpy units because the reduced donor capability of the N-oxide containing moiety leaves the europium ion more positive. An analogous hypothesis may account for the increase of $k_{nr}(T)$ caused by the presence in the ligand of the electron-withdrawing carboxymethyl substituent. In the case of the Tb^{3+} complexes, the thermally activated nonradiative decay, which consists of the deactivation of the metal emitting state via the lowest triplet excited state of the ligand, is important for all complexes containing bpy or phen units with the exception of Tb**37** and **Tb78–Tb81**, where the lowest triplet excited state of the ligand lies at higher energy than in the other Tb^{3+} complexes (Section IV.B). For complexes of the calixarene ligand (**53**), $k_{nr}(T)$ is negligible showing that this ligand does not introduce excited states that can be thermally populated from the metal emitting states.

Finally, we discuss the importance of the nonradiative decay via vibronic coupling with the OH oscillators of the solvent, by considering the number of solvent (water or methanol) molecules coordinated to the metal ion obtained from the decay rate constants (Table 6). First, it is interesting to notice that the numbers of coordinated water molecules, varying from about

0.5 to about 3, indicate efficient shielding of the metal ion by the encapsulating ligands. The Eu^{3+} and Tb^{3+} cryptates containing only bpy units have about 2.5 water molecules in the first coordination sphere of the metal ion. The presence of N-oxides in the bpy cryptands causes a decrease of the number of coordinated water molecules, suggesting that steric hindrance of the N-oxide groups and the charge of the oxygen atoms may hamper water coordination. The branched macrocyclic ligands incorporating the macrocycle **26** shield the metal ion very efficiently, which may be attributed to the presence of four coordinating units and the flexible structure of these ligands. Steric hindrance due to methyl groups in the branches leads to less efficient shielding. In fact, in the complexes of ligand **31** about three water molecules are coordinated to the metal ion and Eu**28** decomposes in water. As to the branched-macrocyclic ligands incorporating the triazacyclononane ring, their shielding ability toward water molecules strongly depends on the

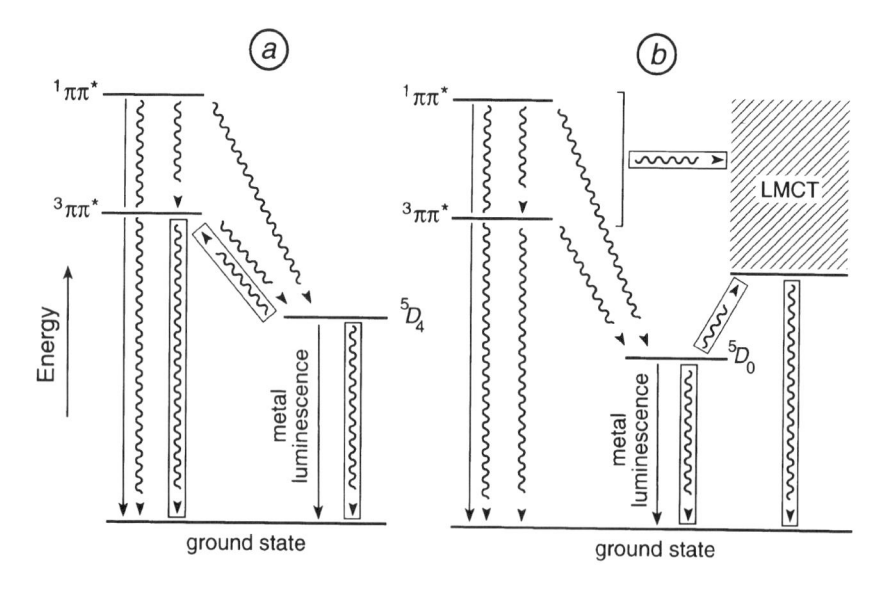

Figure 14. Scheme of the deactivation processes involved in the conversion of the light absorbed by the ligand into light emitted by the metal ion in (*a*) Tb^{3+} and (*b*) Eu^{3+} complexes. The $^1\pi\pi^*$ state is the ligand excited state populated upon excitation and the $^3\pi\pi^*$ state is the ligand excited state populated upon intersystem crossing from the $^1\pi\pi^*$ state. The $^5D_4$ and $^5D_0$ excited states are the luminescent states for Tb^{3+} and Eu^{3+}, respectively. The LMCT excited states are the ligand metal charge transfer states. The pathways responsible for the most important nonradiative losses are boxed.

position of the attachment of the bpy to the ring. The different attachment positions may give rise to different structures of the complexes, as was suggested by the CPK models. As a matter of fact, in the Eu^{3+} and Tb^{3+} complexes of ligand **37** no water molecule is present in the first coordination sphere of the metal ion while the analogous complexes of ligand **38** decompose in water. The podands **48–51** are so inefficient in competing with water molecules for lanthanide complexation that their complexes decompose in water. Ligand **53** shields the metal ion from water molecules relatively well but its complexes are rather unstable in water.

The scheme presented in Figure 14 summarizes the process involved in the conversion of light absorbed by the ligands into light emitted by the metal ions for the complexes discussed. The pathways responsible for the most important nonradiative losses are boxed.

D. Metal Luminescence Intensity

In this section, we discuss the intensity of the metal luminescence obtained upon ligand excitation on the basis of the product of the metal luminescence quantum yield upon excitation at a certain wavelength and the molar absorption coefficient of the ligand in the complex at the same wavelength (Table 7). We would like to recall that the metal luminescence intensity is the photophysical property of main interest in this research dealing with the antenna effect in Eu^{3+} and Tb^{3+} complexes. Furthermore, this quantity is determining for some applications of these compounds (Section III).

First, complexes containing only the bpy chromophore will be considered. The prototypes **Eu2** and **Tb2** showed strong absorption but gave

TABLE 7. Metal Luminescence Intensities[a]

Compound	ε_{max} (M^{-1} cm^{-1})	Φ	$\varepsilon_{max} \times \Phi$
Eu2	28,000	0.02	560
Eu7	26,000	0.02	520
Eu14	20,000	0.15	3,000
Eu15	17,000	0.09	1,500
Eu16	20,000	0.20	4,000
Eu27	36,000	0.10	3,600
Eu28	39,000	0.01	390
Eu29	15,500	0.015	230
Eu30	34,000	0.01	340
Eu31	46,000	0.013	600

TABLE 7. (*Continued*)

Compound	ε_{max} (M^{-1}cm^{-1})	Φ	$\varepsilon_{max} \times \Phi$
Eu32	15,100	0.08	1,200
Eu37	24,500	0.05	1,200
Eu38	30,500	0.36	11,000
Eu40	38,000	0.05	1,900
Eu41	50,400	0.05	2,500
Eu47	49,000	0.02	980
Eu48	32,000	0.06	1,900
Eu49	42,000	0.07	2,900
Eu50	26,500	0.04	1,100
Eu51	43,500	0.04	1,700
Eu53	1,100	0.0002	0.2
Eu67	28,000	0.04	1,100
Eu68	50,000	<0.01	<500
Eu70	22,500	0.03	680
Eu71	26,700	0.14	3,700
Eu72	36,900	0.16	5,900
Eu76	2,100	0.0002	0.5
Eu79	13,200	0.18	2,400
Eu79	14,200	0.32	4,500
Eu80	22,700	0.23	5,200
Eu81	21,000	0.28	5,600
Tb2	29,000	0.03	870
Tb27	44,000	0.14	6,200
Tb31	43,000	0.007	300
Tb32	15,300	0.08	1,200
Tb37	20,400	0.37	7,500
Tb38	31,200	0.08	2,500
Tb47	60,400	0.015	900
Tb53	1,100	0.20	220
Tb67	29,000	0.12	3,500
Tb68	46,700	0.02	930
Tb70	23,400	0.001	20
Tb71	30,000	0.0013	40
Tb72	42,800	0.007	300
Tb75	2,000	0.15	300
Tb76	2,100	0.01	20
Tb78	13,000	0.32	4,200
Tb79	13,000	0.35	4,600
Tb80	23,400	0.39	9,100
Tb81	22,700	0.37	8,400

[a]The ε_{max} and Φ values (in hydrogenated solvents) are those reported in Tables 1–4.

rather low quantum yields so that high values were not obtained for the luminescence intensity. This quantity was not affected by the presence of substituents, as found in Eu7 and Eu8. The Eu^{3+} and Tb^{3+} complexes of the branched-macrocyclic ligand 27 containing one bpy more than the cryptand 2 gave much more intense luminescence thanks to the higher molar absorption coefficients and the good quantum yields. As to the analogous complex Eu28, the luminescence intensity is one order of magnitude lower than that of Eu27. This result is accounted for by the lower quantum yield caused, most likely, by worse metal-ligand interaction due to the methyl groups. The complexes Eu31 and Tb31 gave low luminescence intensities in spite of the high molar absorption coefficients because of the low quantum yields. Concerning Eu32 and Tb32, the presence of only two chromophores results in relatively low molar absorption coefficients. The quantum yields are good so that the luminescence intensities are still rather high. Substitution of the weakly absorbing phenyl units in the phosphinate groups by efficient chromophores is expected to give rise to more intense luminescence if absorption in these chromophores is followed by efficient ligand-metal energy transfer [93]. The Eu^{3+} and Tb^{3+} complexes of the branched macrocyclic ligands 37 and 38, where the triazacyclononane ring carries three bpy units have molar absorption coefficients similar to those of the Eu2 and Tb2 prototypes. Most interestingly, Eu38 and Tb37 gave some of the highest luminescence intensities among the complexes examined thanks to their high quantum yields. It is worthwhile pointing out that for these complexes the LMCT and ligand triplet excited states most likely lie at energy levels sufficiently high so as not to interfere with the ligand-metal energy transfers nor to deactivate the metal emitting states. The Tb38 and Eu37 complexes show less intense luminescence because they have much lower quantum yields. The luminescence intensities of the binuclear Eu^{3+} and Tb^{3+} complexes of ligand 42 are rather low in spite of the high molar absorption coefficients resulting from the presence of six bpy units in the ligand, because of the low quantum yields. The Eu^{3+} and Tb^{3+} complexes of the podands 48–51 show rather intense luminescence because the molar absorption coefficients are high and the quantum yields relatively good.

Complexes of some ligands analogous to those mentioned above but containing one to three $bpyO_2$ and one $biqO_2$ chromophores instead of the bpy chromophore will be discussed. The Eu^{3+} complexes of the cryptands 14 and 15 containing two bpy units and one $bpyO_2$, and 16 containing two bpy units and one $biqO_2$ chromophore gave rather good molar absorption coefficients as well as a high quantum yield, so that good luminescence intensities were obtained. In comparison with the Eu2 prototype, these values are significantly higher. As discussed previously, this different behavior is mainly due to a more efficient ligand-metal energy transfer. The gain of luminescence intensity upon introduction of N-oxides was not observed for Eu**29** and Eu**30**, the values being similar to that of Eu**28**. Most likely, for these three complexes bad interaction between the metal ion and the branches containing methyl groups is responsible for the low quantum yields. The complexes Eu**40** and Eu**41** containing three and four bpyO$_2$ chromophores, respectively, gave good luminescence intensity thanks to high molar absorption coefficients and good quantum yields.

Finally, we discuss the Eu^{3+} and Tb^{3+} complexes of ligands based on substituted calix[4]arenes. The complexes of ligand **53** show low luminescence intensities due to the weak absorption of the calixarene. Most interestingly, the introduction of two to four bpy units in the calix[4]arene moiety as in the complexes of ligands **67**, **68**, and **70–72** gives rise in most cases to a substantial increase of the luminescence intensity thanks to the significant enhancement of the molar absorption coefficients and, for the Eu^{3+} complexes, of the quantum yields. Most interesting are the values obtained for the complexes of the ligands **78–81** ranging from 2400 to 9100 M^{-1} cm^{-1}. These values are mainly due to the high quantum yield values of these complexes.

VI. STATE OF THE ART

First, we draw some general conclusions on the role played by the ligand and the metal ion in luminescent devices consisting of complexes of Eu^{3+} and Tb^{3+} with encapsulating ligands. The results obtained for the complexes overviewed demonstrate that the nature of the ligand is determining in obtaining metal luminescence. The ligand is responsible for the absorption of the complex, and the absorption spectrum observed corresponding to that of the free ligand is slightly modified upon complexation. Furthermore, the complexation ability of the ligand influences the ligand–metal energy transfer. This influence depends also on the energies of the ligand excited states and, in the case of the Eu^{3+} complexes, of the LMCT excited states related to the redox properties of the ligand. Finally, the ligand influences the deactivation processes of the metal emitting states. Shielding of the metal ion by the ligand accounts for the nonradiative deactivation involving solvent molecules coordinated to the metal ion. The energies of the ligand excited states and, for the Eu^{3+} complexes, the LMCT excited states determine the nonradiative deactivations via these states. Concerning the role played by the metal ion, in the Eu^{3+} complexes the thermally activated nonradiative decay of the Eu^{3+} $^5D_0$ emitting state occurs mainly via low-lying LMCT excited states, which are also responsible for some low

$\eta_{\text{en.tr.}}$ values (Section V.B). For the Tb^{3+} complexes, low-lying excited states different from the ligand and metal excited states are not present and $\eta_{\text{en.tr.}}$ values higher than those of the Eu^{3+} complexes are usually obtained (Section V.B). The energy of the Tb^{3+} emitting state, higher than that of the Eu^{3+} emitting state, renders more probable nonradiative deactivation of the former state via the lowest ligand triplet excited state. This process is present in almost all the Tb^{3+} complexes containing the bpy and phen chromophores and is most likely responsible for the lack of metal luminescence at room temperature of several Tb^{3+} complexes with ligands incorporating heterocyclic N-oxide chromophores.

Now we report on the few Eu^{3+} and Tb^{3+} complexes of encapsulating ligands used in TR-FIAs and DNA hybridization assays exploiting their function of luminescent device. Derivatives of the prototypes Eu2 and Tb2 were tested as luminescent labels for TR-FIAs, in spite of their rather low metal luminescence intensities, because they were stable in water and did not decompose upon attachment to the biomolecule. Recently, the application of Eu8, similar to Eu2, in homogeneous TR-FIAs has been reported [Fig. 15(a)] [73, 118]. The assay consists of a method for the quantitative determination of the antigen, prolactin. During the immunoreaction, this antigen reacts with two monoclonal antibodies, one labeled with Eu8 and the other with allophycocyanin (APC). If the formation of the sandwich immunocomplex occurs, excitation of the bpy chromophore in Eu8 is followed by an intramolecular energy transfer to the Eu^{3+} ion and then by intermolecular energy transfer to APC. Sensitized APC luminescence is observed, characterized by a lifetime of 0.25 ms and a broad band with maximum around 665 nm. This signal can be easily distinguished from the luminescence signals of Eu8 and APC attached to the antibodies present in excess. The intensity of the sensitized APC luminescence is proportional to the amount of immunocomplex present in the sample.

The complex Eu8 was used as a luminescent label for a heterogeneous DNA hybridization assay [Fig. 15(b)] [119]. The analysis is preformed as follows. A protein, streptavidin, is labeled with Eu8 and this conjugate, characterized by a 1:4 streptavidin/Eu8 stoichiometry is used as a detection reagent. The target DNA is denatured and immobilized on a nitrocellulose membrane. Hybridization is performed with complementary biotinylated probe DNA. Then the membrane carrying the hybridization product is incubated with a solution of the labeled streptavidin, known to have a high affinity for biotin. Finally, detection of the Eu8 luminescence allows us to quantify the target DNA by comparing the signal with a standard curve.

The use of Eu8 as a luminescent label for the detection of polymerase chain reaction DNA products was reported [Fig. 15(c)] [120]. An antibody recognizing 2,4-dinitrophenol is labeled with Eu8 and this conjugate is used

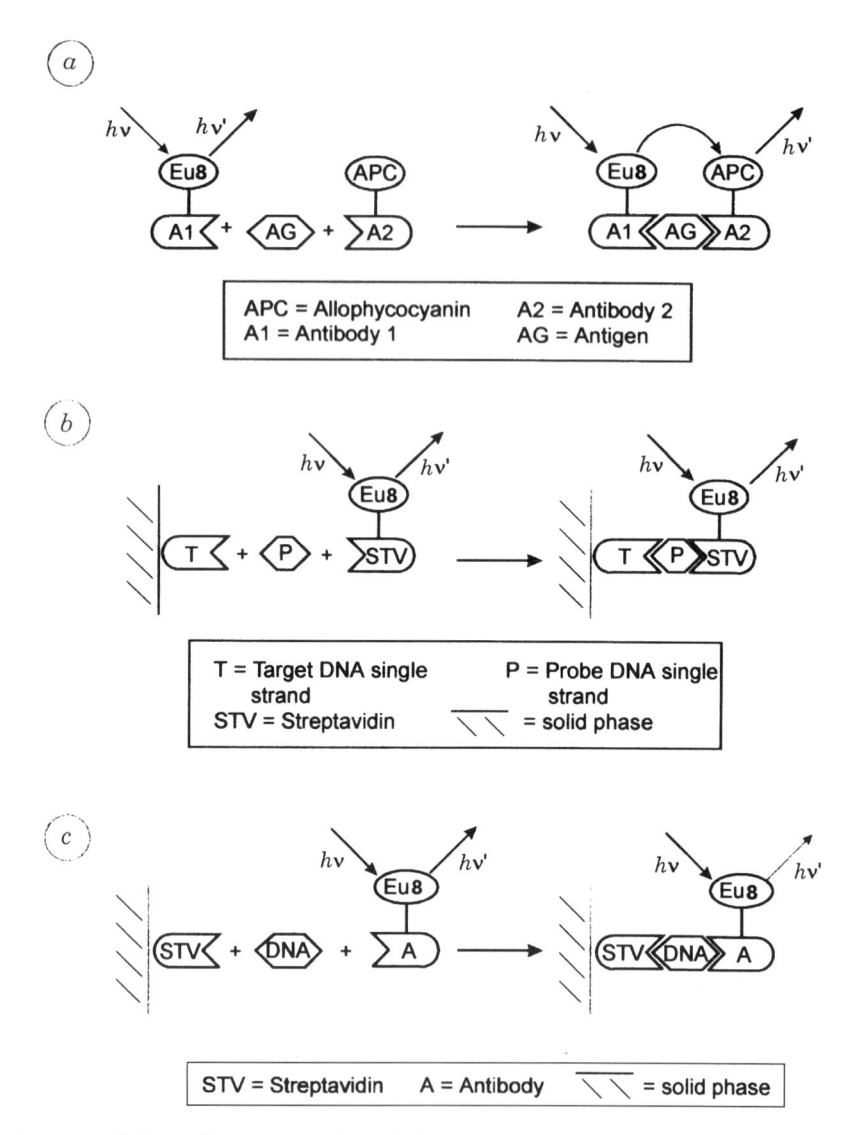

Figure 15. Schematic representation of the three assays using the cryptate Eu8 as label.

as a detection reagent. The procedure consists of DNA amplification via a standard polymerase chain reaction with outer DNA primers. The products of this reaction undergo a second amplification with two types of inner DNA primers, the first biotinylated and the other labeled with 2,4-dinitrophenol. The final duplex DNA is immobilized via the biotinylated end onto a streptavidin-coated microwell. Next, the microwell carrying the duplex DNA is incubated with the solution of the detection reagent recognizing the 2,4-dinitrophenol end of the duplex DNA. Then, detection of the metal luminescence is carried out.

Nevertheless, the demand for complexes characterized by higher metal luminescence intensities remained. Actually, other complexes that were subsequently studied (Section IV) exhibited satisfactory metal luminescence intensities but, up to now, drawbacks related to instability of the complexes in water or their decomposition upon attachment to the biomolecule rendered their application impossible.

ACKNOWLEDGMENTS

This work was supported by M.U.R.S.T. (Ministero dell'Università e della Ricerca Scientifica e Tecnologica) and by the Progetto Strategico "Tecnologie Chimiche Innovative (Sottoprogetto A)" of C.N.R. (Consiglio Nazionale delle Ricerche).

REFERENCES

1. J.-M. Lehn, *Science*, **227**, 849 (1985).
2. J.-M. Lehn, *Angew. Chem. Int. Ed. Engl.*, **27**, 89 (1988).
3. V. Balzani and F. Scandola, *Supramolecular Photochemistry*, Ellis Horwood Limited, Chichester, England 1991.
4. N. Sabbatini, M. Guardigli, and J.-M. Lehn, *Coord. Chem. Rev.*, **123**, 201 (1993).
5. N. Sabbatini, M. Guardigli, and I. Manet, in K. A. Gschneider and L. Eyring, Eds., *Handbook on the Physics, and Chemistry of Rare Earths*, Elsevier Science Publishers B.V., Amsterdam, 1996, Vol. 23, pp. 69–119.
6. R. Reisfeld, *Structure Bonding* (Berlin), **22**, 123 (1975).
7. W. T. Carnall, in K. A. Gschneider and L. Eyring, Eds., *Handbook on the Physics, and Chemistry of Rare Earths*, North-Holland, Amsterdam, 1979, Vol. 3, pp. 171–208.
8. W. Strek, *J. Chem. Phys.*, **76**, 5856 (1982).
9. F. K. Freed, *Top. Appl. Phys.*, **15**, 23 (1976).
10. R. Reisfeld and C. K. Jørgensen, *Inorg. Chem. Concepts*, **1**, 1 (1977).
11. Y. Haas and G. Stein, *J. Phys. Chem.* **75**, 3677 (1971).
12. G. Stein and E. Wurzberg, *J. Chem. Phys.*, **62**, 208 (1975).
13. W. De W. Horrocks, Jr. and D. R. Sudnick, *J. Am. Chem. Soc.*, **101**, 334 (1979).
14. W. De W. Horrocks, Jr. and D. R. Sudnick, *Acc. Chem. Res.*, **14**, 384 (1981).

15. R. C. Holz, C. Allen Chang, and W. De W. Horrocks, Jr., *Inorg. Chem.*, **30**, 3270 (1991).
16. G. Blasse, in K. A. Gschneider and L. Eyring, Eds., *Handbook on the Physics, and Chemistry of Rare Earths*, North-Holland, Amsterdam, 1979, Vol. 4, pp. 237–274.
17. P. G. Sammes and G. Yahioglu, *Natural Product Reports*, 1. (1996).
18. I. A. Hemmilä, in J. D. Winefordner and I. M. Kolthoff, Eds., *Chemical Analysis*, Wiley, New York, 1991, Vol. 117.
19. L. Stryer, *Biochemistry*, Freeman, New York (1988).
20. L. E. M. Miles and C. N. Hales, *Nature (London)*, **219**, 186 (1968).
21. I. Weeks, M. Sturgess, R. C. Brown, and J. S. Woodhead, *Methods Enzymol.*, **133**, 366 (1986).
22. K. E. Rubenstein, R. S. Schneider, and E. F. Ullman, *Biochem. Biophys. Res. Commun.*, **47**, 846 (1972).
23. A. Meyer and S. Neuenhofer, *Angew. Chem. Int. Ed. Engl.*, **33**, 1044 (1994).
24. H. R. Schroeder, P. O. Vogelhut, R. J. Carrico, R. C. Boguslaski, and R. T. Buckler, *Anal. Chem.*, **48**, 1933 (1976).
25. U. Landegren, R. Kaiser, C. T. Caskey, and L. Hood, *Science*, **242**, 229 (1988).
26. A. Oser and G. Valet, *Angew. Chem. Int. Ed. Engl.*, **29**, 1167 (1990).
27. J. Coates, P. G. Sammes, G. Yahioglu, R. M. West, and A. J. Garman, *J. Chem. Soc. Chem. Commun.*, 2311 (1994).
28. J. Coates, P. G. Sammes, R. M. West, and A. J. Garman, *J. Chem. Soc. Chem. Commun.*, 1107 (1995).
29. P. R. Selvin, T. M. Rana, and J. E. Hearst, *J. Amer. Chem. Soc.*, **116**, 6029 (1994).
30. I. A. Hemmilä, S. Dakubu, V.-M. Mukkala, H. Siitari, and T. Lövgren, *Anal. Biochem.*, **137**, 335 (1984).
31. I. A. Hemmilä and S. Dakubu, US Patent No. 4,565,790. (1986).
32. E. P. Diamandis, *Clin. Biochem.*, **21**, 139 (1988).
33. R. A. Evangelista, A. Pollack, B. Allore, E. F. Templeton, R. C. Morton, and E. P. Diamandis, *Clin. Biochem.*, **21**, 173 (1988).
34. H. Takalo, E. Hänninen, and J. Kankare, *Helv. Chim. Acta*, **76**, 877 (1993).
35. V.-M. Mukkala and J. J. Kankare, *Helv. Chim. Acta*, **75**, 1578 (1992).
36. E. F. Templeton Gudgin and A. Pollak, *J. Lumin.*, **43**, 195 (1989).
37. P.G. Sammes, G. Yahioglu, and G. D. Yearwood, *J. Chem. Soc. Chem. Commun.*, 1282 (1992).
38. J. L. Toner, in *Inclusion Phenomena and Molecular Recognition*, J. Atwood, Ed., Plenum, New York, 1990, pp. 185–197.
39. V.-M. Mukkala, M. Helenius, I. Hemmilä, J. Kankare, and H. Takalo, *Helv. Chim. Acta*, **76**, 1361 (1993).
40. C. J. Pedersen, *Angew. Chem. Int. Ed. Engl.*, **27**, 1021 (1988).

41. J.-M. Lehn and J. P. Sauvage, *J. Am. Chem. Soc.*, **97**, 6700 (1975).
42. D. J. Cram, *Angew. Chem. Int. Ed. Engl.*, **27**, 1009 (1988).
43. R. Weiss, B. Metz, and D. Moras, *Proc. 13th Int. Conf. Coord. Chem.*, **2**, 85 (1970).
44. B. Metz, D. Moras, and R. Weiss, *Acta Crystallogr. Sect. B*, **29**, 1377 (1973).
45. M. Dobler and R. P. Phizackerly, *Acta Crystallogr. Sect. B*, **30**, 2746 (1974).
46. M. Dobler and R. P. Phizackerly, *Acta Crystallogr. Sect. B*, **30**, 2748 (1974).
47. M. Dobler, J. D. Dunitz, and P. Seiler, *Acta Crystallogr. Sect. B*, **30**, 2741 (1974).
48. J. D. Dunitz and P. Seiler, *Acta Crystallogr. Sect. B*, **30**, 2739 (1974).
49. J. D. Dunitz, M. Dobler, P. Seiler, and R. P. Phizackerly, *Acta Crystallogr. Sect. B*, **30**, 2733 (1974).
50. P. Seiler, M. Dobler, and J. D. Dunitz, *Acta Crystallogr. Sect. B*, **30**, 2744 (1974).
51. K. N. Trueblood, C. B. Knobler, E. Maverick, R. C. Helgeson, S. B. Brown, and D. J. Cram, *J. Am. Chem. Soc.*, **103**, 5594 (1981).
52. N. Sabbatini, S. Dellonte, M. Ciano, A. Bonazzi, and V. Balzani, *Chem. Phys. Lett.*, **107**, 212 (1984).
53. N. Sabbatini, S. Dellonte, and G. Blasse, *Chem. Phys. Lett.*, **129**, 541 (1986).
54. B. Alpha, R. Ballardini, V. Balzani, J.-M. Lehn, S. Perathoner, and N. Sabbatini, *Photochem. Photobiol.*, **52**, 299 (1990).
55. B. Alpha, J.-M. Lehn, and G. Mathis, *Angew. Chem. Int. Ed. Engl.*, **26**, 266 (1987).
56. M. Pietraszkiewicz, J. Karpiuk, and A. K. Rout, *Pure Appl. Chem.*, **65**, 563 (1993).
57. G. Blasse, M. Buys, and N. Sabbatini, *Chem. Phys. Lett.*, **124**, 538 (1986).
58. N. Sabbatini, S. Perathoner, G. Lattanzi, S. Dellonte, and V. Balzani, *J. Phys. Chem.*, **91**, 6136 (1987).
59. N. Sabbatini, M. Guardigli, J.-M. Lehn, and G. Mathis, *J. All. Comp.*, **180**, 363 (1992).
60. J.-C. Rodriguez-Ubis, B. Alpha, D. Plancheral, and J.-M. Lehn, *Helv. Chim. Acta*, **67**, 2264 (1984).
61. B. Alpha, Ph.D. Thesis, "Cryptates photoactives de lanthanides-de nouveau marqueurs luminescents" University of Strasbourg, Strasbourg, France (1987).
62. B. Alpha, V. Balzani, J.-M. Lehn, S. Perathoner, and N. Sabbatini, *Angew. Chem. Int. Ed. Engl.*, **26**, 1266 (1987).
63. I. Bkouche-Waksman, J. Guilhem, C. Pascard, B. Alpha, R. Deschenaux, and J.-M. Lehn, *Helv. Chim. Acta*, **74**, 1163 (1991).
64. J.-M. Lehn, in V. Balzani, Ed. *Supramolecular Photochemistry*, Reidel, Dordrecht, (1987), pp. 29–43.

65. N. Sabbatini, S. Perathoner, V. Balzani, B. Alpha, and J.-M. Lehn, in *Supramolecular Photochemistry*, V. Balzani, Ed., Reidel, Dordrecht, (1987), pp. 187–206.
66. G. Blasse, G. J. Dirksen, N. Sabbatini, S. Perathoner, J.-M. Lehn, and B. Alpha, *J. Phys. Chem.*, **92**, 24(19) (1988).
67. G. Blasse, G. J. Dirksen, D. van der Voort, N. Sabbatini, S. Perathoner, J.-M. Lehn, and B. Alpha, *Chem. Phys. Lett.*, **146**, 347 (1988).
68. H. Takalo and J. Kankare, *Finn. Chem. Lett.*, **15**, 95 (1988).
69. A. Musumeci, R. P. Bonomo, V. Cucinotta, and A. Seminara, *Inorg. Chim. Acta*, **59**, 133 (1982).
70. J.-M. Lehn, M. Pietraszkiewicz, and J. Karpiuk, *Helv. Chim. Acta*, **73**, 106 (1990).
71. B. Alpha, E. Anklam, R. Deschenaux, J.-M. Lehn, and M. Pietraskiewicz, *Helv. Chim. Acta*, **71**, 1042 (1988).
72. M. Guardigli, Ph.D. Thesis, "Fotofisica di complessi a gabbia di lantanidi" University of Bologna, Bologna, Italy 1993.
73. G. Mathis, *Clin. Chem.*, **39**, 1953 (1993).
74. C. O. Roth, Ph.D. Thesis, "Cryptates luminescents de ligands N-oxides" University of Strasbourg, Strasbourg, France 1992.
75. A. Caron, J. Guilhelm, C. Riche, C. Pascard, B. Alpha, J.-M. Lehn, and J.-C. Rodriguez-Ubis, *Helv. Chim. Acta*, **68**, 1577 (1985).
76. J.-M. Lehn and J.-B. Regnouf de Vains, *Helv. Chim. Acta*, **75**, 1221 (1992).
77. J.-M. Lehn and C. O. Roth, *Helv. Chim. Acta*, **74**, 572 (1991).
78. L. Prodi, M. Maestri, V. Balzani, J.-M. Lehn, and C. Roth, *Chem. Phys. Lett.*, **180**, 45 (1991).
79. C. O. Paul-Roth, J.-M. Lehn, J. Guilhem, and C. Pascard, *Helv. Chim. Acta*, **78**, 1995.
80. J.-C. Bünzli, in Gschneider, K. A. and L. Eyring, Eds., *Handbook on the Physics, and Chemistry of Rare Earths*, Elsevier Science Publishers B. V., Amsterdam, 1987, Vol. 9, pp. 321–394.
81. S. T. Frey, C. Allen Chang, J. F. Carvalho, A. Varadarajan, L. M. Schultze, K. L. Pounds, and W. De W. Horrocks, Jr., *Inorg. Chem.*, **33**, 2882 (1994).
82. V. Alexander, *Chem. Rev.*, **95**, 273 (1995).
83. N. Sabbatini, L. De Cola, L. M. Vallarino, and G. Blasse, *J. Phys. Chem.*, **91**, 4681 (1987).
84. V. Balzani, J.-M. Lehn, J. van de Loosdrecht, A. Mecati, N. Sabbatini, and R. Ziessel, *Angew. Chem. Int. Ed. Engl.*, **30**, 190 (1991).
85. N. Sabbatini, M. Guardigli, I. Manet, F. Bolletta, and R. Ziessel, *Inorg. Chem.*, **33**, 955 (1994).
86. C. J. Pedersen, *J. Am. Chem. Soc.*, **89**, 2495 (1967).
87. M. de B. Costa, M. M. Queimado, and J. J. R. Frausto da Silva, *J. Photochem.*,

12, 31 (1980).
88. S. M. T. Alonso, E. Brunet, C. Hernandez, and J. C. Rodriguez-Ubis, *Tetrahedron Lett.*, **34**, 7465 (1993).
89. L. De Cola, D. L. Smailes, and L. M. Vallarino, *Inorg. Chem.*, **25**, 1729 (1986).
90. F. Benetollo, G. Bombieri, K. K. Fonda, A. Polo, J. R. Quagliano, and L. M. Vallarino, *Inorg. Chem.*, **30**, 1345 (1991).
91. M. Pietraszkiewicz, S. Pappalardo, P. Finocchiaro, A. Mamo, and J. Karpiuk, *J. Chem. Soc. Chem. Commun.*, 1907 (1989).
92. R. Ziessel and J.-M. Lehn, *Helv. Chim. Acta*, **73**, 1149 (1990).
93. N. Sabbatini, M. Guardigli, F. Bolletta, I. Manet, and R. Ziessel, *Angew. Chem. Int. Ed. Engl.*, **33**, 1501 (1994).
94. M. Murru, J. A. G. Williams, A. Beeby, and D. Parker, *J. Chem. Soc. Chem. Commun.*, 1116 (1993).
95. N. Sato, I. Yoshida, and S. Shinkai, *Chem. Lett.*, 1261 (1993).
96. E. Cole, C. J. Broan, K. J. Jankowski, P. K. Pulukkody, D. Parker, A. T. Millikan, N. R. A. Beeley, K. Millar, and B. A. Boyce, *Synthesis*, 67 (1992).
97. G. Ulrich, M. Hissler, R. Ziessel, I. Manet, G. Sarti, and N. Sabbatini, *New J. Chem.*, **21**, 147 (1997).
98. R. Ziessel, M. Maestri, L. Prodi, V. Balzani, J.-M. Lehn, and A. Van Dorsselaer, *Inorg. Chem.*, **32**, 1237 (1993).
99. D. Parker and J. A. G. Williams, J. A. Williams, *J. Chem. Soc. Perkin Trans. 2*, 1305 (1995).
100. A. Beeby, D. Parker, and J. A. G. Williams, *J. Chem. Soc. Perkin Trans. 2*, 1565 (1996).
101. D. Parker and J. A. G. Williams, *J. Chem. Soc. Perkin Trans. 2*, 1581 (1996).
102. V. Balzani, E. Berghmans, J.-M. Lehn, N. Sabbatini, R. Teröde, and R. Ziessel, *Helv. Chim. Acta*, **73**, 2083 (1990).
103. J.-M. Lehn and R. Ziessel, *J. Chem. Soc. Chem. Commun.*, 1292 (1987).
104. O. Carugo, and C. Bisi-Castellani, *Monatsh. Chem.*, **125**, 647 (1994).
105. N. Sabbatini, M. Guardigli, A. Mecati, V. Balzani, R. Ungaro, E. Ghidini, A. Casnati, and A. Pochini, *J. Chem. Soc. Chem. Commun.*, 878 (1990).
106. N. Sato, M. Goto, S. Matsumoto, and S. Shinkai, *Tetrahedron Lett.*, **34**, 4847 (1993).
107. A. Casnati, C. Fischer, M. Guardigli, A. Isernia, I. Manet, N. Sabbatini, and R. Ungaro, *J. Chem. Soc. Perkin Trans. 2*, 395 (1996).
108. N. Sabbatini, M. Guardigli, I. Manet, R. Ungaro, A. Casnati, C. Fischer, R. Ziessel, and G. Ulrich, *New J. Chem.*, **19**, 137 (1995).
109. F. J. Steemers, W. Verboom, D. N. Reinhoudt, E. B. van der Tol, and J. Verhoeven, *J. Am. Chem. Soc.*, **117**, 9408 (1995).

110. N. Sabbatini, A. Casnati, C. Fischer, R. Girardini, M. Guardigli, I. Manet, G. Sarti, and R. Ungaro, *Inorg. Chim. Acta*, **252**, 19 (1996).
111. E. M. Georgiev, J. Clymire, G. L. McPherson, and D. M. Roundhill, *Inorg. Chim. Acta*, **227**, 293 (1994).
112. S. Pappalardo, F. Bottino, L. Giunta, M. Pietraszkiewicz, and J. Karpiuk, *J. Incl. Phenom.*, **10**, 387 (1991).
113. N. Sato and S. Shinkai, *J. Chem. Soc. Perkin Trans. 2*, 621 (1993).
114. H. Matsumoto and S. Shinkai, *Chem. Lett.*, 901 (1994).
115. G. Ulrich and R. Ziessel, *Tetrahedron Lett.*, **35**, 6299 (1994).
116. R. Ungaro, unpublished results.
117. G. Sarti, Studio di complessi luminescenti di ioni Eu^{3+} e Tb^{3+}, MS Thesis, University of Bologna, Bologna, Italy, 1996.
118. J.-M. Lehn, *Supramolecular Chemistry*, VCH, Weinheim (1995).
119. O. Prat, E. Lopez, and G. Mathis, *Anal. Biochem.*, **195**, 283 (1991).
120. E. Lopez, C. Chypre, B. Alpha, and G. Mathis, *Clin. Chem.*, **39**, 196 (1993).

ADVANCES IN THE MEASUREMENT OF CORRELATION IN PHOTOPRODUCT MOTION

Christopher G. Morgan, Marcel Drabbels* and Alec M. Wodtke
Department of Chemistry, University of California, Santa Barbara,
Santa Barbara, CA 93106

CONTENTS

I. Introduction, 280
II. Overview, 284
    A. Uncorrelated measurements of product energy distributions, 284
    B. Vector correlations, 288
    C. Scalar correlation measurements, 293
III. New correlation measurements, 294
    A. Frequency modulation doppler spectroscopy, 295
        1. Introduction, 295
        2. Acrylonitrile photolysis: minor channel, 299
        3. NCCN photolysis, 299
    B. Extraction time-of-flight and ion imaging methods, 302

*Present address: FOM Institute for Atomic and Molecular Physics, Kruislaan 407, 1098 SJ Amsterdam, The Netherlands.

Advances in Photochemistry, Volume 23, Edited by Douglas C. Neckers, David H. Volman, and Günther von Bünau
ISBN 0-471-19289-9 © 1997 by John Wiley & Sons, Inc.

1. Introduction: the Wiley-McClaren time-of-flight mass spectrometer, 302
   a. Extraction time-of-flight methods, 303
   b. Ion imaging, 305
   c. Other extraction time-of-flight and ion imaging Methods, 306
2. Pulsed-extraction studies of the photolysis of triatomics, 309
   a. Ozone photodissociation, 309
   b. Studies of other triatomics, OCS, N_2O, ClNO, BrNO, and NO_2, 312
3. HNCO photodissociation, 317
4. Ion imaging of CH_3I, CD_3I, CH_3Br, and CD_3Br, 318
5. Fragmentation of larger molecules using pulsed-extraction methods, 321
6. Ion imaging secondary dissociation, 325
C. Neutral time-of-flight methods, 326
   1. H-atom rydberg tagging experiments, 326
   2. CO Metastable experiments, 327
      a. Ketene photodissociation at 308 nm, 330
      b. Ketene dissociation at 351 nm, 338
IV. Future outlook, 340
Acknowledgments, 342
References, 342

I. INTRODUCTION

For many years, physical chemists have sought to understand the nature of the transition state in a chemical reaction. Among the questions that need to be answered are What forces act upon the molecule as it passes through the transition state? What vibrational modes and geometries are important? What electronic states play a role? The least complicated chemical reaction might be considered the dissociation reaction, in which a molecule breaks apart into two fragments. These fragments can be atoms, diatomic or polyatomic molecules, or any combination thereof. Molecules can be induced to dissociate by collisions with other molecules or by absorption of many infrared (IR) photons, but in these approaches, it can be difficult to control how much energy is actually deposited into the molecule. Absorption of a single photon of a specific wavelength supplied by a laser provides a specific amount of energy into the molecule at a nearly exact point in time. Dissociation by this method allows a chemist to control how much energy is deposited into the molecule before it breaks apart, and therefore how much energy is left over for the translational and internal energy of the fragments.

The simplest way to model a photodissociation reaction is by using a one-dimensional picture, such as the dissociation of a diatomic molecule into two atoms. The molecule starts out in the bound ground-state potential with the bond distance constrained near the equilibrium separation. One can also consider the dissociation of a polyatomic molecule in this way; instead of using an interatomic distance, one can use either the length of the breaking bond or the distance between the centers of mass (the line of centers) of the two fragments as the dissociation coordinate. Two examples of such a model are shown in Figure 1 for methyl iodide and ketene photodissociation.

$$H_3C-I + h\nu \rightarrow H_3C^{\cdot} + {}^{\cdot}I \qquad (1a)$$

$$H_2C=C=O + h\nu \rightarrow H_2C: + CO \qquad (1b)$$

In both cases, photodissociation results from excitation of a molecule to an excited electronic state above the dissociation limit. Hence, the characteristics of this excited state are essential to understanding the photodissociation process. In a direct dissociation, as seen for methyl iodide dissociation in Figure 1(a), the upper surface is a repulsive antibonding state. The dissociation occurs very quickly, normally within one vibrational period of the dissociation coordinate. For indirect dissociation, often referred to as predissociation, an example of which is shown in Figure 1(b), the molecule is initially excited to an electronic state and cannot dissociate without nonadiabatic electronic transitions. These "radiationless transitions" include intersystem crossing to a dissociative state, which is the mechanism for production of triplet methylene in ketene photolysis [Fig. 1(b)]. Alternatively, "internal conversion" or vibronic coupling to the ground or other lower lying electronic state of the same electron spin multiplicity is another possibility and is the mechanism for singlet methylene formation in ketene photolysis.

One should not be left with the impression that electronically nonadiabatic processes are limited to predissociation. Figure 1(a) shows a crossing between two excited repulsive curves in the photolysis of methyl iodide. If the surface hopping process is efficient enough, it can even influence a dissociating molecule that passes the curve crossing within a few femtoseconds, as is the case for methyl iodide photodissociation.

Clearly, the simplicity of Figure 1(a) and (b) is too great to completely describe these photodissociation processes. As seen above, even within this

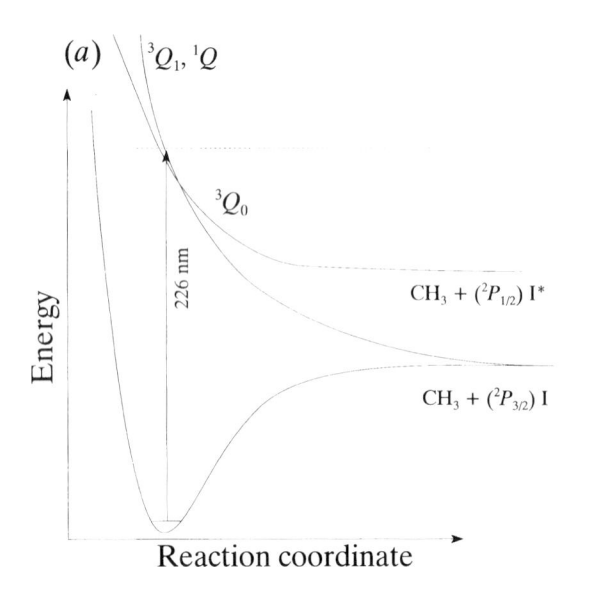

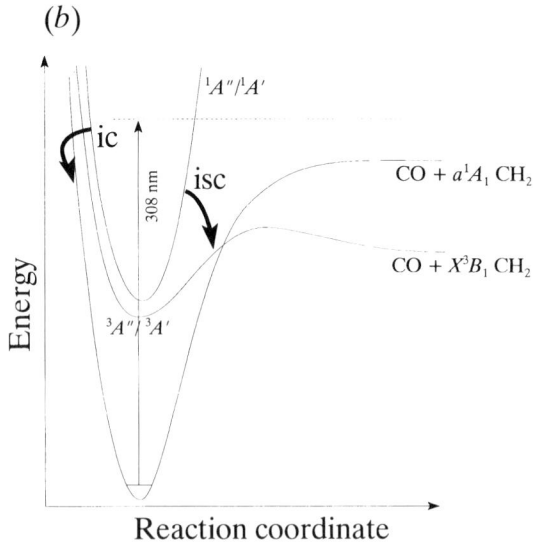

Figure 1. (*a*) One-dimensional potential energy curves for CH_3I photodissociation. Upon excitation by 266-nm light to the $^3Q_0$ state, the molecule can dissociate directly or can undergo curve crossing to the $^1Q, ^3Q_1$ manifold. Direct dissociation to the $^3Q_1, ^1Q$ manifold is also possible. (*b*) One-dimensional potential energy curves for CH_2CO photodissociation. Dissociation occurs through an indirect process of radiationless transfer; first, absorption occurs to the $1A''/1A'$ manifold, then either intersystem crossing (isc) or internal conversion (ic) to the dissociating states.

one-dimension picture there may be a host of electronic states important to the photochemistry. Also, we must consider what effect different vibrational modes in the parent molecule have on the coupling of these potential hypersurfaces, for these other vibrations of the polyatomic molecule play a role in the dissociation and are not merely spectators. While in a diatomic dissociation only one vibrational mode disappears, in the dissociation of a nonlinear polyatomic species, there can be up to five vibrational bending modes that will "disappear" into translation and rotation of the fragments.

Chemists have tried to unravel the mysteries of polyatomic photochemistry in many ways. Theoretical chemists have attempted to calculate excited-state potential hypersurfaces in large, if not complete, dimensionality. The subsequent application of quantum wave packet, variational stationary state calculations or even quasiclassical trajectory calculations using these hypersurfaces allows theoreticians to make predictions of experimental observables. These observables most often fall under the category of product energy distributions, that is, electronic, vibrational, rotational, or translational energy distributions. By comparison with experimental observation, one may gauge the accuracy of the theoretical approach. When good agreement is found between theory and experiment, the theoretical calculations ultimately yield the deepest mechanistic and dynamical insight into the photochemistry. Many of these calculations are discussed in a recently published overview on the subject of photodissociation dynamics [1].

In many photodissociation reactions, there is a lack of information regarding the potential energy surface (PES) involved in the dissociation of the molecule. Several "first-order" theoretical treatments have been devised to model the product state distributions given the absence of information about the PES. Statistical theories model dissociations with very long lifetimes and no significant barrier to dissociation; they assume that all product states have an equal chance of being populated. Therefore, the internal energy distribution of a fragment merely reflects the number of states that correspond to that energy, taking into consideration conservation of both energy and angular momentum. The Prior distribution ignores angular momentum contributions [2], while phase space theory (PST) includes them [3, 4]. Several variations of PST have been devised, including the separate adiabatic channel model (SACM) [5–8] and separate statistical ensembles (SSE) [9]. For dissociation with a barrier, a simple impulsive model can be used [10, 11]. The basic idea of the impulsive model is that the force needed to break apart the molecule is solely directed along the axis of the dissociating bond; any rotational energy resulting from the dissociation is due to the torque created by this repulsive force. However, this model is severely limited because it assumes there is no angular anisotropy in the dissociation process [12]. Because of this limitation, more sophisticated models such as Franck–Condon mapping and infinite order sudden approximation (IOSA) have been developed [13, 14].

Clearly, this interaction between experiment and theory depends on the quality of the experiments that can be carried out. The traditional means to obtain the quantum state distribution of one of the photofragments are slowly being augmented by new methods that yield correlated product information. Such techniques give a more complete and detailed picture of the relationships between the motion of two recoiling polyatomic fragments. This topic is the subject of this chapter and is structured as follows. Section II gives an overview of the basic ideas that are used to implement the various types of experiments that have been devised. These methods include Doppler spectroscopy, field free ion time-of-flight (TOF) and neutral TOF techniques. The body of this chapter (Section III) discusses the "latest and greatest" examples of the applications of these three general approaches.

II. OVERVIEW

A. Uncorrelated Measurements of Product Energy Distributions

One of the most time-tested and reliable means for probing the dynamics of photochemistry is the so-called "PUMP and PROBE" experiment [15]. With this technique, the parent molecule is dissociated with a PUMP laser pulse and the internal state distributions of nascent photofragments is obtained by resonant excitation with a visible or ultraviolet (UV) PROBE laser to a electronically excited state. For many diatomic and triatomic molecules, rotationally resolved spectra have been assigned and transition probabilities can be calculated [16]. After electronic excitation, one can then monitor the fluorescence yield as a function of excitation wavelength. This method, known as laser-induced fluorescence (LIF), has been used to measure product state distributions of numerous photoreactions. In resonant-enhanced multiphoton ionization (REMPI) the molecules are resonantly excited to an electronic excited state and subsequently ionized by the absorption of one or more additional photons. The ion current can then be detected as a function of the wavelength of the first photon. In both REMPI and LIF, the amount of signal for a specific resonant transition will depend on the population of the quantum state being excited the transition probability used to probe the specific quantum state, the alignment or orientation of the ensemble of molecular transition dipole moments, and the polarization as well as the power of the laser. For LIF detection, the fluoresource yield and polarization as well as the precise experimental geometry may need to be considered [17, 18].

Careful determination of the internal state distributions of one of the photoproducts can lead to an understanding of the nature of the forces that act upon the molecule as it breaks apart. One of the nicest examples of this kind of work is water photolysis through the $\tilde{A}$ state. Photolysis via the $\tilde{A}$ state produces little rotational excitation as has been observed by LIF of the OH product [19, 20]. Ab initio studies of the $\tilde{A}$ state have shown that it is purely repulsive, that is, a direct dissociation similar to methyl iodide in Figure 1(a). There is calculated to be little or no variation in the PES as a function of the HOH bending angle [21]. This result means that the OH recoils from the dissociation as if from a central force field and little angular momentum is imparted. This example of photodissociation is an example that nearly perfectly follows the simple one-dimensional model.

Another useful quantity that experimentalists have known how to measure for many years is the translational energy release of the products, E_{trans}. For a simple photodissociation in which the parent breaks apart into just two fragments, one needs only to measure one of the two fragments, since the recoil, or center-of-mass velocities of the two fragments are constrained to one another by the conservation of linear momentum. Measuring the signal as a function of arrival time gives a TOF distribution for one of the photofragments, which can be converted into a center-of-mass frame total translational energy release distribution, $P(E_{trans})$. One particularly successful approach to this measurement has been applied to H atoms recoiling from water photolysis via the $\tilde{B}$ state. With this kind of experiment, photodissociation of molecules prepared in the collision free environment of a molecular beam produces H-atom photofragments. These are then excited with two-color laser excitation to a high-lying, long-lived Rydberg state, which is allowed to travel a measured flight length to a field ionization detector. The TOF spectrum is obtained with very high resolution. Indeed, individual peaks in the TOF spectrum appear that are a measure of the internal energy of the partner OH molecule formed with the H atom. This technique gives high enough resolution to resolve individual rotational states of the OH fragment in the translational energy distribution [22].

These experiments gave very different results than those for photodissociation of water in the $\tilde{A}$ state. Photodissociation via the $\tilde{B}$ state is an example of predissociation and the observed internal energy distribution of the OH is found to peak near 20 kcal mol^{-1} of rotational energy [22]. Ab initio studies have found that the bound $\tilde{B}$ state possesses a deep potential well associated with a linear HOH minimum [23–26]. Excitation from the well-known bent ground state to the linear $\tilde{B}$ state imparts large amounts of bending excitation into the electronically excited water molecule. The

HOH bending coordinate correlates to rotation of the OH fragment at infinite separation of the products, leading to the high rotational excitation of the OH fragment. These two studies show that the measurement of the internal energy distribution of one of the fragments can be a sensitive probe of the mechanism and dynamics of the photochemistry.

In a similar TOF setup that can be applied to a much wider variety of photofragments, the photofragments are allowed to fly a predetermined flight distance into a mass spectrometer where they are ionized by electron bombardment [27] or vacuum ultraviolet (VUV) synchrotron radiation [28]. The energy resolution is good enought in such experiments to resolve different electronic (and sometimes vibrational) states of photoproducts that may be produced in the photolysis. The presence of barriers along the dissociation coordinate is often apparent, and the internal energy of the fragments is found, albeit with relatively low-energy resolution. Under favorable circumstances, the lifetime and symmetry of the excited state may be obtained by use of a polarized photolysis laser. A good review of these kinds of measurements can be found in [29].

With a simple molecule such as H_2O, it is fairly straightforward to obtain an understanding of the dissociation dynamics by looking at the internal state distribution of one of the fragments, since the H atom can only be formed in a single state. However, even for tetra-atomic photodissociation, internal states can be spread out over many vibrational and rotational states of *both* fragments. In this case, one must be careful in interpreting information about the dissociation dynamics from the internal state distribution of one of the fragments.

For the photodissociation reaction

$$ABCD + h\nu \rightarrow AB + CD \tag{2}$$

in which both products have more than one atom, one must consider the *correlated* internal state distribution of the products, which is a matrix of probabilities. An illustration of this is shown in Figure 2. The rows of each matrix shown here refer to the internal state of fragment AB while the columns refer to the internal state of state CD. The ijth element of the matrix gives the probability that state $|j\rangle$ of AB is formed together with state $|i\rangle$ of CD. Let us imagine that this distribution is completely statistical, that is, every combination of AB and CD states is just as likely as any other. In other words is, all the matrix elements are the same. If one measures the uncorrelated distribution of one or another of the fragments, one will obtain the result that all internal states of AB and CD are equally populated, as seen in Figure 2(a). Now let us imagine a second case, one in which *only* the diagonal elements of the correlation matrix are populated, as shown in

$P(|AB_i\rangle, |CD_j\rangle)$'s for reaction $ABCD + h\nu \rightarrow AB + CD$

(a) Statistical product distribution

$|CD_i\rangle$ \ $|AB_j\rangle$	$P(|AB_1\rangle)$ = (1/n)	$P(|AB_2\rangle)$ = (1/n)	...	$P(|AB_n\rangle)$ = (1/n)
$P(|CD_1\rangle) = (1/n)$	$(1/n^2)$	$(1/n^2)$	------	$(1/n^2)$
$P(|CD_2\rangle) = (1/n)$	$(1/n^2)$	$(1/n^2)$	------	$(1/n^2)$
⋮	⋮	⋮		⋮
$P(|CD_n\rangle) = (1/n)$	$(1/n^2)$	$(1/n^2)$	------	$(1/n^2)$

(b) Nonstatistical (highly correlated) distribution

$|CD_i\rangle$ \ $|AB_j\rangle$	$P(|AB_1\rangle)$ = (1/n)	$P(|AB_2\rangle)$ = (1/n)	...	$P(|AB_n\rangle)$ = (1/n)
$P(|CD_1\rangle) = (1/n)$	$(1/n)$	0	------	0
$P(|CD_2\rangle) = (1/n)$	0	$(1/n)$	------	0
⋮	⋮	⋮		⋮
$P(|CD_n\rangle) = (1/n)$	0	0	------	$(1/n)$

Figure 2. Two examples of possible correlated product state distributions from a photodissociation reaction. (a) A completely statistical distribution is shown. (b) A highly correlated distribution is shown.

Figure 2(b), which is indicative of a dynamically controlled photodissociation process. Measuring the uncorrelated distribution yields the *same result* as in the first case. Clearly, we miss important information concerning the dynamics of dissociation if we look at the uncorrelated distribution.

Many methods have been developed to try to obtain this kind of product correlation information. One of the most successful has been that of vector correlation methods, which are described in Section II.B.

B. Vector Correlations

Correlations between populations of quantum states of the two photofragments can be considered as scalar correlations. Vector correlations, which are relationships between different directional properties of the photofragments, are equally important. Indeed, analysis of scalar correlations in the absence of information about vector correlations is sometimes impossible. One of the first vector correlations to be measured was the relationship between the transition dipole moment of the parent molecule and the recoil velocity of the fragments. This $\mathbf{\mu} \cdot \mathbf{v}$ correlation can give information about what type of electronic transition is excited in the parent molecule and its magnitude can be related to the dissociation lifetime of the molecule. If one uses linearly polarized light for photodissociation, the angular distribution of the fragments will be axially symmetric about the electric vector, $\hat{\varepsilon}_D$, of the photolysis light. The angular distribution of the fragments can always be described as follows [30].

$$W(\theta) = \frac{1}{4\pi}[1 + \beta P_2(\cos\theta)] \tag{3}$$

Here, $P_2(\cos\theta)$ is the second Legendre polynomial and θ is the angle between the recoil direction of fragmentation and $\hat{\varepsilon}_D$. The anisotropy parameter β can range from $+2$ (a $\cos^2\theta$ distribution) to -1 (a $\sin^2\theta$ distribution). This correlation will be reduced by the rotations of the parent in the excited state. If the excited-state lifetime is much longer than the rotational period of the parent molecule, the absolute value of β diminishes, indicating more isotropic angular distribution. This correlation can also be washed out by the mixing of electronic states with various symmetries. If the absolute value of β is noticeably greater than zero, then the dissociation is faster than the rotational time of the parent.

Another important vector correlation in photodissociation is called $\mathbf{v} \cdot \mathbf{j}$ correlation, which describes the relationship between the recoil velocity and the angular momentum of one of the fragments. This correlation can give important indications about the kinds of torques in effect between the recoiling products during dissociation. These torques can arise from bending and torsional modes in the parent molecule and from exit channel dynamics.

For example, let us contrast the dissociations of OCS and H_2O_2 in a molecular beam. Because of the strong rotational cooling present in molecular beam expansions, the parent angular momentum will be very small. Thus, in the case of OCS, the dissociation can be considered to occur in a single plane. The OCS bending motion, which provides the torque to give

CO its rotation, must occur such that the angular momentum vector of CO is always perpendicular to the OCS plane, while the recoil direction is always in plane. Therefore, **v** is perpendicular to **j**; this correlation in OCS photodissociation has in fact been seen [31]. The departing CO resembles a tumbling cartwheel. Such **v**·**j** correlations are very commonly observed. They are expected from the simple impulsive model or indeed from any model where the forces all lie within a plane. Any impulsive force imparted to a diatomic fragment that is directed at a position displaced from the center of mass of the diatomic molecule will lead to both translational and rotational energy and a cartwheeling motion will result.

Photolysis of H_2O_2 at 266 nm is the most well known example of torsional control of the **v**·**j** correlations. In the ground state of this molecule, the torsional (or dihedral) angle between the two OH bonds (as viewed end on) is about 110°. In the excited state, however, it is near 180°. Excitation of the molecule results in rapidly developing torsional motion about the disintegrating O—O bond. This torsional motion correlates eventually to OH rotation, but in this case the recoil velocity (along the direction of the original O—O bond) is parallel to the rotational angular momentum vectors. The two newly formed OH molecules appear as departing disrotatory "helicopters," repelling one another [32, 33]. The analysis of **v**·**j** correlation will not only depend on $\hat{\varepsilon}_D$, but also on $\hat{\varepsilon}_{PR}$, the electric vector of the laser used to probe the nascent fragments. It will also depend on the rotational branch (P, Q, or R) used to probe the fragment, along with the direction of the probe laser propagation.

An extension of both classical and quantum PST calculations has been devised in order to model **v**·**j** correlations [34, 35]; the quantum PST uses a helcity basis set to express the distribution of **j** about **v**. These calculations have been used to model several of the experimentally determined **v**·**j** correlations described in this chapter.

Several papers have been written on the subject of analyzing the anisotropic distribution of **v** and **j** vectors of molecules produced by a photodissociation reaction [17, 36–42]. The most satisfying scheme is the renormalized bipolar moment analysis proposed by Dixon [18]. The basic premise is to use bipolar harmonics as the basis set to describe the angular distribution of a single recoil velocity **v** and fragment angular momentum **j**:

$$P(\theta'_t, \phi'_t, \theta'_r, \phi'_r) = \sum_K \sum_Q \sum_{k_1} \sum_{k_2} \frac{\sqrt{2k_1+1}\sqrt{2k_2+1}}{16\pi^2} b_Q^K(k_1 k_2) B_{KQ}(k_1 k_2; \theta'_t, \phi'_t, \theta'_r, \phi'_r) \quad (4)$$

Here, (θ'_t, ϕ'_t) and (θ'_r, ϕ'_r) refer to polar coordinates for the orientation of the vectors **v** and **j**, respectively. The $B_{KQ}(k_1k_2;\theta'_t,\phi'_t,\theta'_r,\phi'_r)$ are the bipolar harmonics, and $b_Q^K(k_1k_2)$ are the expansion coefficients, referred to as bipolar moments. All of these quantities are in the body-fixed frame. This distribution is rotated to the space-fixed axis considering the relationship between the photolysis laser's polarization (space-fixed frame) and the transition dipole of the molecule (body-fixed frame). This transformation from the body-fixed to the space-fixed frame will limit the possible values of K and Q. For example, in photodissociation by one photon of linearly polarized light only B_{00} and B_{20} are nonzero and the following angular distribution will be produced:

$$P(\theta_t,\phi_t,\theta_r,\phi_r) = \frac{1}{16\pi^2}\sum_k (2k+1)b_0^0(kk)B_{00}(kk,\theta_t,\phi_t,\theta_r,\phi_r)$$

$$+\frac{2}{5}\sum_{k_1}\sum_{k_2}\frac{\sqrt{2k_1+1}\sqrt{2k_2+1}}{16\pi^2}b_0^2(k_1k_2)B_{20}(k_1k_2;\theta_t,\phi_t,\theta_r,\phi_r)$$

(5)

Now, (θ_t,ϕ_t) and (θ_r,ϕ_r) refer to polar coordinates in the space-fixed frame for the orientation of the vectors **v** and **j**, respectively. The absorption intensity of photofragments with an anisotropic distribution of **j** has been described by Fano and Macek [43] by using the spherical tensor moments $A_q^k(j)$. For example, the intensity of a one-photon, dipole-allowed transition of the photofragment in state j is described by the following equation:

$$I(j) = C_{\text{det}}\{A_0^0(j) - \tfrac{1}{2}h^{(2)}[A_0^2(j) - 3\cos(2\beta)A_{2+}^2(j)] + \tfrac{3}{2}\sin(2\beta)h^{(1)}A_0^1(j)\}$$

(6)

In this equation, C_{det} can be considered a proportionality constant and β describes the degree of ellipticity of the probe laser. The probe light of arbitrary polarization is described by a vector $(\cos\beta, e^{i\delta}\sin\beta, 0)$. Linearly polarized light is described when $\beta = 0$ and circularly polarized light is described when $\delta = \pi/2$ and $\beta = \pm\pi/4$. The depolarization factors, $h^{(i)}$, depend on which branch is being used in the probe step.

The anisotropies we have described above ($\boldsymbol{\mu}\cdot\mathbf{v}$ and $\mathbf{v}\cdot\mathbf{j}$ correlation) are measurements of the alignment of one physically significant vector with respect to another. Another type of anisotropy is orientation. The alignment tells one whether two vectors will tend to be, for example, parallel or perpendicular to one another. The orientation will tell if these vectors are pointing in the same or the opposite direction. A chiral parent molecule and/or the use of elliptically polarized light is necessary in order to obtain information about orientation. From Eq. 6, it is easily seen that the "odd" spherical tensor moment $A_0^1(j)$, which describes orientations, will not be detected unless elliptically polarized light is used.

Expressions for $A_q^k(j)$ can be derived from the equation above in the following way. The rotational moments $\rho_q^k(j)$ can be obtained by projecting $P(\theta_t, \phi_t, \theta_r, \phi_r)$ onto a basis set of modified spherical harmonics, $C_{kq}^*(\theta_r, \phi_r)$.

$$\rho_q^k(j) = \int C_{kq}^*(\theta_r, \phi_r) P(\theta_t, \phi_t, \theta_r, \phi_r) \sin\theta_r d\theta_r d\phi_r \tag{7}$$

The geometry of the experiment is taken into account by rotating $\rho_q^k(j)$ onto the probe laser frame. The rotated $\rho_q^k(j)'$ moments are proportional to the $A_q^k(j)$ moments. In order to obtain the proper form for $A_q^k(j)$, the $b_Q^K(k_1, k_2)$ moments must be renormalized to $\beta_Q^K(k_1, k_2)$ moments according to Dixon [18]; the renormalization constants and the ranges of $\beta_Q^K(k_1, k_2)$ are shown in Tables 1a and 1b.

As can be seen from this introduction, there are many vector and scalar quantities that are important to correlated product state measurements. In order to aid the reader in keeping track of the quantities important to these measurements, Table 2 defines and summarizes all the scalar and vector quantities used in this chapter.

TABLE 1a Renormalized Bipolar Moments with Even k_1 and k_2[a].

Moment	Equivalence	Argument of expectation value for renormalized moment	Range
$\beta_0^0(00)$	$b_0^0(00)$	1	1
$\beta_0^2(20)$	$b_0^2(20)$	$P_2(\cos\theta_t')$	$-\frac{1}{2}$ to 1
$\beta_0^2(02)$	$b_0^2(02)$	$P_2(\cos\theta_r')$	$-\frac{1}{2}$ to 1
$\beta_0^0(22)$	$\sqrt{5}b_0^0(22)$	$P_2(\cos\omega_{rt}')$[b] $= P_2(\cos\theta_t')P_2(\cos\theta_r')$ $+ \frac{1}{3}P_2^1(\cos\theta_t')P_2^1(\cos\theta_r')\cos(\phi_t' - \phi_r')$ $+ \frac{1}{12}P_2^2(\cos\theta_t')P_2^2(\cos\theta_r')\cos[2(\phi_t' - \phi_r')]$	$-\frac{1}{2}$ to 1
$\beta_0^2(22)$	$\sqrt{\frac{7}{2}}b_0^2(22)$	$-P_2(\cos\theta_t')P_2(\cos\theta_r')$ $- \frac{1}{6}P_2^1(\cos\theta_t')P_2^1(\cos\theta_r')\cos(\phi_t' - \phi_r')$ $+ \frac{1}{12}P_2^2(\cos\theta_t')P_2^2(\cos\theta_r')\cos[2(\phi_t' - \phi_r')]$	-1 to $\frac{1}{2}$
$\beta_0^2(42)$	$\sqrt{\frac{7}{2}}b_0^2(42)$	$P_4(\cos\theta_t')P_2(\cos\theta_r')$ $+ \frac{1}{6}P_4^1(\cos\theta_t')P_2^1(\cos\theta_r')\cos(\phi_t' - \phi_r')$ $+ \frac{1}{72}P_4^2(\cos\theta_t')P_2^2(\cos\theta_r')\cos[2(\phi_t' - \phi_r')]$	$-\frac{1}{2}$ to 1

[a]Reference [18].
[b]ω_{rt}' is angle between **v** and **j**.

TABLE 1b Renormalized Bipolar Moments with Either k_1 and k_2 odd[a].

Moment	Equivalence	Argument of expectation value for renormalized moment	Range
$\beta_0^0(11)$	$\sqrt{3}b_0^0(11)$	$-\hat{\mathbf{v}}\cdot\hat{\mathbf{j}} = -\cos\theta'_t\cos\theta'_r - \sin\theta'_t\sin\theta'_r\cos(\phi'_t - \phi'_r)$	-1 to 1
$\beta_0^2(11)$	$\sqrt{\frac{3}{2}}b_0^2(11)$	$\frac{1}{2}(3\hat{j}_{z'}\cdot\hat{v}_{z'} - \hat{\mathbf{v}}\cdot\hat{\mathbf{j}}) = \cos\theta'_t\cos\theta'_r$ $-\frac{1}{2}\sin\theta'_t\sin\theta'_r\cos(\phi'_t - \phi'_r)$	-1 to 1
$\beta_0^2(12)$	$2i\sqrt{\frac{2}{3}}b_0^2(02)$	$2\hat{j}_{z'}\cdot(\hat{\mathbf{v}}\times\hat{\mathbf{j}})_{z'} = -2\cos\theta'_r\sin\theta'_t\sin\theta'_r\sin(\phi'_t - \phi'_r)$	-1 to 1
$\beta_0^2(21)$	$2i\sqrt{\frac{2}{3}}b_0^2(21)$	$2\hat{v}_{z'}\cdot(\hat{\mathbf{v}}\times\hat{\mathbf{j}})_{z'} = -2\cos\theta'_t\sin\theta'_t\sin\theta'_r\sin(\phi'_t - \phi'_r)$	-1 to 1
$\beta_0^2(31)$	$\sqrt{\frac{7}{3}}b_0^2(31)$	$-P_3(\cos\theta'_t)\cos\theta'_r - \frac{1}{3}P_3^1\sin\theta'_r\cos(\phi'_t - \phi'_r)$	-1 to 1
$\beta_0^2(32)$	$i\sqrt{\frac{35}{6}}b_0^2(32)$	$\sqrt{15/72}[4P_3^1(\cos\theta'_t)P_2^1(\cos\theta'_r)\sin(\phi'_t - \phi'_r)$ $+ P_3^2(\cos\theta'_t)P_2^2(\cos\theta'_r)\sin 2(\phi'_t - \phi'_r)$	-1 to 1

[a]Reference [42].

TABLE 2 Description of Commonly Used Symbols in This Chapter

Symbol	Description
$\hat{\varepsilon}_D$	Polarization vector of photolysis light
$\hat{\varepsilon}_{PR}$	Polarization vector of probe light
$\boldsymbol{\mu}$	Transition dipole moment of patent molecule
$\mathbf{v}, v$ or $\mathbf{v}_i, v_i$	Recoil, or center-of-mass, velocity of fragment i being probed
θ	Angle between $\mathbf{E}_{ph}$ and $\mathbf{v}_i$
$\mathbf{j}, j$ or $\mathbf{j}_i, j_i$	Angular momentum of fragment i being probed
β	Alignment parameter of $\mathbf{v}_i$ about $\boldsymbol{\mu}$. Also known as $\beta_0^2(20) = \frac{1}{2}\beta$
β_{eff}	Effective alignment parameter of $\mathbf{v}_i$ about $\boldsymbol{\mu}$. Used when other vector correlations, for example, $\mathbf{v}\cdot\mathbf{j}$ correlation, cannot be separated from $\boldsymbol{\mu}\cdot\mathbf{v}$ correlation
$A_0^{(2)}, A_0^{(4)}$	Alignment of $\mathbf{E}_{ph}$ about $\mathbf{j}$
$\beta_0^0(22)$	Bipolar moment describing $\mathbf{v}\cdot\mathbf{j}$ correlation
$h^{(i)}$	Depolarization factors[a]. $h^{(1)} = (j+1)/[j(j+1)]^{1/2}$, $h^{(2)} = -(j+1)/(2j+1)$ for P branch. $h^{(1)} = 1/[j(j+1)]^{1/2}$, $h^{(2)} = 1$ for Q branch. $h^{(1)} = -j/[j(j+1)]^{1/2}$, $h^{(2)} = -j/(2j+3)$ for R branch.

[a]Reference [42] and [43].

C. Scalar Correlation Measurements

In addition to vector correlations and the wealth of information they provide, an essential question to ask is If state $|\chi_1\rangle$ of fragment 1 is formed, what is the probability to form state $|\chi_2\rangle$ of fragment 2? These *scalar* correlations can be obtained from the quantum state specific translational energy distribution of one of the fragments. How this works becomes clear if one considers the energy balance equation of the photoreaction:

$$E_{h\nu} + E_{\text{int}}(\text{ABCD}) = E_{\text{int}}(\text{AB}) + E_{\text{int}}(\text{CD}) + E_{\text{trans}} + D_0(\text{AB} - \text{CD}) \quad (8)$$

When performing one of these experiments, the photolysis energy, $E_{h\nu}$, and the bond dissociation energy, $D_0(\text{AB} - \text{CD})$, are already known. The internal energy of the parent $E_{\text{int}}(\text{ABCD})$ can either be minimized using molecular beam methods, or its distribution can be calculated by knowing the temperature of the experiment. The remaining unknown quantities in Eqs. 8 are the internal energies of the fragments, $E_{\text{int}}(\text{AB})$ and $E_{\text{int}}(\text{CD})$, and E_{trans}. Let us run a laser spectroscopy experiment probing a specific internal state of AB. Now $E_{\text{int}}(\text{AB})$ is known from the spectroscopy. If we can measure $P(E_{\text{trans}})$ solely for this probed internal state of AB, it would be directly related to the internal energy distribution of CD, the other fragment.

One way to accomplish this is by way of the Doppler effect. The wavelength at which an atom or molecule absorbs light will depend on the velocity of the molecule with respect to the propagation direction of the photon. If ν_0 is the rest absorption frequency, then ν, the frequency at which the molecule with velocity $\mathbf{v}$ absorbs light with propagation direction $\mathbf{k}_{\text{pr}}$ and speed c, is given by

$$\nu = \nu_0[1 - (\mathbf{v} \cdot \mathbf{k}_{\text{pr}})/c] \quad (9)$$

The probed nascent fragment population traveling with a distribution of velocities will give rise to a Doppler profile about ν_0. Analysis of this profile can lead to a state-specific $P(E_{\text{trans}})$ for the photodissociation process [31, 44–50]. A variation of Doppler spectroscopy can be accomplished by measuring a series of Doppler profiles and changing the delay time between the 'PUMP' and the 'PROBE' steps between each measurement. This technique, known as velocity-aligned Doppler spectroscopy (VADS), is a more direct observation of the speed distribution of the probed molecule along $\mathbf{k}_{\text{pr}}$, rather than just a measurement of the velocity components in this direction [51]. As is clear from the above theoretical discussion, complications arise when one considers how the various vector correlations play a role in the Doppler profile. However, these problems can be solved.

Another approach to the determination of state specific translational energy distributions uses REMPI to ionize photofragments in a state-specific manner, and then allows these ions to travel in a field-free region before being accelerated onto a detector [52–55]. This approach allows the ions to be separated in time and space before being detected. A disadvantage of this method is that the ions tend to "smear out" in time due to space charge and stray field effects.

A way around this problem is not to excite the atoms or molecules to the ionization continuum, rather, one promotes them to a long-lived high Rydberg state or a metastable state. This method allows the molecules to travel as a neutral entity through the flight path. They possess sufficient electronic energy that they may be ionized by a small electric field in the case of high Rydberg states or surface ionization in the case of metastable states. These three ideas: Doppler, field-free ion TOF, and neutral metastable TOF comprise the essential ideas of all the methods being tested presently.

III. NEW CORRELATION MEASUREMENTS

Recent advances have been made in methods of correlated product state measurements and these advances will be descussed in the remainder of this chapter. Generally, are improvements to the correlated measurements we have just outlined. Doppler spectroscopy has been improved by using frequency modulation techniques to carefully measure the Doppler profile with astoundingly high signal-to-noise (S/N) ratios. The extraction TOF method, first developed by Mons and Dimicoli [56], and Chandler and Houston [57], have been used since the late 1980s. These methods are presented in this chapter because there is still new and fascinating work being done with them, and because several experimental groups have developed variations of these techniques that merit discussion. Finally, advances in neutral TOF methods will be discussed, including the metastable TOF methods used in our laboratory. These neutral methods have been shown to produce the most detail for correlation measurements, including the extraction of vector correlations as a function of kinetic energy release.

A. Frequency Modulation Doppler Spectroscopy

1. Introduction. In an effort to get the most detail out of absorption profiles in electronic transitions with a high S/N ratio, frequency modulation (FM) Doppler absorption spectroscopy has been used. The reason that FM spectroscopy is especially suited for this task are manifold. First, it is an absorption measurement, as opposed to LIF, where the Doppler profile is determined by measuring the fluorescence from an absorption feature. Therefore, it is a more general method that yields data that is more easily analyzed. This method also used single-mode continuous wave (CW) lasers with a very narrow bandwidth (3 MHz) as the probing light source. In Doppler measurements, a deconvolution of the probe laser line width needs to be carried out in order to obtain information on the velocity dependency of the Doppler profile. However, if the bandwidth is sufficiently small, as it is with single-mode CW lasers, the frequency profile of the laser can be treated simply as a delta function and no deconvolution is necessary.

Frequency modulation spectroscopy as a method to detect first-order susceptibilities in visible spectroscopy was first described by Bjorklund [58] and Hall et al. [59], and used by Bjorklund and Levenson [60] to observe hyperfine splitting in I_2. Since then, it has been used to obtain spectra of transient species such as HCO, ND_4, and HO_2 [61-64]. Photodissociation studies using FM and absorption spectroscopy have been done by the group at Brookhaven National Laboratory [65-68]. This chapter will concentrate on their particular methodology.

In the setup of an FM spectroscopy experiment, shown in Figure 3, the CW laser light with a carrier frequency ω_c is fed into an electrooptic modulator (EOM) driven at the modulation frequency ω_m, typically 200 MHz. The output of the EOM is the carrier frequency and two sidebands with frequency $\omega_c \pm \omega_m$. When these three frequencies hit a photodetector, the signal from the detector will have a direct current (dc) component and an alternating current (ac) component with a frequency that results from the beating of the optical frequencies. The two leading terms of the ac component will have frequency ω_m, and they result from the beating of $\omega_c + \omega_m$ with ω_c, and $\omega_c - \omega_m$ with ω_c. However, in the absence of any amplitude or phase modification, these two leading terms will be equal in amplitude and 180° out of phase; therefore, a null signal will be recorded. The null is broken when the attenuation of amplitude or the phase shift at $\omega_c + \omega_m$ is different than at $\omega_c - \omega_m$. The ac component is fed into a mixer driven by reference frequency ω_m. The mixer works as the demodulator of the signal. It extracts the signal arriving with beat frequency ω_m, sends it

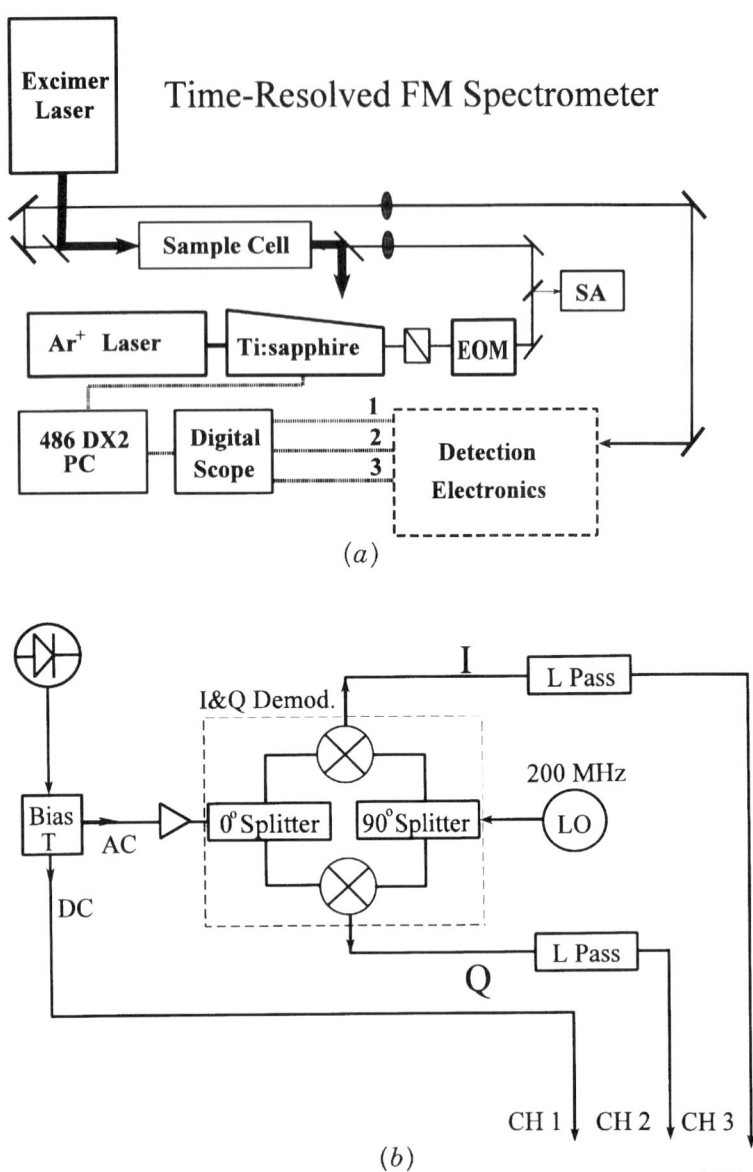

Figure 3. (*a*) Schematic diagram of the Brookhaven transient FM spectrometer, in which EOM = electrooptic modulator and SA = spectrum analyzer. (*b*) Schematic diagram of the FM detection electronics, in which LO = local oscillator, L pass = low pass filter. [Reprinted with permission from S. W. North, X. S. Zheng, R. Fei, and G. E. Hall, *J. Chem. Phys.*, **104**(6), 2129–5. Copyright © 1996 American Institute of Physics.]

first through a low pass filter in order to remove $2\omega_m$ components, and then to the signal processing electronics.

There are two regimes of FM spectroscopy, the first is when the modulation frequency is greater than the line width of the absorption feature, and the second is when it is less. It is in the latter regime that the Brookhaven group performs their experiments, which is sometimes called "wavelength modulation." The Brookhaven group scans an absorption profile by changing the carrier frequency. The output, $I(\omega_c)$, has both absorption and dispersion elements that depend on amplitude or phase modification, respectively, as is shown in Eq. 10,

$$I(\omega_c) \propto M[A_{FM}(\omega_c)\cos\theta_{FM} + D_{FM}(\omega_c)\sin\theta_{FM}] \tag{10}$$

where M is the sideband modulation index, $A_{FM}(\omega_c)$ is the absorption component, and $D_{FM}(\omega)$ is the dispersion component of the FM signal, and θ_{FM} is the absolute phase angle between the beat signal and the reference signal in the mixer. The absorption and dispersion components are given by

$$A_{FM} = \delta(\omega_c + \omega_m) - \delta(\omega_c - \omega_m) \tag{11}$$

$$D_{FM} = \phi(\omega_c - \omega_m) - 2\phi(\omega_c) + \phi(\omega_c + \omega_m) \tag{12}$$

where $\delta(\omega)$ and $\phi(\omega)$ are the amplitude attenuation and optical phase shift, respectively, at frequency ω. Since the absorption elements are needed in order to determine the Doppler profile, these two components must be separated from one another, which requires a determination of the absolute phase angle θ_{FM}. This determination can be done by measuring the FM signal from thermalized transient molecules whose absorption and dispersion components can be calculated as a function of temperature. Then, instead of a simple mixer, the FM signal is fed into an I & Q demodulator that produces an in-phase $I(\omega_c)$ and a quadrature $Q(\omega_c)$ output. The $I(\omega_c)$ output is similar to Eq. 10, and $Q(\omega_c)$ output will differ from $I(\omega_c)$ in its absolute phase angle by $\pi/2$. From these two outputs, the absolute phase angle can be found, and once this angle is determined, the absorption and dispersion elements of the signal can be separated.

Since the absorption component of the FM signal is recorded as a derivative spectrum, a method for extraction of the Doppler profiles form the FM absorption signal is also needed. North et al. [69] first suggested using a resursive relationship in order to extract this information. In a more recent paper, they settled on a method that expands the FM absorption signal in a Taylor series [67], and then used both intergation and differentiation with respect to $d\omega$ in order to obtain two solutions, one for $\delta(\omega_0)$, and the other for $d^2\delta/d\omega^2$. These two solutions can be combined, and the combination can be truncated to give the following relation:

$$\delta(\omega_0) = \frac{1}{2\omega_m}\int_{-\infty}^{\omega_0} A\,d\omega - c\left(\frac{dA}{d\omega}\right)_{\omega_0} \tag{13}$$

The coefficient c is adjusted by the least-squares method to obtain the most consistent result for $\delta(\omega_0)$.

Recent work by the Brookhaven group has shown the success of this method. This group has focused mainly upon photoreactions that produce

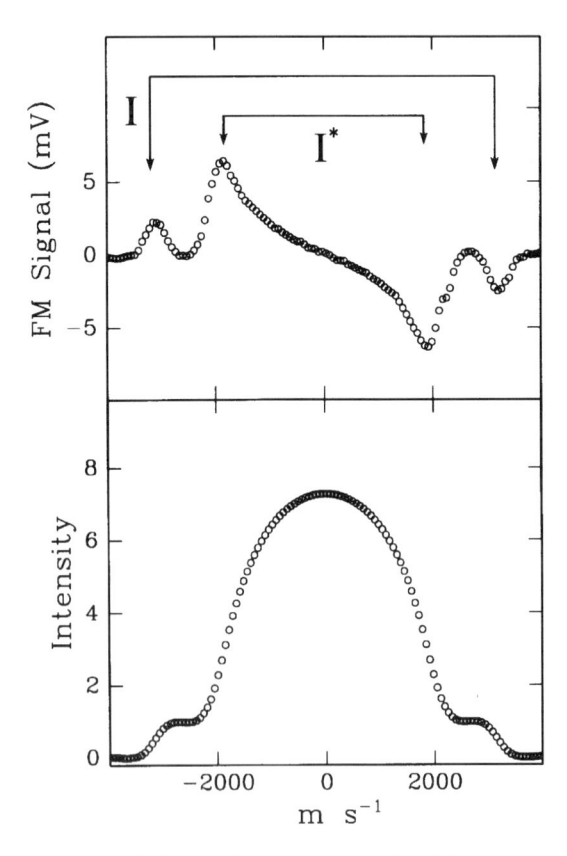

Figure 4. The upper panel shows frequency modulated lineshapes of nascent CN ($N'' = 24$) in its ground electronic and vibrational state, which was produced from the photolysis of ICN with 248-nm light. The lower panel shows the resulting Doppler profile. [This figure was provided courtesy of S. W. North and G. E. Hall.]

CN radicals, photolysis of NCCN and CH$_2$=CHCN [67, 68]. The group has also performed absorption spectroscopy on CH$_2$ radicals formed by ketene photolysis [66] and spectroscopy to study CHCl radicals [70]. All these species are detectable with their Ti:sapphire CW laser system.

The resolving power of FM Doppler spectroscopy is demonstrated in the dissociation of ICN at 248 nm. North and Hall [71] were clearly able to resolve the two spin–orbit states of an I atom in the Doppler profile of the CN fragment. As seen in Figure 4, there are two distinct components in both the raw FM signal and the Doppler profile. Further experiments are currently being performed on this system.

2. Acrylonitrile Photolysis: Minor Channel. The major channel of acrylonitrile photodissociation at 193 nm forms HCN and either acetylene or vinylidene [72]. The CN channel is a very minor one, measured to have a quantum yield of only (0.003 ± 0.001) [73].

$$H_2C=CHCN + hv \rightarrow H_2C=CH + CN \text{ (minor)} \quad (14)$$

$$H_2C=CHCN + hv \rightarrow HCCH + HCN \quad H_2C=C + HCN \text{ (major)} \quad (15)$$

The rotational distribution of CN produced by this reaction at 193 nm has been measured and fitted to a Boltzmann distribution with $T = 1450$ K [74]. Combined with photofragmentation translational spectroscopy, it was initially concluded that this channel is formed by an impulsive dissociation through a repulsive excited state [72].

North and Hall [68] used FM spectroscopy to obtain Doppler profiles of the nascent CN produced by this reaction. A feature of their experimental results is that the Q and R branches for the same rotational state of CN exhibit nearly identical Doppler profiles. This result indicates an absence of vector correlations in the dissociation; these correlations should occur in an impulsive reaction, where one would expect $\mathbf{v} \perp \mathbf{j}$. Their results suggest that the dissociation occurs as a statistical fragmentation after internal conversion to the ground state rather than as an impulsive one via an excited state.

3. NCCN Photolysis. Cyanogen (NCCN) photodissociation has also been considered a prototypical statistical dissociation reaction. After absorption to the $^1\Sigma_u^-$, $^1\Delta_u$ manifolds, the dissociation is thought to proceed via internal conversion to the ground-state surface. Previous studies have measured the rotational state distribution of the CN fragments from photolysis of cyanogen at 193.3 nm [75–77]. The threshold photon energy for CN formation was found to be 47,000 ± 200 cm^{-1} [78].

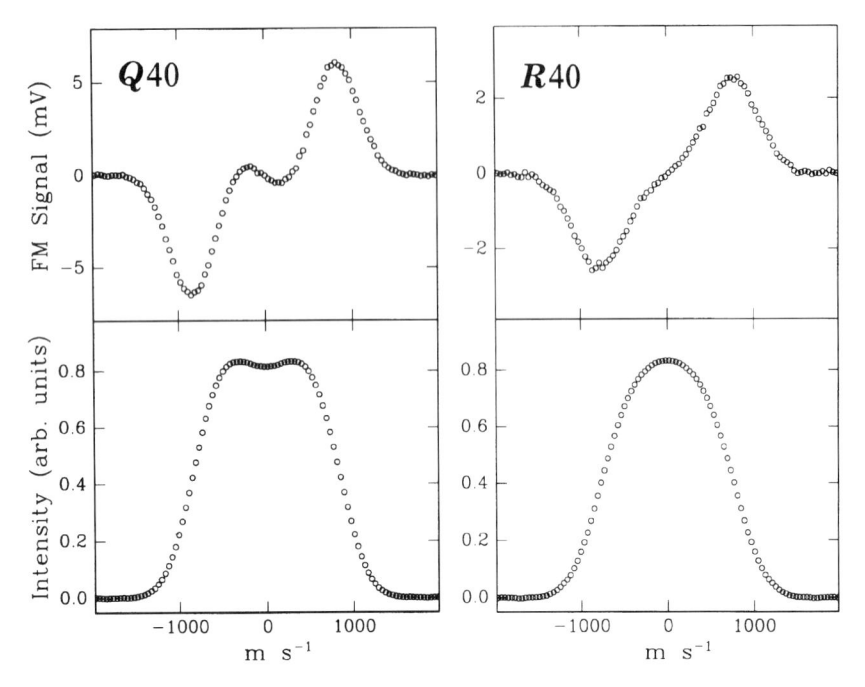

Figure 5. The upper panels show the frequency modulated lineshapes of nascent CN ($N'' = 40$) in its ground electronic and vibrational state, which was produced from the photolysis of NCCN with 193.3-nm light. The lineshapes for both the Q and R branches are shown. The lower panels show the resulting Doppler profiles.

As in the case of ketene, the correlated product-state measurements of NCCN at 193.3 nm by FM spectroscopy were also performed in a cell [67]. Once again, differences in Doppler profiles between Q and R branches were used to determine vector correlations; only $\mathbf{v} \cdot \mathbf{j}$ correlations were considered to contribute to changes between Q and R branches. As can be seen in Figure 5, now there are distinct differences in the Doppler profile between Q and R branches for NCCN photolysis. The angular distribution of the photoproducts is isotropic [76, 79], therefore the Doppler profile in the laboratory frame is given by

$$D'(w) = \int_{|w|}^{\infty} \frac{1}{2v_{\text{lab}}} [1 - h^{(2)} < \beta_0^0(22)_{\text{lab}} > P_2(w/v_{\text{lab}})] v_{\text{lab}}^2 f'(v_{\text{lab}}) dv_{\text{lab}} \quad (16)$$

where v_{lab} is the laboratory frame velocity, w is the component of v_{lab} along the direction of the probe laser, $P_2(x)$ is the second Legendre polynomial,

$\langle \beta_0^0(22)_{\text{lab}} \rangle$ is the average value of the velocity dependent bipolar moment that describes $\mathbf{v} \cdot \mathbf{j}$ correlation, and $v_{\text{lab}}^2 f'(v_{\text{lab}})$ is the distribution of speeds in the laboratory frame. The depolarization factor $h^{(2)}$ is equal to $+1$ if a Q branch is probed and $-j/(2j+3)$ if an R branch is probed.

Vector correlation effects can be removed from the Doppler profile by normalizing each profile to unity and then adding the profiles of a Q and R branch for a particular j state with the Q/R weighing factor of $1/(2+3/j)$. From this composite profile, the laboratory speed distribution can be found through the following:

$$v_{\text{lab}}^2 f'(v_{\text{lab}}) = -2v_{\text{lab}} \left(\frac{dD'_0(w)}{dw} \right)_{w=v_{\text{lab}}} \tag{17}$$

A comparison was made between the experimentally determined Doppler profiles and those profiles predicted by PST [35] for each detected state. In order to make the correct comparison, the PST velocity distribution, which is in the center-of-mass frame, must be first averaged over the room temperature Boltzmann distribution of the parent, and then convoluted with the parent velocity distribution. This calculated laboratory distribution can then be converted to a Doppler profile and compared to experimental data. A comparison can also be made between the experimentally determined $\langle \beta_0^0(22)_{\text{lab}} \rangle$ values and those calculated by PST. The transformation between the two frames is given by [80]

$$\beta_0^0(22)_{\text{lab}} = \frac{\beta_0^0(22)_v I_{5/2}(mv_{\text{lab}}v/k_B T)}{I_{1/2}(mv_{\text{lab}}v/k_B T)} \tag{18}$$

where m is the mass of the parent, k_B is Boltzmann's constant, T is the temperature of the parent, $\beta_0^0(22)_v$ is the velocity dependent $\mathbf{v} \cdot \mathbf{j}$ correlation in the center-of-mass frame, v_{lab} is the fragment velocity in the laboratory frame, and $I_{1/2}$ and $I_{5/2}$ are modified spherical Bessel functions.

It was found that some physically interesting modifications of PST calculations were necessary in order to explain the experimental Doppler profiles. In the helicity state count used to calculate the PST predicted $\mathbf{v} \cdot \mathbf{j}$ correlation, the total helicity Λ was found not to be restricted by the total angular momentum but by quantum number K, the projection of the total angular momentum on the axis of the linear molecule. This result can be explained as follows. Ground-state cyanogen is a linear molecule, therefore $K = 0$. A one quanta transfer of vibrational angular momentum to states of π vibronic symmetry will occur when the molecule is excited to $^1\Sigma_u^-$ or $^1\Delta_u$, so K will go from 0 to 1. Matrix elements that mediate internal conversion should be nonzero only if $\Delta K = 0$, K scrambling due to Coriolis mixing takes place in many nonoseconds [81, 83], while the cyanogen lifetime in the ground state is only around a few picoseconds. Restricting the total helicity to $\Lambda = 1$ increases the perpendicular character of the $\mathbf{v} \cdot \mathbf{j}$ correlation. As the rotational quantum number of CN in its ground vibrational state increases from $N = 13$ to $N = 40$, the experimental $\langle \beta_0^0(22)_{\text{lab}} \rangle$ decreases from -0.04 to -0.21. Total helicity restricted PST keeps up with experimental $\langle \beta_0^0(22)_{\text{lab}} \rangle$, while the unrestricted PST calculation only goes up to $\langle \beta_0^0(22)_{\text{lab}} \rangle = -0.13$ for $N = 40$. The experimental $\mathbf{v} \cdot \mathbf{j}$ correlation is similar to the PST calculated $\mathbf{v} \cdot \mathbf{j}$ correlation with $J = 0$, as if the parent molecule was cooled in a beam. Although a decrease in the tendency for the recoil velocity and the angular momentum of the fragment to be perpendicular should be evident for dissociations from thermalized parent molecules, in the case of NCCN, this decrease is not seen due to the geometry of the ground state and the nature of the electronic transition.

Experimental Doppler profiles could be matched to PST synthesized Doppler profiles only if the available energy $(E_{\text{ph}} - D_0)$ was set to 4700 cm^{-1} and an excess of 600 cm^{-1} of translational energy was added. This finding indicates that the total available energy for the fragments is 5300 cm^{-1}, and that there is a 600-cm^{-1} barrier to the dissociation, which will show up exclusively as translational energy. This barrier has not been detected in previous work, and the Brookhaven group would like to do more experiments on NCCN, this time at lower photolysis energies, in order to confirm the existence of the barrier. Their explanation as to the nature of the barrier-to-ground state NCCN dissociation involves the dipole–dipole repulsion of the two CN fragments. The electrostatic interaction energy of the two CN fragments (1.45 D) at a distance of 3.5 Å from each other is about 585 cm^{-1}, roughly equivalent to the estimated barrier height. Of course, this dipole–dipole interaction energy will change if NCCN isomerizes into NCNC or CNNC, but if the large amplitude motions that will produce these isomers occurs on a time scale much longer than that of dissociation, this electrostatic barrier will have to be surmounted in order for dissociation to occur.

Clearly the use of high S/N FM absorption spectroscopy has a bright future in the investigation of correlation in photochemistry.

B. Extraction Time-of-Flight and Ion Imaging Methods

1. Introduction: The Wiley-McClaren Time-of-Flight Mass-Spectrometer.
In an effort to obtain the best velocity resolution in TOF spectra of photofragments ionized in a REMPI scheme, several groups have used variations of

the Wiley–McClaren TOF mass spectrometer (TOF–MS) [84]. In the standard magnetic or quadrupole mass spectrometer, ionized molecules are separated by mass using either electric or magnetic fields. The Wiley–McClaren TOF-MS uses differences in ion flight times to separate masses. The TOF-MS consists of an electrically isolated metal repeller plate and an electrically isolated accelerator grid, placed on either side of the region where the photofragments are ionized by REMPI. A flight tube is placed normal to the plane of the repeller and accelerator, and a second grid, electrically grounded, is placed between the flight tube and the accelerator to ensure that the flight tube is free of electric fields. An ion detector is placed at the end of the flight tube. As described in a paper by Houston and coworkers [85], one of two things can happen depending on what voltages are applied to the grids. At certain settings of the repeller and acceleration voltages, the space focusing condition is fulfilled and the ions arrive at the detector independent of their initial position, which results in an increased mass resolution. If other conditions are used, the ions retain the spatial separation, and the net result is that instead of performing as mass analyzer, the TOF–MS will perform as a velocity analyzer.

The voltages on the acceleration and repeller grids can either be constant or pulsed. In the pulsed extraction TOF methods, the repeller and accelerator are kept at ground potential for a few microseconds after the photolysis and ionizing lasers have fired in order to let the ions expand in a field-free region. In the center-of-mass frame, these ions expand like a balloon being blown up. That is, molecules with the same speed expand on the surface of a "Newton sphere". At a certain delay, the grids are switched on, and the electric fields around the interaction region extract the photon-produced ions into the flight tube.

Two basic types of experiments can be preformed with the TOF–MS that will reveal information about the correlated photoproduct distribution. One set of experiments involves the use of the TOF–MS as a velocity analyzer. In this chapter we call this technique as the *extraction time-of-flight method*. Another type of experiment uses the TOF–MS in its mass resolution mode in conjunction with a two-dimensional ion detector. This technique is known as *ion imaging*. Variations of these techniques have been developed and are discussed in Section III.B.1.c. Experimentalists have used these methods to measure scalar and vector correlations in many types of photodissociation reactions; their results are presented in Section III.B.2.

a. Extraction Time-of-Flight Methods. Photofragments which have been ionized by a REMPI scheme can be extracted by the acceleration and repeller voltages through the flight path and onto the detector in one of two ways. If the ions are extracted onto the detector without hindrances in the flight path, they will separate as a function of the projection of their recoil velocities on the flight axis, producing a one-dimensional Doppler-type profile [86]. Since the projection of recoil velocities is measured, we call this method *projection TOF*. More information about the Newton sphere can be obtained by changing the geometric features of the experiment. The directions of the probe and photolysis lasers' polarization vector can be varied in order to sample different portions of the Newton sphere. Alternatively, an aperture can be placed in front of the flight tube in order to allow only a small core of the Newton sphere with a very specific recoil direction to reach the detector. This can also be accomplished by using a very long flight tube relative to the active area of the detector [85, 87]. Since only a small core of the Newton sphere is sampled, this method is known as *core extraction* or *core sampling*. In both cases, a one dimensional image of the photofragmentation as a function of arrival time at the detector is then obtained.

The extraction TOF method can be done using either pulsed or constant voltage sources to the acceleration and repeller grids of a Wiley–McClaren TOF–MS. The first extraction TOF experiments were Mons and Dimicoli [56] on NO_2 photodissociation using constant voltage sources. Later, Black and Powis [88] used a similar extraction TOF technique, also with constant voltage sources, in order to study CH_3I photolysis reactions. Most subsequent experimentalists have used pulsed voltage sources.

Uberna, Hinchliffe and Cline (referred to now as UHC) have used Dixon's formalism to obtain expressions for the geometric and bipolar moment dependency on the signal of a $1 + n'$ REMPI scheme, in which the ionization step is saturated [42]. In this way they have analyzed how the core extraction method can be used to probe vector correlations. For one-photon photolysis by linearly polarized light and $1 + n'$ REMPI, the number of bipolar moment's $\beta_Q^K(k_1, k_2)$ will be limited to $K = 2$ and 0, $Q = 0$, $k_1 \leq 4$, and $k_2 \leq 2$. Bipolar moments with even k_1 and k_2 give information about the alignment of **μ**, **v**, and **j** with respect to one another, and these can be probed by using linearly polarized light in the probe step. Since the polarization of both the photolysis laser and the probe laser can be in two different orthogonal directions with respect to the flight axis, and since the propagation directions of these two lasers can also be placed in two orthogonal positions (either parallel or perpendicular to each other), eight different experimentally practical geometries can be obtained for this experiment. A difference TOF spectrum can be obtained by subtracting the TOF signal obtained for one geometry from the TOF signal obtained for another. Uberna, Hinchliffe, and Cline have shown that this difference spectrum will be more sensitive to certain bipolar moments and less sensitive to others. Therefore, all the bipolar moments can be extracted by taking difference spectra for a variety of experimental geometries.

The UHC treatment of bipolar analysis has also been extended into looking at orientation moments. These bipolar moments with either k_1 or k_2 odd are for the most part probed by using circularly polarized light. All but one of these moments will be zero unless the system is chiral; the exception is the $\beta_0^2(21)$ moment, which describes the orientation of **j** about **v** × **μ**. If we imagine a prompt dissociation of a nonlinear triatomic ABC + $h\nu \to$ AB + C with **v** having components not perpendicular to **μ**, then depending on which end of AB gets the impulse of the dissociation, the **j** vector of AB will either be oriented on one side or the other of the plane of the triatomic molecule. Therefore, determination of the $\beta_0^2(21)$ moment is a measure of which end of a diatomic fragment received the impulsive recoil torque.

b. Ion Imaging. Correlated photoproduct distributions can also be measured while using the TOF–MS in its mass resolution mode in a method known as ion imaging [57]. In this method, the displacement of the photofragments perpendicular to the flight path is measured directly by using a two-dimensional imaging detector. As in the extraction TOF method, the Newton sphere is allowed to expand before the ion extraction voltages are turned on, but not the TOF–MS is in the mass resolution mode so that the ions will no longer be separated by their displacement parallel to the ion flight path. However, they will retain their displacement in the plane perpendicular to the flight path. What remains for the experimentalist to do is to record the positions of the ions on this two-dimensional plane with some sort of position sensitive ion detector. Such a two-dimensional, or ion imaging, detector consists of two microchannel plates (MCPs) in a chevron assembly with a fast phosphor screen as the anode, which is viewed by a CCD camera.

One inconvenience associated with this method arises for the portions of the two-dimensional image that lie along the propagation direction of the ionizing laser. Due to the Doppler effect, which is described by Eq. 9, the velocity component along this propagation direction will cause the fragment to absorb a slightly different frequency of light. Depending on the frequency and bandwidth of the probe laser, there will be a greater probability of ionizing certain parts of the Newton sphere at the expense of others. This velocity-selective effect can be minimized by scanning the laser through the entire absorption profile. If there is sufficient sparseness in the REMPI spectrum of the fragment, this should not affect the state-specific nature of the experiment. Another nontrivial aspect of the experiment results from the fact that the ion imaging detector is actually seeing a three-dimensional image "crushed" onto a two-dimensional screen. Mathematical routines were developed in order to take the raw two-dimensional image and transform it into a three-dimensional map of the velocity distribution, using an inverse Abel transform [89]. Several routines have been suggested and devised by experimentalists to transform the two-dimensional data. The inverse Abel transform assumes that the image is axially symmetric, but raw images will not always have this symmetry due to experimental imperfections. Therefore, the data processing routines must take this into consideration, along with the fact that the raw images will have noise in them [90]. These symmetrizing and noise reduction routines tend to blur results, hence efforts have been made to either obtain two-dimensional images through other means or to obtain three-dimensional images directly.

c. Other Extraction Time-of-Flight and Ion Imaging Methods. Variations of the two methods discussed above have been successfully performed. One method can be considered a variation of the one-dimensional TOF–MS method in which fragments are detected as a velocity-dependent signal. However, in this method the dissociation and probe lasers are separated in space from one another [91,92]. A schematic diagram of this approach is shown in Figure 6, as implemented by the Reisler group at USC. The parent molecules are prepared in a supersonic beam traveling along the x axis. The photolysis laser dissociates the parent molecules and a short distance further along the x axis, the position of the focus of a REMPI probe laser beam is scanned perpendicular to the molecular beam direction, that is along the z axis. Scanning along the z axis is essentially the same as recording the state selective lab frame angular distribution, which can depend strongly on the translational energy release of the photofragment. This method has been used in the dissociation of NO_2 to observe the $O(^3P_J)$ atom electronic branching ratio as a function of the rotational state of NO [91]. In another similar study, NO_3 dissociation near the threshold of the O_2 + NO channel was investigated with this method [92].

Another variation of the core-extraction method is the technique developed by Syage and co-worker [93–95]. In this variant of the core-extraction method, deflector plates are placed at the beginning of the field-free region, and the aperture is placed directly in front of the MCPs. As in the core-extraction experiment, the time-resolved signal profile will depend on the velocity component along the flight time axis. However, depending on what voltages are applied to the deflector plates, the dimension of the Newton sphere perpendicular to the ion flight path may be varied. In this way, a one-dimensional slice of the Newton sphere is seen by the detector. Rastering through each slice can be done by changing the voltages on the deflector plates, and hence, a two-dimensional image can be constructed. Syage used this method to obtain two-dimensional images of the dissociation of ozone and CH_3I in great detail, and also for the study of the photodynamics of CH_3I clusters [96].

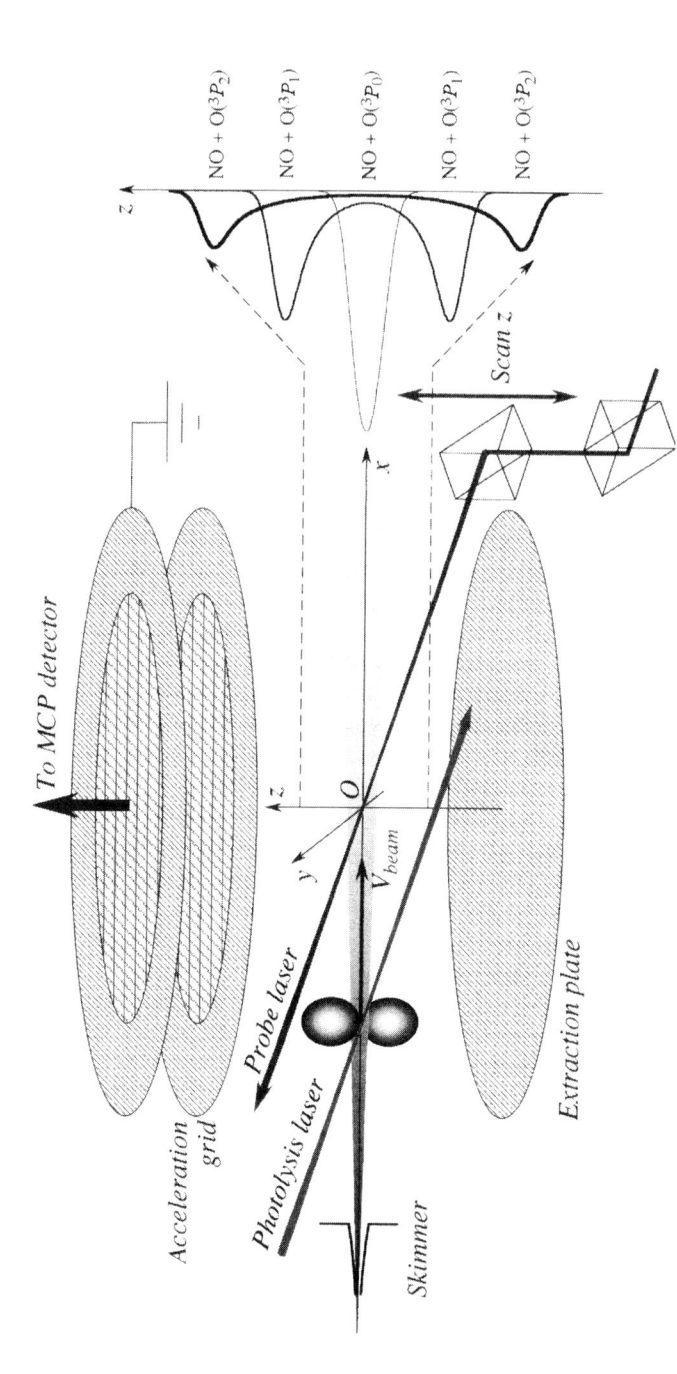

Figure 6. Schematic of the position-sensitive TOF–MS arrangement used by the Reisler group (see text). On the right are broadened single-mode distributions of NO correlated with different states of O(^3P_j). These distributions are calculated assuming the anisotropy parameter $\beta = 1.35$ and $^3P_2:^3P_1:^3P_0 = 1:1:0.1$. The O($^3P_2$), O($^3P_1$), and ($^3P_0$) channels have 227, 69, and 1 cm^{-1} available as relative translational energies, respectively. The positions of the O($^3P_2$) peaks correspond to $z \approx \pm 4$ mm at a photolysis-probe laser delay of 17.1 μs. [Reprinted with permission from A. Sanov, C. R. Bieler, and H. Reisler, *J. Phys. Chem.*, 1995, **99**, 13637 (1995). Copyright © 1995 American Chemical Society.]

Another variation of the ion TOF–MS techniques yields three-dimensional imaging of photofragments, using the method of laser sheet ionization. Instead of using a laser beam focused to a spot to ionize the fragments in a state-specific manner, the beam is focused into a sheet of light with a cylindrical lens, thereby ionizing the fragments that lie in the plane of the laser sheet. By changing the delay between photolysis and probe lasers, a complete three-dimensional image of the fragmentation can be obtained. This imaging technique has been demonstrated by Tonokura and Suzuki [97] for NO_2 dissociation at 355 nm in which three-dimensional images using the laser sheet ionization method matched the reconstructed Abel transform images from the two-dimensional experiment, showing that this method can measure the three-dimensional image directly without the need to process the raw data. A similar experiment has been carried out by Grünefeld and Andresen [98] on the photodissociation of HNO_3 at 193 nm and NO_2 at 308 nm. Instead of imaging the ionized photofragments, they imaged the LIF resulting from the excitation of the nascent OH products to an electronically excited state.

Since one of the axes of a two-dimensional ion image needs to lie along the propagation direction of the probe laser, one of the features of the ion imaging method is that care must be taken to either include or minimize the Doppler effect, as has been previously discussed. This effect can be turned to the experimentalist's advantage. Fragments with a velocity component along the probe laser propagation direction can be selectively ionized by the probe laser according to the Doppler shift. Probe laser ionization will select fragments in a two-dimensional slice out of the three-dimensional Newton sphere; therefore, by tuning the probe laser to see only this Doppler selected slice, a three-dimensional mapping can be achieved. The two-dimensional slice can be resolved by ion imaging, which requires that the probe laser points almost directly at the detector. The laser can be placed at a slight inclination ($\sim 6°$) to avoid directly hitting the detector. This method has been shown to be able to image both channels of the HBr photolysis at 243 nm,

$$HBr + h\nu(243\,nm) \rightarrow H + Br(^2P_{3/2}) \qquad (19)$$

$$HBr + h\nu(243\,nm) \rightarrow H + Br(^2P_{1/2}) \qquad (20)$$

in which the H atom is detected [99].

This slice in velocity space can also be resolved by core-extraction methods. Changing the polarization of the probe laser will "core out" a different part of the two-dimensional slice. These core-extraction experiments were recently done by Lai et al. [100] on acetylene photolysis at

121.6 nm, once again probing the H atom. At this photolysis wavelength, C_2H is not formed in its ground electronic state ($\tilde{x}^2\Sigma$). There are two distinct dissociation channels; the first correlates to C_2H in its ($\tilde{A}^2\Pi$) state. This channel has a negative β parameter, indicating a perpendicular transition, and it also has distinct peaks in the $P(E_{trans})$ coincident with C—H stretching vibrational energies. The second channel is less structured and has an isotropic angular distribution. Recently, this technique has also been successfully applied to the study of the chemical reaction between the CN radical and the deuterium molecule [101]

$$CN + D_2 \rightarrow DCN + D \qquad (21)$$

In another Doppler-selective method, the position of the probe laser beam is kept parallel to that of the photolysis laser, but is moved in one dimension with respect to it in order to spatially scan the Newton sphere [102, 103]. This method can be considered a two-dimensional version of Reisler's experiment shown in Figure 6. Work by the Valentini group has shown that this method works well for probing the velocity distributions of systems with large kinetic energy release, as in the case of HI dissociation at 266 nm, and also systems with small kinetic energy release, for example, the vibrational predissociation of $(HCl)_2$.

2. Pulsed-Extraction Studies of the Photolysis of Triatomics

a. Ozone Photodissociation. The photolysis of ozone by UV light is one of the most important reactions in atmospheric chemistry [104]. For wavelengths around 240 nm near the peak of the Hartley bands, two reaction channels are thought to be important; the so-called "triplet channel": which produces ground electronic state products, and the "singlet channel."

$$O_3 + h\nu \rightarrow O(^3P_{2,1,0}) + O_2(^3\Sigma_g^-) \text{ "triplet"} \qquad (22)$$

$$O_3 + h\nu \rightarrow O(^1D_2) + O_2(^1\Delta_g) \text{ "singlet"} \qquad (23)$$

Photofragmentation translational spectroscopy showed that the triplet channel represents about 10% of the total yield for photolysis at 266 nm [105, 106]. The rotational distribution of $O_2(^1\Delta_g)$ is consistent with this branching ratio [107].

The Houston group at Cornell used the ion imaging technique to study ozone dissociation through the Hartley band. They first looked at the singlet channel by performing (2 + 1) REMPI on the nascent $a^1\Delta_g O_2$ photofragment and imaging the ions [108]. The triplet channel was characterized with

(2 + 1) REMPI on the O(^3P) atom [109]. The singlet channel experiment was performed with a photolysis laser at 248 nm. The first step of the REMPI scheme was a two-photon $^1\Pi_g \leftarrow \leftarrow {}^1\Delta_g$ transition at either 303.19 nm (the 4–0 band) or 310.70 nm (the 4–1 band). The upper state is a Rydberg state whose molecular constants have not been characterized. The ionization photon is supplied by the same laser. The polarizations of both the dissociation, $\hat{\varepsilon}_D$, and probe, $\hat{\varepsilon}_{PR}$, lasers could be independently varied with respect to the plane of the image. Four different configurations of pump and probe geometries are possible, which can be designated as cases A–D. The images produced by these four cases are seen in Figure 7 (see color insert). Case A had $\hat{\varepsilon}_D$ and $\hat{\varepsilon}_{PR}$ in the plane of the image, while case B had $\hat{\varepsilon}_{PR}$ perpendicular to the plane. In case C, $\hat{\varepsilon}_D$ is perpendicular to the plane, while $\hat{\varepsilon}_{PR}$ is in the plane. Case D has $\hat{\varepsilon}_D$ and $\hat{\varepsilon}_{PR}$ perpendicular to the plane of the image.

Ion images were collected and symmetrized for all four cases in both the (4–0) and (4–1) bands of O_2. When $\hat{\varepsilon}_D$ is parallel to the plane of the image, it points to the top and bottom (north and south) of the image. In the images from the (4–0) and (4–1) bands, noticeable changes can be seen between each of the cases. Note that in cases A and B, there are two distinct lobes about the north and south poles, while in cases C and D, the ion counts are distributed throughout the image. This distribution is indicative of a parallel transition in which the detector sees the photofragments recoil in the direction of $\hat{\varepsilon}_D$. When $\hat{\varepsilon}_D$ is perpendicular to the plane, the detector sees the fragments travel toward or away from it, hence the entire image can be filled with ion counts. Changing the direction of $\hat{\varepsilon}_{PR}$ changes the ion image. In cases C and D, one would expect a more uniform distribution, peaking at the center of the image, but instead photofragments are concentrated in particular areas. These changes are attributable to the perpendicular correlation between the recoil velocity and the angular momentum of the O_2 fragment.

A Monte Carlo simulation was used to model the results of the experiment. The parameters required for this simulation were the anisotropy parameter of the fragmentation (β), and the rotational quantum number and branch used in the first step in the REMPI scheme. The two-photon probabilities were calculated by taking the product of two one-photon probabilities via a virtual intermediate allowed transition. Five branches are possible; O, P, Q, R, and S. The dissociation process was assumed to be impulsive, therefore $\mathbf{v} \perp \mathbf{j}$. Finally, it was assumed that no $\mathbf{\mu}$–$\mathbf{v}$–$\mathbf{j}$ correlation exists. This finding is not strictly true, as pointed out by Hall et al. [38], but it was found that this correlation can be ignored in the calculation.

Successful simulations of the ion images were made for this experiment using the model just described. It was found that an anisotropy parameter

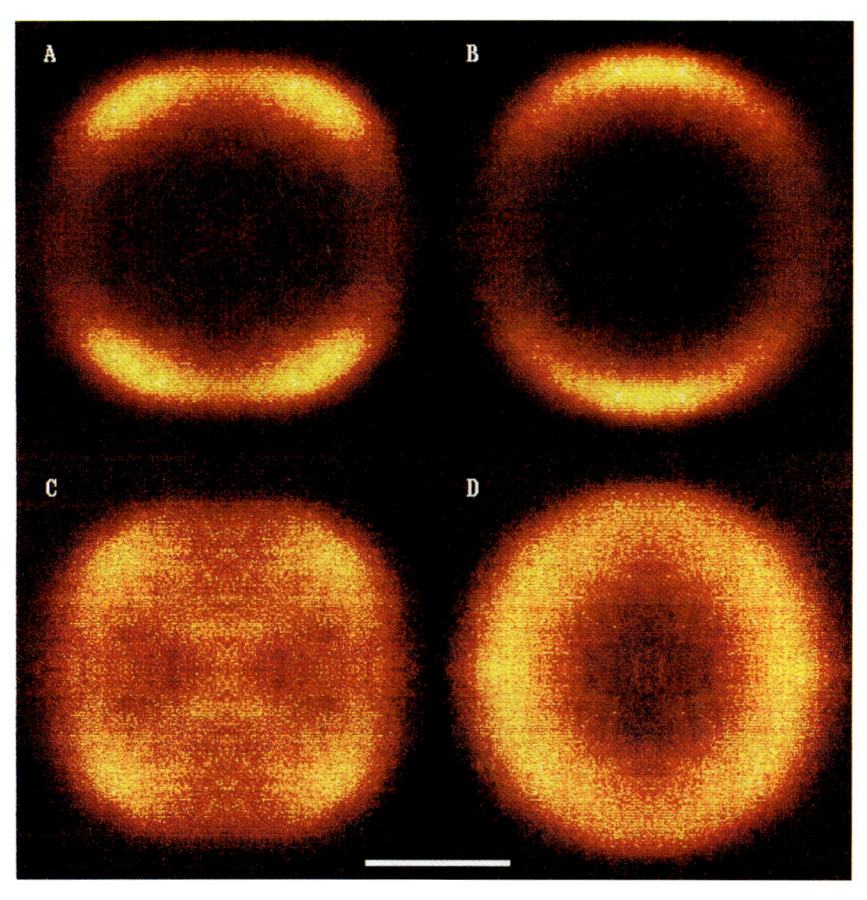

Figure 7. Symmetrized photofragment images of $O_2(a^1\Delta_g)$ produced by photolysis of ozone at 248 nm and recorded on the (4-0) band. The four images correspond to different experimental geometries as follows: (A) both ϵ_D and ϵ_{PR} are parallel to the plane of the image. (B) ϵ_D is parallel to the plane, ϵ_{PR} is perpendicular to it. (C) ϵ_D is perpendicular to the plane, ϵ_{PR} is parallel to it. (D) Both ϵ_D and ϵ_{PR} are perpendicular to the plane of the image. Reprinted from A.G. Suits, R.L. Miller, L.S. Bontuyan, and P.L. Houston, *J. Chem. Soc. Faraday Trans.* **89,** 1443 (1993). Reprinted by permission of The Royal Society of Chemistry.

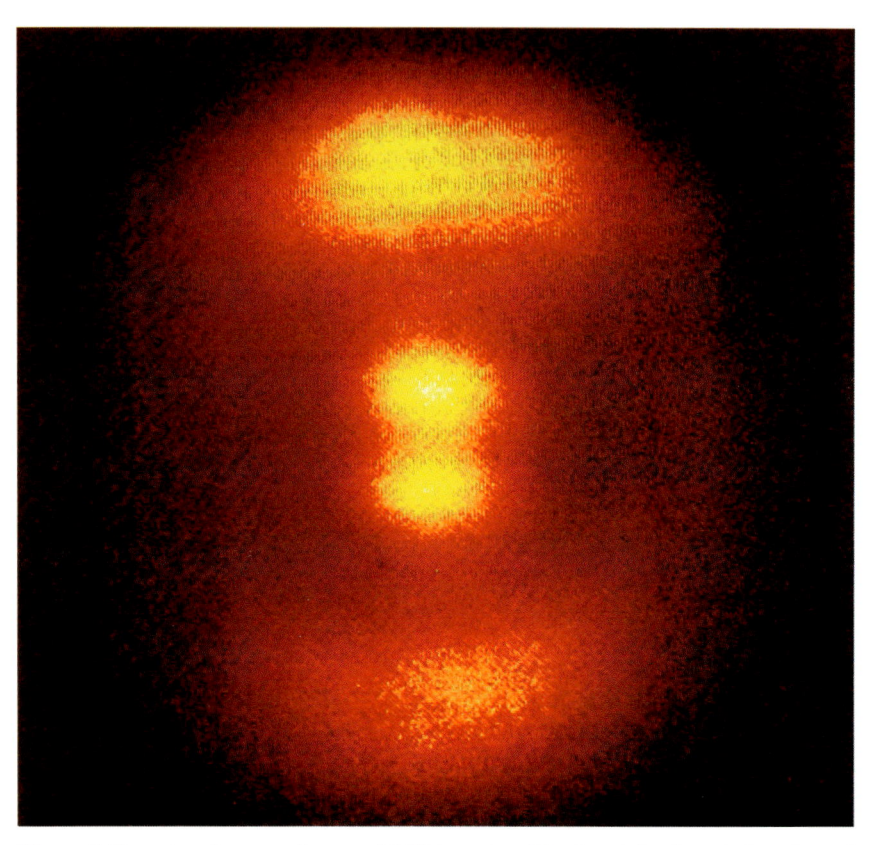

Figure 8. Raw photofragment image of $O(^3P_2)$ produced by the triplet dissociation channel of ozone photolyzed by 226 nm light. The laser polarization is aligned vertically in the figure. Reprinted with permission from R.L. Miller, A.G. Suits, P.L. Houston, R. Toumi, J.A. Mack, and A.M. Wodtke, *Science* **265,** 1831-1838 (1994). Copyright 1994 American Association for the Advancement of Science.

of $\beta = 1.18$ provided the best fit of the data. The other adjustable parameters used in the simulation were $j = 38$, P branch for the (4–0) band, and $j = 34$, O branch for the (4–1) band. The difference between the (4–0) and (4–1) images are attributable to the fact that different branches were used to ionize the O_2. This experiment clearly shows that ion imaging can be used to obtain not only scalar and angular velocity distribution but also $\mathbf{v} \cdot \mathbf{j}$ correlation in a state-specific manner.

The ion images of the triplet channel from the Hartley bands yielded results of interest to both physical and atmospheric chemists. Theoretical calculations of the amount of stratospheric ozone have often "come up short" [110] and several hypotheses have been proposed and tested to explain this "ozone deficit" [111, 112]. One of the more recent calculations proposes that highly vibrationally excited ground-state oxygen molecules can contribute to ozone formation through the following mechanism [113]:

$$O_3 + hv(\lambda < 243\,\text{nm}) \rightarrow O_2(X^3\Sigma_g^-, v \geq 26) + O(^3P) \tag{24}$$

$$O_2(X^3\Sigma_g^-, v \geq 26) + O_2 \rightarrow O_3 + O(^3P) \tag{25}$$

$$2 \times [O(^3P) + O_2 + M \rightarrow O_3 + M] \tag{26}$$

$$\text{Net} \qquad 3O_2 + hv(\lambda < 243\,\text{nm}) \rightarrow 2O_3 \tag{27}$$

Although work remains to be done on this problem, stimulated emission pumping (SEP) experiments strongly suggests the possibility of its occurrence. The SEP studies of highly vibrationally excited oxygen show a sharp increase in the disappearance of $O_2(v \geq 26)$ when $O_2(v = 0)$ is the collision partner [114] and the existence of a "dark channel" above $v = 25$, where molecules prepared in a given vibrational state do not cascade down the vibrational ladder [115]. A second SEP experiment demonstrated that $O_2(v = 0)$ removes $O_2(v \geq 26)$ at a rate 25–150 times faster than N_2 [113]. Subsequent studies revealed that other common atmospheric constituents such as CO_2 cannot compete with self-relaxation [116].

Can we show that the dissociation of ozone produces highly vibrationally excited ozone in significant amounts? To answer this question, ion imaging experiments were carried out on the ozone molecule with a 226-nm polarized laser providing the photolysis light [109]. This photolysis laser also ionized the $O(^3P_j)$ fragment via a (2 + 1) REMPI scheme using the $O(3p^3P_j \leftarrow \leftarrow 2p^3P_j)$ transition. The three $O(^3P_j)$ transitions were at 226.23 nm for $j = 0$, 226.06 for $j = 1$, and 226.25 for $j = 2$. The laser was scanned over the entire Doppler profile to ensure a complete image was recorded.

The image collected from this reaction, seen in Figure 8 (see color insert), is striking. By probing $O(^3P_2)$, we see a double dumbbell shape with the ion counts situated about the direction of the polarization axis of the laser. The outer dumbbell corresponds to O atoms with high kinetic energy release (~2 eV); this corresponds to O_2 vibrational energy with maximum probability around $v = 14-15$. The second, inner dumbbell corresponds to very low kinetic energy release (<0.4 eV), produced in coincidence with O_2 being formed in a high vibrational state ($v \geqslant 26$). The LIF experiments performed in a cell on ozone photolysis at this wavelength, in which vibrational populations up to $v = 26$ were quantified, confirmed that the double lobes in the ion image correspond to O_2 being formed with a significant fraction of vibrational energy above $v = 26$ [109].

Work by Syage [94, 95] using his core-extraction deflector plate method has confirmed the findings of the Houston group for the triplet dissociation channel of ozone. By looking at the O-atom velocity distribution, the double dumbbell shaped image was seen once again. Syage was also able to measue the anisotropy parameter as a function of kinetic energy release, and he found that it decreases with lower E_{trans}.

Recently, the first observations have been made of a spin-forbidden channel $O_3 + hv \rightarrow O(^1D_2) + O_2(^3\Sigma_g^-)$ at photolysis wavelengths of 317–327 nm using Doppler profiles of VUV LIF spectra of the $O(^1D_2)$ fragment [117]. This result may have very important implications for tropospheric chemistry, since new channels for the production of $O(^1D)$ potentially mean new source terms for OH. Ozone photochemistry clearly shows the fascinating complexity possible even for a triatomic molecule, the dynamics of which are only slowly being revealed by application of the newest correlation methods.

b. Studies of Other Triatomics, OCS, N_2O, ClNO, BrNO, and NO_2. The photodissociation dynamics of many triatomic molecules are complicated due to multiple overlapping electronic transitions in the parent leading to many degenerate or near-degenerate dissociation pathways. One classic example is the photodissociation of OCS at 222 nm [31]. The nascent CO rotational distribution has been shown to be bimodal, peaking at $j = 56$ and $j = 67$. Careful examination of the CO Doppler profiles indicated that both dynamically distinct channels lead to formation of $S(^1D_2)$. The bimodal distribution is thought to be due to the molecules dissociating via different excited electronic surfaces in the parent. These different surfaces are due to a Renner–Teller splitting between the two nominally degenerate bending modes of linear OCS.

Following the Doppler spectroscopy work of Sivakumar et al. [31], several ion imaging experiments have been done on OCS. As mentioned previously, when OCS is excited by a UV photon around 220 nm, the major dissociation channel is ground-state CO and $S(^1D_2)$. A minor channel produces ground-state CO and $S(^3P)$ [50]. No vibrational excitation of CO occurs from this process. Excitation from the linear ground-state $^1\Sigma^+(1^1A')$ OCS molecule accesses the nominally forbidden $^1\Delta(2^1A')$ and $^1\Sigma^-(1^1A'')$ through vibronic coupling. Ion imaging experiments near 217 and 230 nm conducted by the Sato et al. group [118] have confirmed the results of the Doppler experiments. The CO was resonantly ionized using a (2 + 1) REMPI scheme via the $B^1\Sigma^+$, $C^1\Sigma^+ \leftarrow \leftarrow X^1\Sigma^+$ transitions. The CO rotational distribution is bimodal and β increases with increasing rotational quantum numbers of CO. These results are consistent with the results of Sivakumar et al. [31] and recent PES calculations of Nanbu et al. [119]. For low rotational states of CO, there are direct dissociation pathways of both A' and A'' symmetry; the low β values for these rotational states are actually the incoherent average of the β values for both surfaces. For higher rotational states, an additional pathway opens up on the $2^1A'$ surface, indicated by the bimodal rotational distribution. This channel is characterized by larger β values.

Ion imaging of nascent S atoms from OCS photodissociation can be achieved with a (2 + 1) REMPI scheme using the $^1F_3 \leftarrow \leftarrow {}^1D_2$ and the $^3P_{1,2} \leftarrow \leftarrow {}^3P_2$ resonances. A bimodal velocity distribution of $S(^1D_2)$ atomic photofragments was seen in such experiments [130]. The low recoil velocities, corresponding to high rotational states of CO ($j \approx 67$), had a large amount of angular anisotropy, while those with high recoil velocities, correlating with lower rotational states of CO, has a more isotropic distribution. This result is consistent with the previous OCS experiments in which CO was probed. The $S(^3P_2)$ atoms also show a bimodal distribution, fitted by two Gaussian components, one with a wider distribution overlapping the other. The wider component has its peak at slightly larger energy. The β_{eff} parameter for the narrower peak is 1.1, and for the wider peak, $\beta_{\text{eff}} = 0.3$. Since this distribution is fairly similar to the singlet distribution, it indicates that triplet formation occurs via intersystem crossing after singlet–singlet excitation.

Vector correlation of the $S(^1D_2)$ atomic photofragment has also been observed in the OCS photodissociation at 223 nm. Mo et al. [131] determined the alignment of the $S(^1D_2)$ atom angular momentum with the recoil velocity from ion images taken with the polarizations of the probe and photolysis lasers in various relations to one another and relative to the plane of the image, as was done in the O_2 imaging experiments in ozone photolysis (see Fig. 7). The population of the magnetic sublevels ($M_J = -2, -1, 0, 1, 2$) with the recoil velocity as the axis of symmetry was found in order to fit each observed image. This analysis showed that, as expected, $\mathbf{v} \cdot \mathbf{j}$ alignment was mostly perpendicular ($m = 0$) and that the slower channel, with more angular anisotropy ($\beta = 1.8$), also had more strongly polarized perpendicular $\mathbf{v} \cdot \mathbf{j}$ correlation.

In the case of N_2O, the major photodissociation pathway produces ground-state N_2 and $O(^1D_2)$ with a branching ratio of more than 90% [122–124]. Previous studies of the O atom velocity distribution measured $\langle E_{trans} \rangle = 27$ kcal mol^{-1} and $\beta \sim 0.5$ for photolysis around 200 nm [124–127]. However, N_2O should be subject to the same Renner–Teller effects as OCS, since they are isoelectronic. This possibility was recently investigated. A study of N_2O photodissociation at 205 nm using $(2 + 1)$ REMPI of the O atom ($^1P_1 \leftarrow \leftarrow {}^1D_2$) as the probe produced ion images [128]. The ion image of the O atom shows at least two components as seen in Figure 9; one component has a $\cos^2\theta$ angular distribution relative to $\hat{\varepsilon}_D$, while the other component, which has a larger recoil velocity than the first component, has a $\sin^2\theta$ distribution. The scalar kinetic energy release distribution

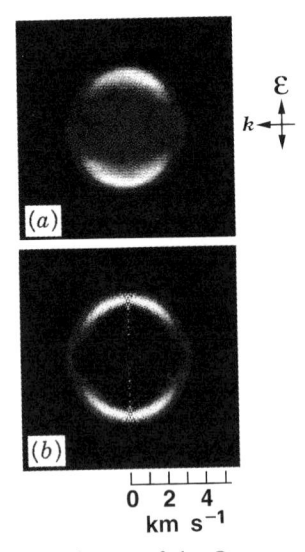

Figure 9. (a) Raw photofragment image of the O atom produced by the photolysis of N_2O with 205-nm light. The laser polarization is aligned vertically in the figure. (b) The inverse Abel transform image of (a). [Reprinted from Chemical Physics Letters, vol. 256 T. Suzuki, H. Katayanagi, Y. Mo, and K. Tonokura, "Evidence for multiple dissociation components and orbital alignment in 205 nm photodissociation of N_2O", 90–95, Copyright © 1996 with kind permission of Elsevier Science-NL, Sara Burgerhartstraat 25, 1055 KV Amsterdam, The Netherlands.]

was fitted with two Gaussians corresponding to the two components, both peaking at about 26 kcal mol^{-1}. The major component ranged in E_{trans} from about 15 to 45 kcal mol^{-1}, while the other component has a much larger width, ranging from 0 to 60 kcal mol^{-1}. The first component is slightly larger in intensity, although the second component seems much less intense since it is spread out over more of velocity space. The angular distribution was found to have three parts; at low E_{trans}, the effective beta parameter, β_{eff}, was measured to be 0.0, and at $E_{trans} \sim 25$ kcal mol^{-1}, β_{eff} was measured to be 1.3. At high E_{trans}, β_{eff} was found to be negative, indicating a perpendicular transition. However, Suzuki et al. [128] were hesitant to say that a third component was present in their measurement due to the possibility that orbital alignment of the O($^1D_2$). They noted that the extreme values for the angular distribution do not fall on $\theta = 0°$ and 90°, and this behavior was also seen when imaging nascent O_2 from ozone photolysis as discussed above and as seen in Figure 7. The average kinetic energy release and anisotropy parameter ($\langle E_{trans} \rangle = 27$ kcal mol^{-1} and $\beta = 0.6$) from this experiment matched previous studies. Their conclusions were that the major component with $\beta > 1.0$ is due to excitation to an A' Renner–Teller component of the $B^1\Delta$ state, while the other component is tentatively assigned to overlapping transitions to the $A'(^1\Delta)$ and $A''(^1\Sigma^-)$ states, in direct analogy to OCS.

Projection TOF techniques can also be used in the study of triatomic photolysis, as seen in the study of branching ratios in ClNO and BrNO dissociation by the Qian group. Nitrosyl chloride was dissociated at various wavelengths of light from 600–355 nm [129]. At these wavelengths, the two spin–orbit states of both NO($^2\Pi_{3/2}$ and $^2\Pi_{1/2}$) and Cl($^2P_{3/2}$ and $^2P_{1/2}$) are energetically accesible. With the use of an ab initio study and their measurements of the Cl($^2P_{3/2}$)/Cl($^2P_{1/2}$) and NO($^2\Pi_{3/2}$)/NO($^2\Pi_{1/2}$) correlated branching ratios, they were able to construct a correlation diagram between the various electronic states of the parent molecule and the NO and Cl fragments. In BrNO photodissociation at 355 nm, from the projection TOF data, the transition dipole moment was found to be along the Br—N bond ($\beta = 1.95 \pm 0.2$) [130]. A high amount of rotational excitation of the NO ($j \approx 50$) was produced by the dissociation; for larger amounts of NO rotation, an increase in the Br($^2P_{1/2}$)/Br*($^2P_{3/2}$) branching ratio was observed.

In NO_2, both the NO and O fragments have several low-lying electronic states that are accessible in the dissociation process. The NO fragment has two spin–orbit states ($^2\Pi_{3/2}$ and $^2\Pi_{1/2}$), while the O atom can have three possible spin–orbit states ($^3P_2$, $^3P_1$, and $^2P_0$). A total of 18 doubly degenerate pathways are possible for NO_2 dissociation [101]. Ion imaging experiments [131, 132] on both the NO and O fragments formed in the photodissociation of NO_2 at 355 nm show that this is a strongly parallel transition ($\beta = 1.2 \pm 0.3$). The NO molecule was probed for a variety of rotational states while for the O atom, only the $^3P_2$ state was probed. In examination of NO fragments, possible effects due to $\mathbf{v} \cdot \mathbf{j}$ correlation were minimized by saturating the NO ($A \leftarrow X$) transition. The image from the O atom shows a bimodal distribution corresponding to the two energetically accessible vibrational states of NO. Earlier room temperature studies of NO_2 dissociation at 355 nm had measured $\beta = 0.6$, a depressed value due to the increased parent rotational motion present in the cell experiments. This difference could be used to estimate the lifetime of the dissociating state of NO_2 using the thermal averaging relation [133]:

$$\beta = \frac{1}{2} P_2(\cos\theta)\left(1 + 3\gamma e^\gamma \int_\gamma^\infty \frac{e^{-t}}{t} dt\right) \quad \gamma = \frac{I}{8k_B T \tau^2} \quad (28)$$

where T and I are the temperature and moment of inertia of the parent, respectively. The dissociative lifetime of the parent is τ, and k_B is Boltzmann's constant. Comparison between the beam and room temperature experiments yield a lifetime for NO_2 dissociation of 200–400 fs.

As mentioned in an earlier part of this section, the Reisler group has performed experiments on NO_2 dissociation, using a spatially resolved velocity probing method to examine the branching ratio of the O atom formed with NO in a specific quantum state [91]. These experiments were performed with photolysis wavelengths close to the dissociation threshold for NO_2, therefore the kinetic energy release of the fragments was kept at a minimum. The $^3P_2$ and $^3P_1$ states of the O atom were clearly resolved, and for some rotational states of NO the $^3P_0$ state could be observed. The results showed definite fluctuations in the $^3P_1:^3P_2$ ratio when different Λ-doublets were probed. This result is presented in Figure 10. These fluctuations are thought to be due to either quantum interference among resonances in the parent molecule or changes in the matrix elements that couple long-range electronic states. For O atoms correlated to both the ground $^2\Pi_{1/2}$ and excited $^2\Pi_{3/2}$ states of NO, the $^3P_1:^3P_2$ ratio is colder than statistically predicted; the $^3P_1:^3P_2$ ratio for an O atom correlated to the $^2\Pi_{1/2}$ state of NO was colder than that for the $^2\Pi_{3/2}$ state. A simple model [144] based on infinite-order sudden approximation and the Franck–Condon approximation has been found to reproduce these results in a qualitative fashion. Long-range, nonadiabatic transitions between the spin–orbit states that occur beyond the transition state are considered the cause of the observed spin–orbit distributions [135].

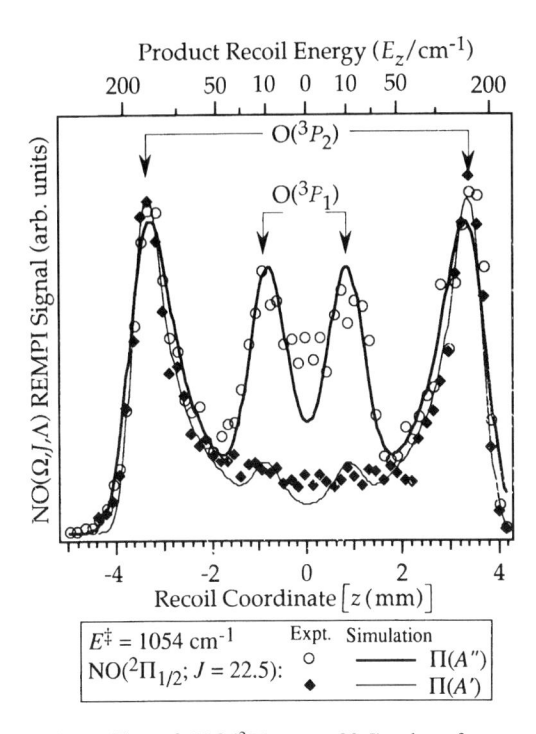

Figure 10. Spatial profiles of NO($^2\Pi_{1/2}; j = 22.5$) photofragment obtained from NO$_2$ photodissociation with the photolysis energy 1054 cm^{-1} above threshold. The $Q_{11} + P_{21}(22.5)$ transition was used to obtain the NO spatial distribution of the $\Pi(A'')\Lambda$-doublet state, while the $R_{11} + Q_{21}(22.5)$ transition was used for the $\Pi(A')$ state. The arrows indicate the peaks correlated to $^3P_1$ and $^3P_2$ states of the O atom, while the top axis indicates NO + O relative translational energy along the z axis. See the text for further details. [Reprinted with permission from A. Sanov, C. R. Bieler, and H. Reisler, *J. Phys. Chem.*, **99**, 13637 (1995). Copyright © 1995 American Chemical Society.]

3. **HNCO Photodissociation.** Isocyanic acid (HNCO) has two possible dissociation channels.

$$\text{HNCO} + h\nu \rightarrow \text{H} + \text{NCO} \tag{29}$$

$$\text{HNCO} + h\nu \rightarrow \text{HN} + \text{CO} \tag{30}$$

Previous studies of the translational energy distribution of REMPI detected CO produced from the photolysis of a thermal beam of HNCO gave a C—N bond dissociation energy of 41530 ± 100 cm^{-1} [136]. However, recent studies by Brown et al. [137, 138] and Zyrianov et al. [139] found that the actual bond energy is 42,620 ± 100 cm^{-1}. The discrepancy is due to the fact that hot band transitions in HNCO, especially bending modes, have a larger probability than transitions starting from the ground vibrational state, because of the bend geometry in the excited electronic state. Excitation of these hot bands can give a false threshold energy for dissociation. Ion imaging experiments of nascent CO reported by Kawasaki et al. [140] showed that dissociation at 217 nm occurs via a perpendicular transition with $\beta = -0.7$. The Reisler group recently performed an extensive survey of the HNCO photodissociation reactions in the 220–260-nm range using ion imaging [139]. Since they performed their experiments in a molecular beam, they greatly reduced the importance of hot-band photolysis, and therefore they were able to derive a more accurate value for the HN—CO bond energy. Their results confirmed the higher value found by the Crim group, who used careful LIF measurements of the NH yield to determine the bond energy [137, 138]. Reisler's group was also able to examine both the H as well as the CO product, and concluded that at these photolysis wavelengths, both channels undergo internal conversion to the ground electronic state.

4. Ion Imaging of CH_3I, CD_3I, CH_3Br, and CD_3Br. The mono-halogenated methanes offer examples of photochemistry that are easily probed and quite fascinating. In many cases, there are two possible electronic channels for the halogen. In CH_3I and CD_3I photodissociation, iodine is produced in both the ground ($^2P_{3/2}$) and the excited state ($^2P_{1/2}$) as seen in Figure 1a. For 266-nm photolysis, it has been found that the majority of the iodine is formed in the excited channel [141]. For photolysis around 266 nm, almost 90% of the available energy goes into translation [142–146], and the only two vibrational modes in CH_3 and CD_3 are excited by the dissociation: the v_1 C—H symmetric stretch [88, 147] and the v_2 umbrella inversion motion [148]. The excited electronic states about 266 nm are referred to as $^3Q_1$, $^3Q_0$, and $^1Q_1$ [149]. The majority of the oscillator strength in the excitation of CH_3I or CD_3I goes into the $^3Q_0$ state [145] which correlates to the formation of I*($^2P_{1/2}$); the other states correlate to I($^2P_{3/2}$). With pulsed extraction TOF methods, one can ascertain the I/I* branching ratios as a function of v_1 and v_2 vibrational excitation probing CH_3 or CD_3 with REMPI, revealing interesting information about the nature of the electronic transition from the velocity vector anisotropy.

One interesting result on CH_3I and CD_3I photodissociation at 266 nm employed the core-extraction method [85, 87]. It was found that the I/I* ratio increases with increasing amounts of v_2 vibration, as shown in Figure 11. It was also found that there was a strong preference for the rotation of

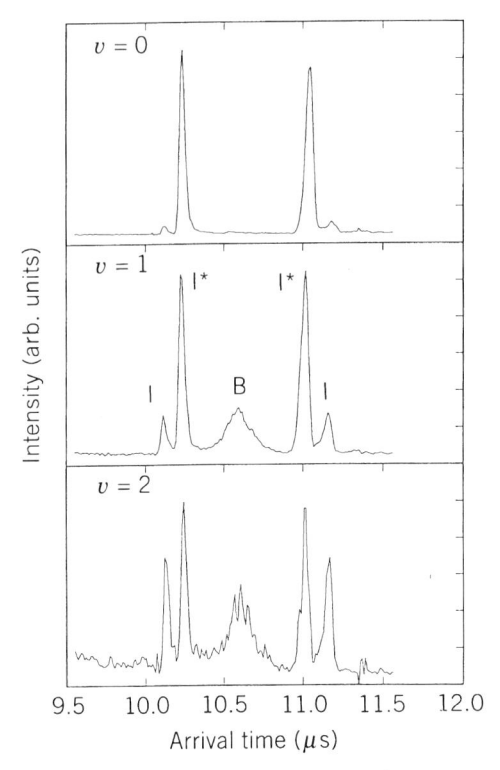

Figure 11. Core-extraction TOF spectra of CH$_3$ from the 266-nm photodissociation of CH$_3$I. The three spectra were collected by probing transitions of the CH$_3$ product with different amounts (from top to bottom $v = 0$, $v = 1$, and $v = 2$) of vibrational energy in the v_2 umbrella mode. In the middle panel, the peaks corresponding to the formation of I($^2P_{1/2}$) and I*($^2P_{3/2}$) are labeled, along with a peak resulting from background (B). [Reprinted with permission from R. Ogorzalek Loo, H.-P. Haerri, G. E. Hall, and P. L. Houston *J. Chem. Phys.*, **90**(8), 4222 (1989). Copyright © 1989 American Institute of Physics.]

the CH$_3$ fragment to be perpendicular to its symmetric top axis ($K \sim 0$), and that the angular distribution of the photofragments is highly anisotropic ($\beta = 1.7$–1.8).

From the results of the 266-nm photolysis of CD$_3$I, the rotational N- and K-state populations of the methyl fragment, where N is the nonelectronic rotational quantum number and K is the symmetric top molecule-fixed projection quantum number, can be fitted by a double exponential function with two temperatures, one of the N-state and the other for the K-state population. These two temperatures, T_N and T_K, were found to be 150 and

20 K, respectively. The N-state rotational temperature reflects the CH_3—I bending modes that turn into rotation during the dissociation. The K-state rotational temperature reflects the initial rotational temperature of the parent molecule.

In a ion imaging experiment of CH_3I dissociation by 266-nm light, the rotational structure of the 0_0^0 band could be measured [150]. When the wavelength of the probe laser was tuned to the Q-branch band head, the ion image shows a distribution of ion counts centered around the polarization axis of the photolysis laser, the result expected for a parallel transition. A cut through the center of the image reveals two peaks of the same recoil velocity: these correlate to the formation of I*. The small shoulder on either side of the main peaks correlates to ground-state iodine. Under these probe conditions, the I/I* ratio for the 0_0^0 band was determined to be 0.12. Probing the $P(4)$ transition of CH_3 reveals something different. Now the cut through the center has three distinct peaks, which is due to blending of the 1_1^1 vibrational band with the $P(4)$ line of the 0_0^0 band. The three peaks can be correlated to I* coincident with ground-state CH_3, and both I and I* coincident with CH_3 with one quanta of excitation in the stretching mode. By using CD_3I instead of CH_3I, a separation of the 1_1^1 band in frequency from the 0_0^0 transitions was achieved [151]. It was found that this band correlated with an I/I* ratio of 1.3 ± 0.2. Similar to earlier results of the core-extraction experiments, where the I/I* ratio increases with increasing quanta of v_2, this experiment gave a similar result for excitation of v_1. According to the Landau–Zener one-dimensional curve-crossing model [152], the logarithm of the branching ratio I*/(I + I*) should scale linearly with the velocity at the curve-crossing region. For each v_2 vibration, the branching ratio is increasing much faster than predicted by Landau–Zener. This bending indicates that the simple one-dimensional model is inadequate to describe the coupling between the I and I* electronic states and the v_2 vibration mode at the curve-crossing region.

In a separate experiment, the Chandler group was able to probe the alignment of the CH_3 fragment with respect to the CH_3—I axis [153]. In this experiment, they replaced the CCD camera with a photomultiplier tube (PMT), and placed a mask between the phosphor screen and the PMT. The mask was placed so that only the center part of the image would be seen by the PMT; the acceptance angle was $15°$. The polarization of the photolysis laser was set to be parallel to the flight path (perpendicular to the plane of the image). The net effect is that the detector will now measure only the intensity of the ion signal for certain velocity groups. Of course, scanning the probe laser frequency will detect different rotational states of CH_3 being excited through different branches. Changing the angle between the polarizations of the probe and photolysis lasers will change the intensities of the

CH$_3$ REMPI transitions; these changes will be caused by the alignment of CH$_3$ rotation relative to photolysis laser polarization. Different branches will depend on the (N, K) levels of the CH$_3$ rotation in different ways. For example, the $P(2)$ line only sees $(N, K) = (2, 1)$, while $R(2)$ sees $(N, K) = (2, 1)$ and $(2, 2)$ and $S(2)$ sees $K = 0, 1$, and 2 sublevels. From these, the $A_0^{(2)}(N, K)$ and $A_0^{(4)}(N, K)$ rotational alignment parameters were obtained.

Hertz and Syage examined CH$_3$I dissociation using their version of the core extraction method to look at the I and I* fragments at photolysis wavelengths of 266 and 304 nm [93, 95]. They detected weak perpendicular transitions at both wavelengths in the I-atom TOF spectra, indicating direct absorption to either the ^1Q excited state, in the case of 266-nm photolysis, or the $^3Q_1$ state, in the case of 304-nm photolysis [see Fig. 1(a)]. From the ratio of the perpendicular transition cross-section to that of the parallel transition and from the I/I* branching ratio, the surface crossing probability was calculated to be 0.25 ± 0.05 for the ^1Q state and 0.43 ± 0.08 for the $^3Q_1$ state.

The Chandler group at Sandia has also examined the photolysis of brominated methanes (CH$_3$Br and CD$_3$Br) by 205-nm light [154]. This time, the bromine atoms, formed in both the ground ($^2P_{3/2}$) and excited states ($^2P_{1/2}$), were probed using a (2 + 1) REMPI scheme at 250.98 and 277.77 nm, respectively. The results were similar to the iodomethanes. The ground-state Br atoms have a distinct $\sin^2\theta$ component in their distribution, while excited state Br* atoms have a $\cos^2\theta$ distribution. Furthermore, the Br atoms have a broader speed distribution than the Br* atoms, indicating that there is more internal excitation of the CH(D)$_3$ fragment correlating to ground-state Br formation.

5. Fragmentation of Larger Molecules Using Pulsed-Extraction Methods.

Several pulsed extraction TOF experiments have also been done on molecules with more than five atoms. Suzuki et al. [165] performed ion imaging experiments on *trans*-dichloroethylene at 193 and 235 nm, examining the speed and angular distribution of both the Cl ($^2P_{3/2}$) and Cl ($^2P_{1/2}$) channels [155]. The images show two components; one corresponding with a Boltzmann-like $P(E_{\text{trans}})$ and the other corresponding with a Gaussian-like one. Strong angular anisotropy is seen in the Gaussian-like distribution, while weak anisotropies are seen for the other channel. The two distributions are manifestations of different dissociation events. Surface crossing between the (π, π^*) and either (n, σ^*) or (n, π^*) state causes rapid C—Cl bond breaking and the highly anisotropic, Gaussian-like distribution, while internal conversion causes the more isotropic, Boltzmann-like behavior.

Hwang and El-Sayed [156] performed analogous experiments on the photodissociation of C$_2$F$_5$I at 304.7 nm using the core-extraction method.

They probed both the ground- and excited-state I atoms. As with methyl halides, a strong parallel transition ($^3Q_0 \leftarrow N$) was observed. The velocity distribution of the excited-state I* atom was more anisotropic than the ground-state I atoms. This result suggests that the ground-state I atoms are formed in two different ways; through both a direct dissociation from the $^3Q_1$ state as well as curve crossing from $^3Q_0$ to ^1Q. The perpendicular transition was found to release more translational energy than the parallel transition. Hwang and El-Sayed [166] suggest the reason to be less vibrational energy redistribution through the direct channel as opposed to the curve-crossing one.

With the use of their core-extraction method to study the photodissociation of iodobenzene, the El-Sayed group found that this molecule could dissociate through two channels. At a photolysis energy of about 304.7-nm one channel was found to be a direct dissociation through a $\sigma^* \leftarrow n$ excitation, and the other was found to be excitation to the predissociative π,π^* states [156, 157]. The direct channel was found to be a parallel transition ($\beta \approx 1.6$), while for the slow channel, β was found to rapidly decrease with decreasing kinetic energy release (E_{trans}). The predissociative iodobenzene apparently has a long enough lifetime to redistribute its energy from the C—I bond into the internal degrees of freedom of the rest of the molecule before it dissociates, leaving less energy available for translation when the molecule finally breaks apart. While this redistribution is occurring, the molecule is also rotating and diminishing the angular anisotropy of the dissociation. However, using an aperture in a TOF experiment which measures vector correlation can introduce artifacts that are difficult to account for [158]. Therefore, the interpretation of vector correlations measured with core-extraction experiments must be treated with a certain amount of caution.

The dissociation dynamics of nitroso compounds in the molecular beam has been studied by Cline and co-workers [86, 159, 160]. These reactions, where R is $(CH_3)_2ClC$, for example, occur with visible

$$\text{RNO} + h\nu(600-650\,\text{nm}) \rightarrow \text{R} + \text{NO} \tag{31}$$

photolysis light. The NO fragment can be probed with (1 + 1) REMPI. The translational energy distribution for several rotational states of NO was measured; changing the polarizations of the photolysis and probe lasers allowed the bipolar moments $\beta_0^2(20)$, $\beta_0^2(02)$, and $\beta_0^0(22)$ to be measured as a function of the rotational state of NO. The scalar results show two components, as seen in Figures 12 and 13, and are consistent with the hypothesis that 2-chloro-2-nititrosopropane (CNP) dissociates along two pathways, a singlet and a triplet channel [86]. The singlet channel has no

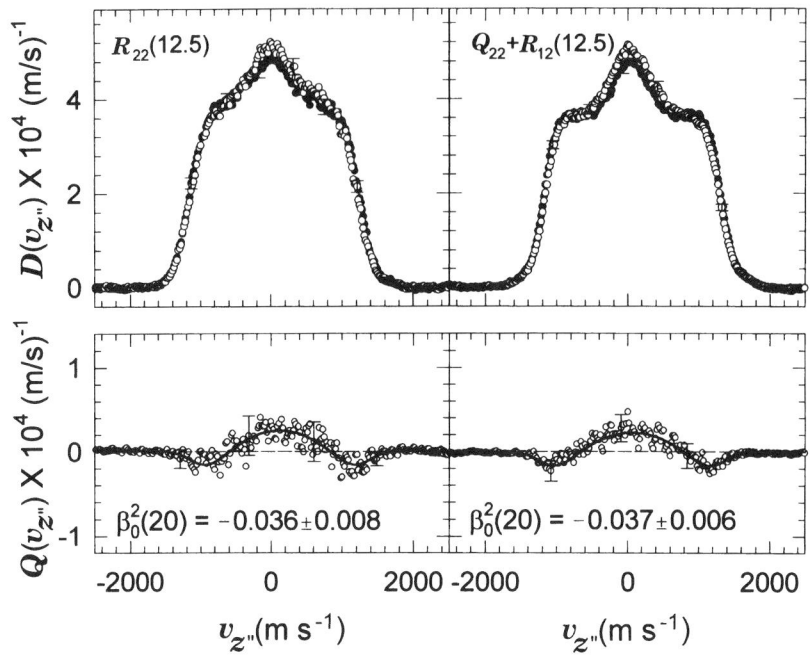

Figure 12. The upper panels show one-dimensional velocity distributions of NO from the photodissociation of 2-chloro-2-nitrosopropane at 650 nm using the projection TOF method. The NO ions were produced by REMPI via the $A^2\Sigma^+ \leftarrow X^2\Pi$ transition using the branches indicated in the figure. The velocity distributions were taken with $\hat{\varepsilon}_D$ and $\hat{\varepsilon}_{PR}$ collinear and both ε_{PR} and the probe laser propagation direction perpendicular the axis of the detector, $\mathbf{k}_D$. The symbols ○ and ● indicate velocity distributions taken with ε_D parallel and perpendicular to $\mathbf{k}_D$, respectively. The lower panel shows the difference spectra of the velocity distributions, which are sensitive to the $\beta_0^0(20)$ bipolar moment. The solid line shows a fit to the difference spectra using equations discussed in [42] and [170]. [Reprinted with permission from R. Uberna, R. D. Hinchliffe, and J. I. Cline, *J. Chem. Phys.*, **105**(22), 9847 (1996). Copyright © 1996 American Institute of Physics.

barrier and the C—N bond energy is about 13,000 cm^{-1}. The triplet channel has a barrier of about 1500 cm^{-1}. The results from the vector correlation measurements are consistent with this picture [160] and are also shown in Figures 12 and 13. The $\beta_0^2(20)$ parameter (indicative of $\mathbf{\mu} \cdot \mathbf{v}$ correlation) is very small, indicating a long-lived excited state. The $\beta_0^0(22)$ parameter ($\mathbf{v} \cdot \mathbf{j}$ correlation) becomes more negative as j increases for NO, meaning the $\mathbf{v}$ and $\mathbf{j}$ vectors tend to be perpendicular to one another. Phase space theory

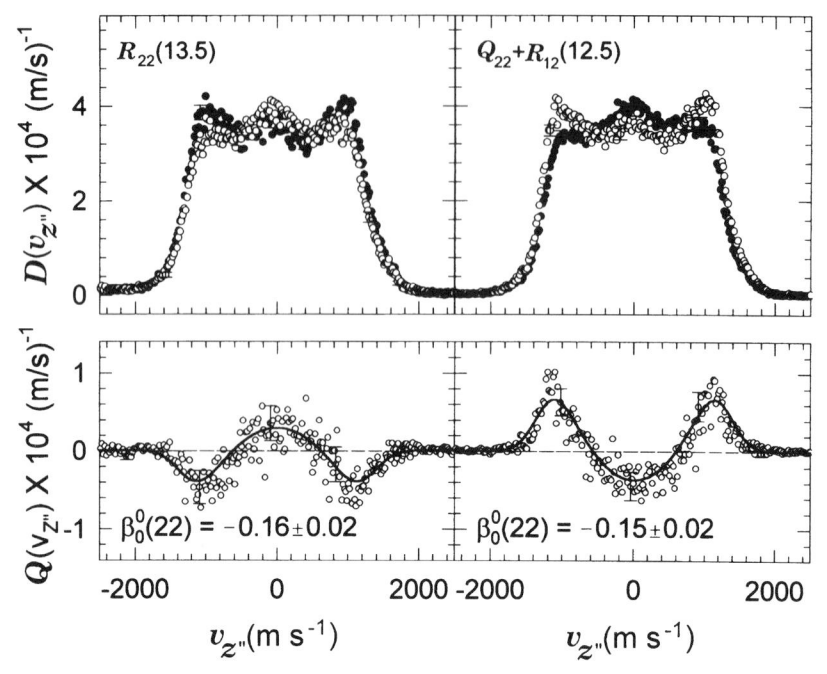

Figure 13. The upper panels show one-dimensional velocity distributions of NO from the photodissociation of 2-chloro-2-nitrosopropane at 650 nm using the projection TOF method. The NO ions were produced by REMPI via the $A^2\Sigma^+ \leftarrow X^2\Pi$ transition using the branches indicated in the figure. The geometry is similar to that in Figure 11, except now ε_D is perpendicular to $\mathbf{k}_D$ and ε_{PR} is either parallel (○) or perpendicular (●) to $\mathbf{k}_D$. The lower panel shows the difference spectra of the velocity distributions, which are sensitive to the $\beta_0^0(22)$ bipolar moment. The solid line shows a fit to the difference spectra using equations discussed in [42] and [170]. [Reprinted with permission from R. Uberna, R. D. Hinchliffe, and J. I. Cline, *J. Chem. Phys.*, 105(22), 9847 (1996). Copyright © 1996 American Institute of Physics.]

calculations underestimate the magnitude of $\beta_0^0(22)$ by a factor of 5. This value is indirect evidence of a large impulsive force at the transition state for dissociation, likely a barrier on the triplet surface.

The photodissociation of larger NO containing molecules has revealed more interesting results. By using a novel delivery system for introducing temperature labile parent molecules that have low volatility into a vacuum system as a beam of thermal molecules, the Cline group was able to study the photodissociation of 1-chloro-1-nitrosocyclohexane (CNCH) and 2-chloro-2-nitroso-6,6-dimethylbicyclo[3.1.1]heptane (CNMH) [161], and

CNP at room temperature. They were able to measure the effect of parent excitation on **v·j** correlation, finding that room temperature CNP has its $\beta_0^0(22)$ moment reduced by a factor of 2 compared to CNP in a molecular beam. One would expect that for the larger molecules, the $\beta_0^0(22)$ moments would be greatly reduced compared to CNP, but instead, they also found that CNP and CNMH have comparable $\beta_0^0(22)$ moments, while for CNCH it is smaller by a factor of 2. They postulate that the dissociation dynamics of these molecules depends more on parent rotational excitation and skeletal rigidity than on structural complexity.

6. Ion Imaging Secondary Dissociation. Recent ion imaging experiments have been able to look at multiple dissociation in a molecule. Two examples of this are acetyl chloride (H_3C—CO—Cl), which after absorbing a photon breaks into a Cl atom, carbon monoxide, and a methyl radical, and oxalyl chloride (Cl—CO—CO—Cl), which falls apart into two Cl atoms and two CO molecules after photon absorption [162]. Photofragment translational spectroscopy experiments of acetyl chloride by Butler and co-workers [163] were used to determine the nature of the excited state dissociation mechanism, namely an $n \rightarrow \pi^*$ photoexcitation followed by α-cleavage of the chloride. In the acetyl chloride photolysis at 236 nm, the Cl atom was found to have an anisotropic distribution ($\beta \approx 0.9 \pm 0.2$), indicating a fast dissociaton of this fragment. The CO and CH_3 were observed to have more isotropic distribution, suggesting that the dissociation occurs in two steps,

$$H_3CCOCl + h\nu \rightarrow (H_3C-CO^{\cdot})^* + Cl^{\cdot} \qquad (32)$$

$$(H_3C-CO^{\cdot})^* \rightarrow H_3C^{\cdot} + CO \qquad (33)$$

where $(H_3C-CO^{\cdot})^*$ indicates internally excited acetyl radicals. Ab initio calculations on this complex and its dissociation pathway predict a 19.1-kcal mol^{-1} barrier between CH_3CO and the CH_3 and CO fragments. Photofragment translational spectroscopy experiments [164] have measured the barrier height to be 17 ± 1 kcal mole^{-1}. The experimentally determined internal energy distribution of the acetyl radical, as determined from the ion images of the Cl atoms, was sufficient for some decomposition above the barrier to occur.

When oxalyl chloride absorbs a 230-nm photon, it breaks apart into four fragments. Ion images of the Cl and CO fragments from this reaction have been recorded [165]. Both Cl and CO have two components in their velocity and angular distribution. The fast component has a great deal of anisotropy, while the slower component is more isotropic. A simple model very similar to acetyl chloride dissociation can describe this dissociation event.

$$(ClCO)_2 + h\nu(230 \text{ nm}) \rightarrow Cl + [Cl(CO)_2^{\cdot}]^* \text{ (fast)} \quad (34)$$

$$[Cl(CO)_2^{\cdot}]^* \rightarrow CO + (ClCO^{\cdot})^* \text{ (fast)} \quad (35)$$

$$(ClCO^{\cdot})^* \rightarrow Cl + CO \text{ (slow)} \quad (36)$$

Here, $[Cl(CO)_2^{\cdot}]^*$ and $(ClCO^{\cdot})^*$ indicate excited radicals.

As is evident, the power of the pulsed-extraction ion TOF–MS techniques derives from their generality, a result of the fact that REMPI may be applied to a large number of molecules. But velocity measurements of low-energy ions is a difficult business. Stray fields are never truly eliminated even in the hands of the cleverest experimentalist. Space charge is another potential problem for which one must be on the look out. In order to carry out the highest resolution measurements of velocity, one must work with neutrals.

C. Neutral Time-of-Flight Methods

1. H-Atom Rydberg Tagging Experiments. From the pioneering work of Schnieder et al. [166], the Rydberg tagging neutral-TOF (nTOF) method has been used by many groups to study photodissociation processes in which one of the fragments is an H atom. The method has also been extended to tagging HD molecules [167], and it has also been used to study bimolecular reactions [168, 169]. With this kind of experiment, photodissociation of molecules prepared in the collision free environment of a molecular beam produces H-atom photofragments. These are then excited with two-color laser excitation to a high-lying, long-lived Rydberg state, which is allowed to travel a measured flight length to a field ionization detector. The polarization of the photolysis laser can be varied with respect to the detection axis. An extensive review of experiments using this method appeared in the last volume of *Advances in Photochemistry* [170]. We therefore feel no need to comment extensively on this method and will instead refer the reader to this excellent review. We would, however, like to draw the readers' attention to a few more experiments that were not covered by that review.

The Wittig group has used this method to study the H-atom channels in HNCO and H_2CO dissociation [171, 172]. They have also studied the predissociation of $HF(v = 3)$ prepared by overtone excitation and photodissociated by 193.3-nm light [173]. The TOF spectrum resulting from HNCO dissociation at 193 nm was found to have two components, the major one was from H atoms produced directly from the dissociation (Eq. 29), and the minor one was from the dissociation of NH

$$NH + h\nu(193\text{ nm}) \rightarrow H + N(^2P) \tag{37}$$

by a second 193-nm photon. The contribution of this minor H-atom source could be modeled by the forward convolution method for secondary dissociation. The measured dissociation energy for H—NCO from this experiment was $D_0 = 110.1 \pm 0.5$ kcal mol^{-1}, in agreement with previous measurements. The anisotropy parameter was found to be negative, $\beta = -0.85 \pm 0.1$, meaning a perpendicular transition ($A'' \leftarrow A'$) was involved in the process. The average translation energy release was found to be very high, $\langle E_{\text{trans}} \rangle \sim 70$ kcal mol^{-1}, in contrast to earlier Doppler work, which measured a much smaller value for $\langle E_{\text{trans}} \rangle$ [174]. The internal energy distribution of NCO obtained from the H-atom TOF shows a progression of v_2 bending modes up to 15 quanta of vibration, suggesting the importance of a bent HN—C—O A'' excited state.

One interesting advance obtained with this experimental approach is the measurement of velocity-resolved **μ**·**v** correlation in photodissociation [175]. The dissociation of NH$_3$ and ND$_3$ at 212 and 216 nm proceeds through an $\tilde{A}^1A_2'' \leftarrow X^1A_1'$ absorption band. The nTOF spectra show peaks that are clearly assignable to the rotational lines of the NH$_2$ (ND$_2$) fragment. By observing the changes in nTOF spectra as a function of the angle between the polarization vector of the photolysis laser and flight axis, the β parameter as a functon of NH$_2$(ND$_2$) rotation was able to be assigned. It was found that H atoms coinciding with (D atoms) rotationally cold NH$_2$(ND$_2$) tended to be ejected perpendicular to the C_3 axis of the parent, which is parallel to the transition moment. Hence, $\beta \sim -1$ in this regime. Hydrogen or deuterium atoms correlated to rotationally hot NH$_2$(ND$_2$) tended to be ejected with recoil velocities parallel to the C_3 axis, so that in this regime $\beta \sim +2$. An explanation of this behavior of β as a function of E_{trans} using the known potential energy surfaces of the X and A states and a classical impact parameter model was presented.

2. CO Metastable Experiments. The H-atom Rydberg tagging experiments of the Ashfold, Welge, and Wittig groups have shown the resolving power of nTOF, but what other molecules can be probed using nTOF methods? Experiments in our group have shown that this can also be accomplished with both CO and NO photofragments [176, 177], and there are certainly other molecules that are candidates. In recent experiments performed in our laboratory, our metastable detector has found previously unseen singlet–triplet optical transitions in acetylene. These transitions were measured to have a lifetime of about 100 μs, and were found to be especially strong in the region overlapping the 3_0^3 singlet transitions of acetylene. Neutral state specific TOF of CO or NO allows correlation measurements to be made in reactions in which both products are molecules. Instead of promotion to a high-lying Rydberg state, which are often predissociative for molecules, one excites the nascent molecules to a metastable state with a high spectral brightness laser, literally overpowering the forbidden transition to a long-lived metastable state. The metastable molecule can then be detected by surface ionization via an Auger-like process. In CO, the $a^3\Pi \leftarrow X^1\Sigma^+$ Cameron system can be used with very high efficiency, while in NO, one can access the $a^4\Pi \leftarrow X^2\Pi$ bands. In both molecules, spin-forbidden transitions are used. This requires the use of a high spectral brightness light source for best results.

Metastable nTOF measurements on CO produced by the photolysis of ketene at 308 and 351 nm have been performed in our laboratory [178–180]. The way these experiments were performed is slightly different than either the ion experiments or the H-atom Rydberg tagging experiments and deserves some comment. A diagram of the metastable TOF experiment is found in Figure 14. The molecules are excited to the $a^3\Pi$ metastable state by the frequency tripled output of a single-mode CW ring dye laser operating at 600 nm. A KD*P and a BBO crystal are used, first to frequency double the light, and second to mix the fundamental and doubled light to produce 200-nm light with 2–3 mJ of power and a 200-MHz Fourier transform bandwidth. The spectral brightness of this laser system is sufficiently high that a large fraction, about 5%, of the nascent CO molecules can be converted into metastables. The polarization of this probe laser can be varied by a Berek's polarization compensator to any desired polarization. The metastable detector itself is deceptively simple. It consists of a heated nickel foil, a set of nonimaging MCPs, and an electron repeller housed in a conducting ion shield. The metastables strike the nickel foil and produce electrons in an Auger-like process. These electrons are directed by the electron repeller onto the MCPs, where they are amplified by 10^8. The resulting pulses are sent into an amplifier discriminator and then to a multichannel scaler for time sorting and analysis.

Unlike the H atom nTOF measurements, the CO products have recoil velocities comparable to the molecular beam velocity of the parent, therefore a Jacobian transformation from center-of-mass (recoil velocity) coordinates to laboratory coordinates must be performed. To extract the $P(E_{\text{trans}})$, TOF spectra must be taken with the beam axis at different angles relative to the flight path of the detected photofragments; this angle is designated θ_{lab}. The

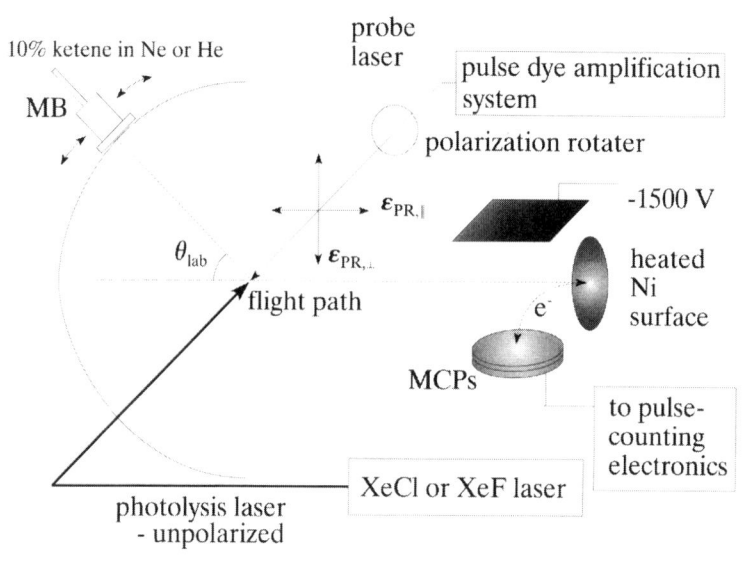

Figure 14. Schematic diagram of the CO metastable TOF experimental apparatus is shown. The molecular beam (MB) containing 10% ketene in neon or helium can be placed at any acute angle (θ_{lab}) relative to the flight path, and it is collimated by an electroformed skimmer (not shown). The photolysis laser is an unpolarized excimer (XeCl or XeF), and the probe laser is a pulse dye amplification system whose polarization can be made either parallel ($\varepsilon_{PR,\parallel}$) or perpendicular ($\varepsilon_{PR,\perp}$) to the flight path. The metastables pass through a 1-cm orifice and deflector plates and grids (both not shown), and they strike a heated Ni surface. Electrons produced from the Ni surface by the metastables are steered by a plate set at -1500 V onto a stack of 3 MCPs; the resulting pulses are then amplified, discriminated against noise from dark current, and counted by a multichannel scaler.

ability to change this angle allows one to obtain TOF spectra that are sensitive to different ranges of recoil velocities in the center-of-mass frame. A small θ_{lab} is sensitive to fragments with low kinetic energy release. As the angle increases, these low kinetic energy fragments will no longer be seen by the detector and sensitivity to the high kinetic energy release fragments will increase. In order to fit a set of TOF data, a $P(E_{trans})$ is guessed and fitted to a particular TOF spectrum at a particular angle, then this guess is compared to the TOF spectrum taken at a different angle. This guess is iteratively improved until all the data is reproduced. This procedure for extracting the $P(E_{trans})$ from TOF spectra is known as the forward convolution method. The Jacobian transformation can be accomplished with the following equation, which gives $I(\theta_{lab}, t)$, the intensity of the TOF signal of ketene photodissociation, for example, as a function of time and θ_{lab} [181]:

$$I(\theta_{lab}, t) = C \frac{v_{lab}^3}{t} \frac{m_{ketene} m_{CO}}{m_{CH_2}} \int \frac{P(E_{trans}, v_{CO}, j_{CO}) S(j_{CO}, \Omega, v; j'_{CO} \Omega')}{|v|} \qquad (38)$$
$$\times N_{beam}(v_{beam}, \omega_{beam}) dv_{beam} d\omega_{beam} d\omega_d$$

In Eq. 38, C is a proportionality constant, v_{lab} is the velocity of CO in the laboratory frame, m_i refers to the masses of molecule i, v_{CO} is the vibration quantum number of the CO fragment, $d\omega_d$ refers to the angle of capture of the detector, t is the flight time, Ω is the body-fixed projection of j_{CO}, and j'_{CO} and Ω' are the total angular momentum and its body-fixed projection for the electronically excited state, respectively. Several factors that affect the experiment can be seen from this equation. Forward scattering will be much more prominent than backward scattering due to the v_{lab}^3/t factor in the Jacobian, in contrast to the pulsed-extraction experiments. The angular and velocity distribution of the parent in the beam, $N_{beam}(v_{beam}, \omega_{beam}) dv_{beam} d\omega_{beam}$, affects the TOF data and needs therefore to be quantified.

a. Ketene Photodissociation at 308 nm. Considered a prototype of unimolecular dissociation reactions, ketene is a molecule whose photochemistry has been studied extensively [182–191]. The potential energy curves associated with ketene photolysis are shown in Figure 1(b). In the wavelength region below 330 nm, there are two dissociation pathways for ketene [185, 187]. The first pathway, called the singlet channel, occurs through an internal conversion to the ground-state continuum. It produces excited-state singlet methylene ($a^1A_1 CH_2$) and its threshold is 30,116 cm^{-1} above the ground state [188]. It has no barrier to dissociation, which makes this pathway an excellent candidate for the testing of statistical theories of unimolecular dissociations. A second pathway occurs through intersystem crossing to an excited triplet state in ketene. Extensive measurements of uncorrelated photoproduct distributions have been obtained by the Moore group at Berkeley, and these measurements have been compared with statistical theories. It was found that the CO rotational distribution is roughly approximated by statistical theory, although their results showed that the CO vibrational distribution is slightly hotter than statistically predicted [187]. However, the CH_2 rotational distribution experimentally observed is much colder than statistical theories predict [189, 190].

The first correlated measurement of ketene photodissociation found in the literature was that of the Brookhaven group, using absorption spectroscopy to measure Doppler profiles of singlet CH_2 fragments in low rotational states [66]. They were able to draw two conclusions from these studies. First, there is no rotational alignment ($\mathbf{v}\cdot\mathbf{j}$ correlation) for low rotational states of the CH_2 fragment. Second, the ratio of CO with one quanta of vibrational energy to CO with zero quanta of vibrational energy is about 0.2:1. This value is twice the ratio predicted by SSE and three times the ratio predicted by PST.

Our experiments on ketene photodissociation at 308 nm were done using an unpolarized XeCl excimer laser for the photolysis step. Typical TOF spectra for this experiment can be seen in Figure 15. For example, Figure 15(b) shows an experiment where the probe laser is tuned to the $Q_3(20)$ line of CO and the polarization of the probe laser is perpendicular to the flight path. The TOF spectrum shows three distinct peaks. The very small peak at 125 μs is due to formation of triplet CH_2. The dominant peak at 160 μs corresponds to singlet CH_2 in its ground vibrational state (000), and, in the 4° data, a third peak is seen at 200 μs, which corresponds to singlet CH_2 with one quantum of bending vibration excited (010). The TOF spectra taken for $j_{CO} = 0, 6, 10, 15, 20, 25, 30$ were all consistent with this assignment. Note that at 14° the (010) peak disappears. This results because there is not enough energy available for translation to cause such a large angular deflection out of the molecular beam when the CH_2 fragment is formed with this much internal energy.

If there were no vector correlations in the ketene photolysis reaction the analysis of the data would be very straightforward, however, this is definitely not the case, as can be seen by comparing Figure 15(b) and (a). In Figure 15(a), the only change is that the probe laser polarization is parallel to the detection axis instead of perpendicular as in Figure 15(b). Not only are the intensities of the two vibrational components of the singlet channel different relative to one another, but the shapes of the peaks are altered. Figure 15(c) and (d) shows the effect of changing the probe branch used. Again drastic changes in the shapes of the TOF spectra arise. This dependence on branch is a well-known fingerprint for the presence of $\mathbf{v}\cdot\mathbf{j}$ correlations.

In order to analyze this data, one must be able to extract both the center-of-mass translational energy distribution as well as the translational energy dependence of the $\mathbf{v}\cdot\mathbf{j}$ correlations, a quantity never measured before. The formalism of UHC [42], presented in the extraction TOF section of our chapter, was therefore modified to describe our experiment and incorporated in the forward convolution program. The photolysis laser used in our experiment is unpolarized but we can treat this as if the direction of polarization is along the propagation direction of the laser. This treatment requires multiplying all b_0^2 moments in Eq. 5 by $-\frac{1}{2}$ [18]. The only angle that varies in the experiment is the angle between the polarization direction of the probe laser and the center-of-mass recoil velocity of CO. The result

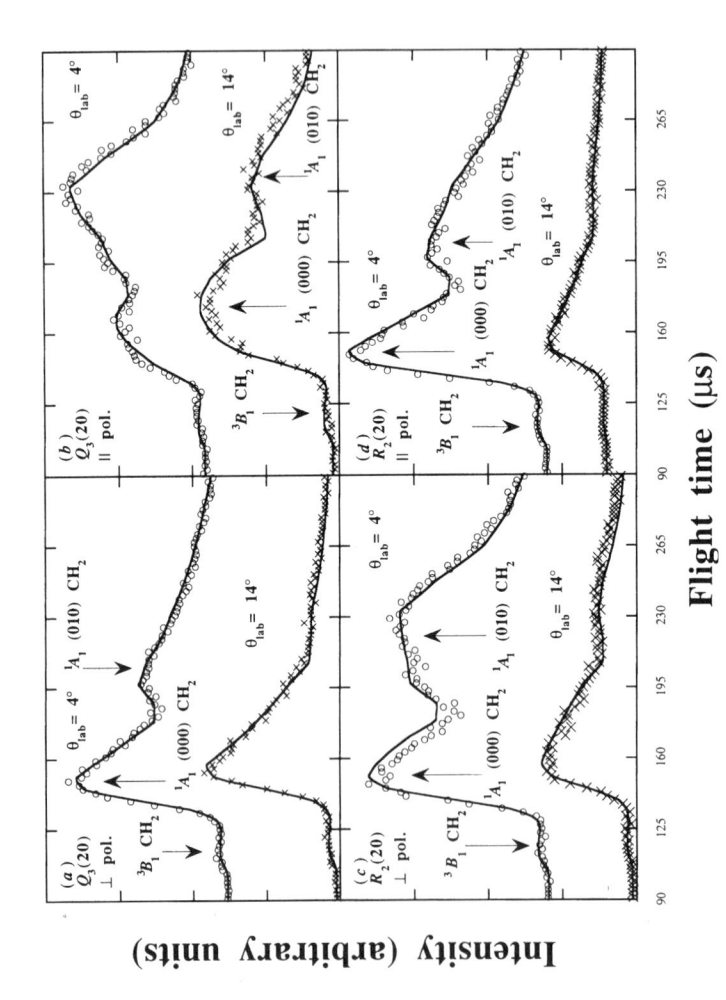

Figure 15. The CO metastable TOF spectra produced from ketene photolysis at 308 nm when $j_{CO} = 20$ is probed. Spectra for both Q and R branches and for both polarizations of the probe laser are shown. The empty circles and x's are data points for $\theta_{lab} = 4°$ and $14°$, respectively. The solid lines are data simulations, which are described in the text. The arrows point to the three vibronic channels of CH_2 produced by this dissociation.

is a recoil velocity dependent line strength factor S:

$$S(j_{CO}, \Omega, v; j'_{CO}, \Omega') = 1 + \frac{1}{2}\beta_0^2(20) + h^{(2)}\left[\frac{1}{5}\beta_0^2(02) + 2\beta_0^0(22)P_2(\cos\chi_p) \right. \\ \left. + \frac{2}{7}\beta_0^2(22)(1 - 2P_2(\cos\chi_p))\right] \quad (39)$$

Where χ_p is defined as the angle between the $\hat{\varepsilon}_{PR}$ and **v**. The direction of $\hat{\varepsilon}_{PR}$ can be either parallel or perpendicular to the axis of the flight path. Only the terms are multiplied by a function of χ_p, which will influence the shape of the TOF spectra. Empirically, it was found that all of the data could be fitted by considering only the $\beta_0^0(22)$ parameter, which describes **v**·**j** correlations. In order to separate the signal dependence on $P(E_{trans})$ versus its dependence on S, we first approximated both the beam and laboratory velocity distributions with single velocity vectors. This process allows us to remove the integral of Eq. 38. With this approximation, we can write an equation for $\beta_0^0(22)$ that is independent of $P(E_{trans})$.

$$\frac{1}{\cos 2\chi_{p,\parallel}} \frac{I_\parallel - I_\perp}{I_\parallel + I_\perp} = \frac{3h^{(2)}\beta_0^0(22, E_{trans})}{2 + h^{(2)}\beta_0^0(22, E_{trans})} \quad (40)$$

Here, $I_\parallel$ and $I_\perp$ refer to the intensity of the signal when the polarization of the probe laser is parallel and perpendicular to the flight path, respectively. The angle $\chi_{p\parallel}$ is the angle between **v** and axis of the flight path. Relative normalization of $I_\parallel$ and $I_\perp$ is accomplished by setting them equal to one another at the arrival time where $\chi_p = 45°$. Plots of this equation (Fig. 16) when $j_{CO} = 15$ and 20 and probed for the Q branch and show that our fitting routine is adequate to fit the TOF data. In the next step, both $\beta_0^0(22, E_{trans})$ and $P(E_{trans})$ are optimized until a satisfactory fit to all data is achieved.

The resulting state-specific $P(E_{trans})$ and $\beta_0^0(22)$ are shown in Figures 17 and 18, respectively, for $j_{CO} = 15$ and 20. They can be compared to statistical calculations (PST). It is found that for each singlet CH_2 vibrational channel, PST overestimates the amount of singlet CH_2 with low rotational energy coincident with a particular rotational state of CO, and, to a greater extent, it overestimates the amount of CH_2 with high rotational energy. Although ketene dissociation through the ground-state singlet channel should behave statistically since it is long lived and there is no barrier to dissociation, it is clear that there are some dynamical restrictions to this dissociation at photolysis energy around 308 nm.

In order to treat this, we modified the PST calculations slightly by restricting the possible values of the impact parameter of the dissociation.

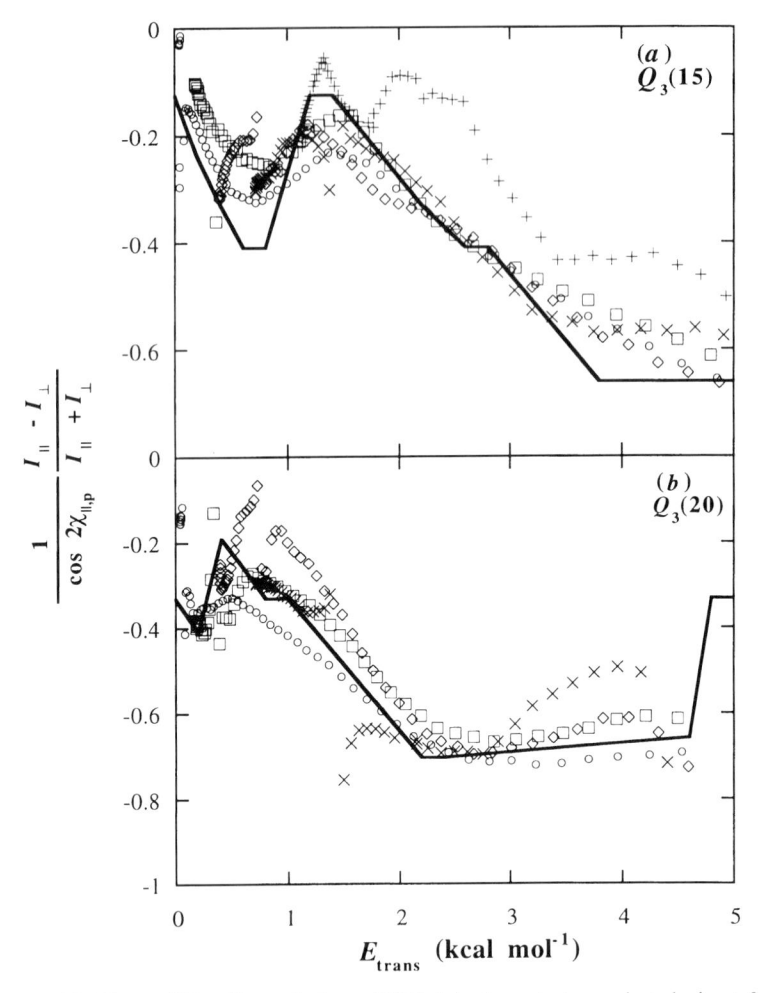

Figure 16. Plots of Eq. 18 made from TOF data from ketene photolysis at 308 nm when (a) $j_{CO} = 15$ and (b) $j_{CO} = 20$ are probed. The TOF data were taken at a variety of laboratory angles ($\bigcirc = 4°$, $\square = 9°$, $\Diamond = 14°$, $\times = 19°$, and $+ = 24°$). Both lines probed were Q transitions. The solid line is the solution to Eq. 18 using the $\beta_0^0(22)$ parameters used to fit the data.

In our PST calculation, the orbital angular momentum **l** of the dissociation must be included in the state count. The impact parameter b is a semiclassical function of the l quantum number in the following way:

$$b = \sqrt{\frac{l(l+1)\hbar^2}{2\mu_m E_{trans}}} \tag{41}$$

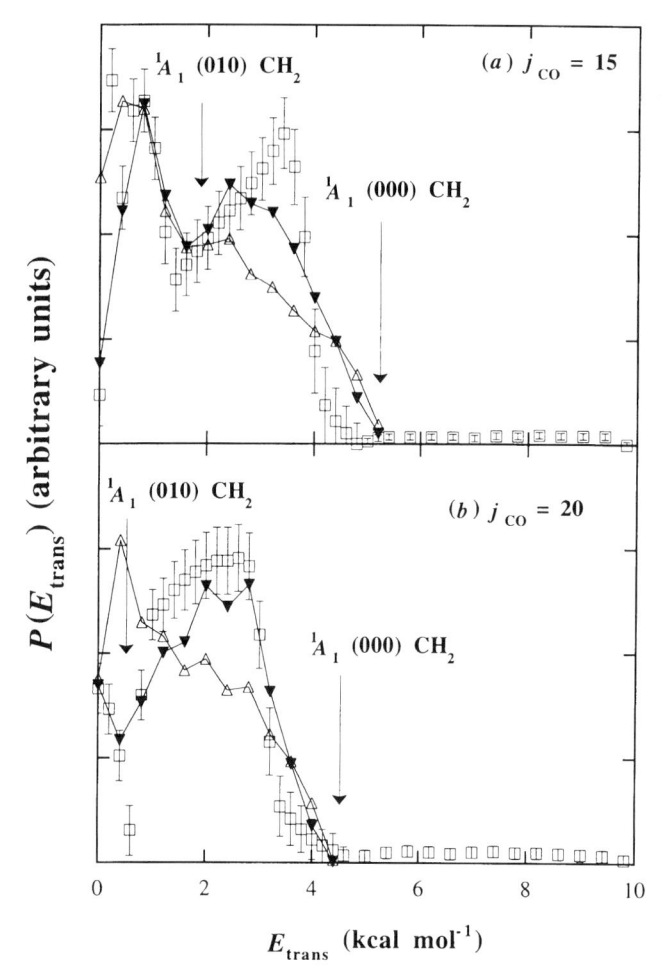

Figure 17. The experimentally determined state-specific $P(E_{trans})$ of CO in its ground vibrational state and (a) $j_{CO} = 15$ and (b) $j_{CO} = 20$ produced by ketene photolysis at 308 nm are shown with error bars as open squares (□). The lines with symbols are calculated state-specific $P(E_{trans})$ using either PST (△) or RPST (▼). See test for details of calculations. The arrows point to energy thresholds for the $^1A_1(000)$ and $^1A_1(010)$ CH$_2$ channels.

where μ_m is the reduced mass of the two fragments. At this point in a counting routine, a box-type opacity function can be included that has two parameters, a minimum and maximum allowed impact parameter, for each rotational state of CO and each vibrational state of CH$_2$. The opacity function will throw out any l state whose calculated impact parameter does

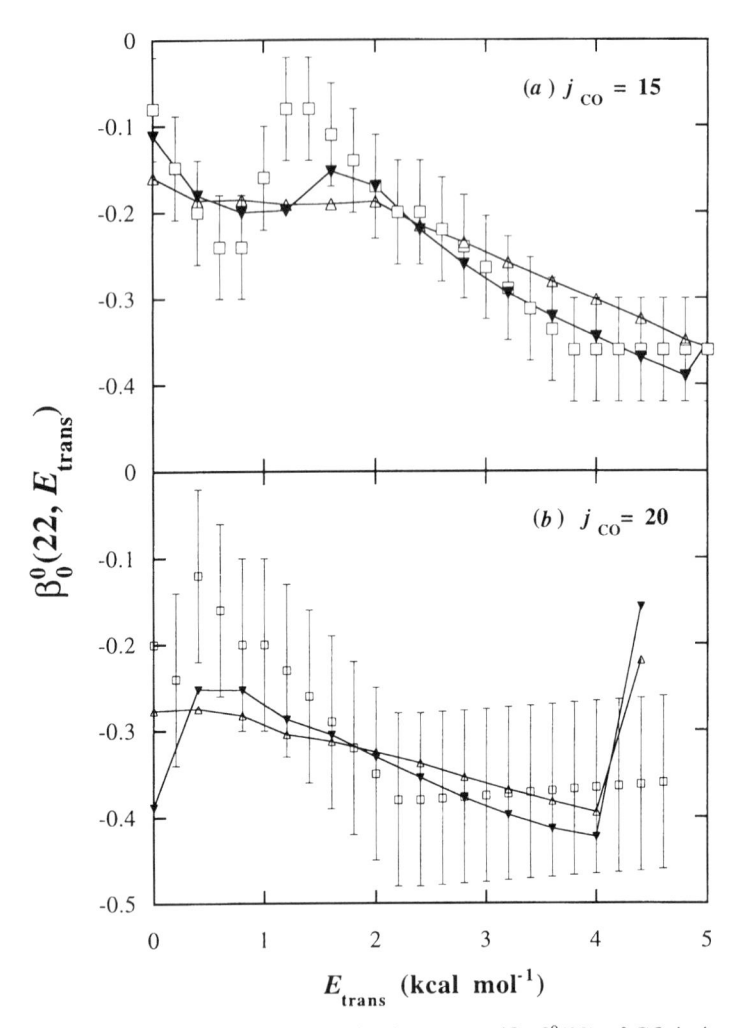

Figure 18. The experimentally determined state-specific $\beta_0^0(22)$ of CO in its ground vibrational state and (a) $j_{CO} = 15$ and (b) $j_{CO} = 20$ produced by ketene photolysis at 308 nm and plotted as a function of E_{trans} are shown with error bars as open squares (□). The lines with symbols are calculated state-specific $\beta_0^0(22)$ using either PST (△) or RPST (▼). See text for details of calculations.

not fall within the two limits. We call this variation restricted phase space theory (RPST). Figures 17 and 18 show the results of both our PST and RPST calculations for $j_{CO} = 15$ and 20, and it is clear that RPST does a much better job modeling the $P(E_{trans})$ data than PST. For $\beta_0^0(22)$, the RPST calculation is marginally better at modeling the experimental results than

PST. The net result of RPST is a shrinking of the phase space in which the dissociation occurs. It was found that this shrinkage explains discrepancies between the experimentally determined global distributions of rotations and vibrations of each fragment and the statistically determined ones.

But does this shrinkage of phase space have any physical basis? For this reaction, the major component of the impact parameter is the distance of a line perpendicular to the recoil velocity axis that intersects the center-of-mass of the CO fragment. The CH_2 fragment will make a much smaller contribution to the impact parameter distance since its center of mass is so close to the carbon atom. The amount of C—C—O bending during dissociation will have the largest effect on the value of the impact parameter. The opacity functions used for our RPST calculations are plotted in Figure 19 for each j_{CO} in the ground vibrational state that was probed in our

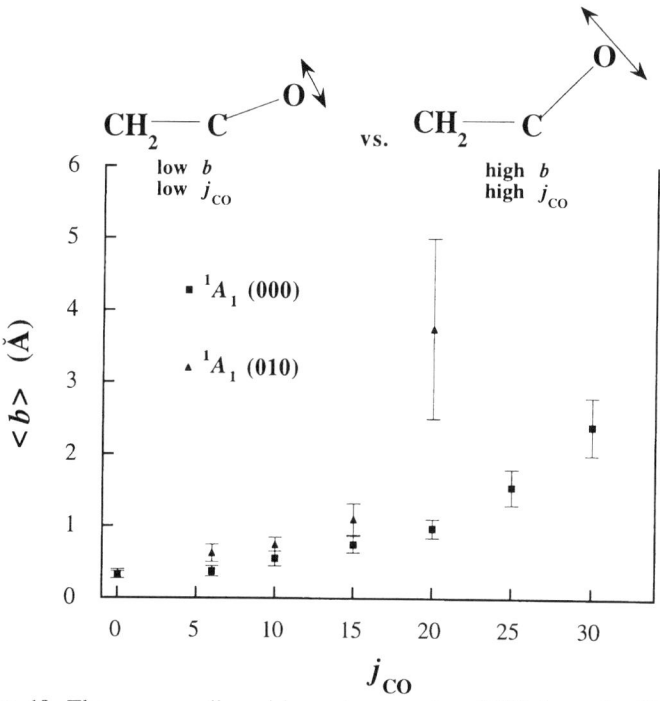

Figure 19. The average allowed impact parameters ($\langle b \rangle$) from the RPST calculations on ketene photolysis at 308 nm are shown as either solid squares (■) for $\langle b \rangle$ corresponding to $^1A_1(000)$ CH_2 or solid triangles (▲) for $\langle b \rangle$ corresponding to $^1A_1(010)$ CH_2 and are plotted as a function of j_{CO}. The extrema of the error bars are the minimum and maximum allowed impact parameters used in the RPST calculations. The drawings above the graph indicate the trend relating the energy of the C—C—O bending mode to the value of the impact parameters and j_{CO}.

experiment. With our correlated product state measurements, which were analyzed with RPST, we can see the general trend of the range of C—C—O bending angles for each CO rotational state probed. The general trend seen is that with increasing j_{CO} there is an increase in the average value of the impact parameter and an increase in the width of allowed impact parameters. More excitation of the C—C—O bending mode leads to access of smaller C—C—O bond angles at the point of dissociation; this causes an increase in the average value of the impact parameter. It also leads to a wider accessible range of impact parameters. From impulsive model arguments, greater amplitude in the bending mode motion leads to more rotational excitation of the CO fragment. Therefore, the relation between the allowed impact parameters and j_{CO} can be explained by the importance of the C—C—O bending motion in the dissociation of singlet ketene.

From this high-resolution experiment, one can begin to see the kind of detailed information that can be obtained experimentally. We are beginning to see the different structures a molecule may have at the transiton state and how they lead to different correlated product pair channels.

b. Ketene Dissociation at 351 nm. Below the singlet dissociation threshold in ketene (~ 332 nm) [186], only the triplet CH_2 channel is accessible. Experimental measurements of the rotational distribution of the CO fragment produced from this dissociation at 351 nm [190] follow predictions made by the impulsive model using the transition state geometry calculated by ab initio methods [180, 181]. The impulsive model used to successfully model the observed CO rotational distribution also makes predictions about the CH_2 rotational distribution. Due to the fact that the center-of-mass of the fragment is so close to the C atom, very little torque can be applied to this fragment. Therefore, according to this model, there should be very little CH_2 rotation from the fragmentation process. Most of the rotational energy of the CH_2 will come from the zero-point vibrational energy of the H—C—O bending modes and the torsional mode of the ketene, therefore, according to the impulsive model, the CH_2 rotational energy distribution should peak at very low energy and slowly decrease as rotational energy increases. A measurement of $(X^3B_1)CH_2$ internal energy by spectroscopic means has not been carried out. Our method provides a direct way to obtain this information.

Carbon monoxide metastable TOF spectra for ketene photodissociation at 351 nm were taken from $j_{CO} = 3$–18. There was no difference in the shapes of the TOF spectra when the polarization of the probe laser was changed with respect to the flight path. Thus, our results are not sensitive to whatever vector correlations may exist in this photodissociation and the line strength factor of Eq. 38 can be ignored. The resulting $P(E_{trans})$ functions reveal that

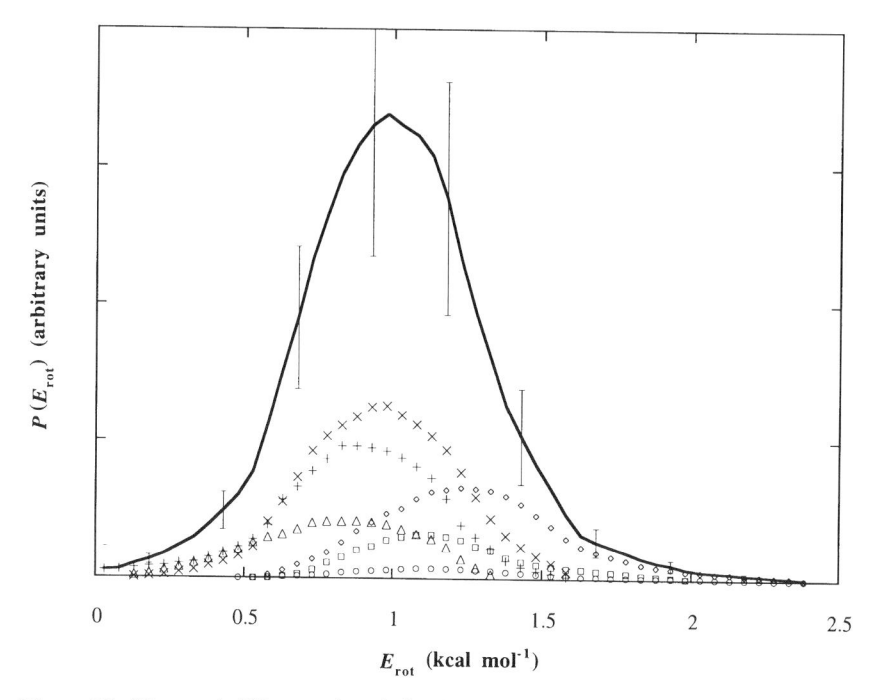

Figure 20. The total CH_2 rotational distribution from ketene photodissociation at 351 nm as measured by the metastable TOF method is shown as a solid line with error bars. The symbols show the contributions to the total CH_2 rotational distribution from each experimentally derived state-specific $P(E_{trans})$ measured ($\bigcirc j_{CO} = 3$, $\square j_{CO} = 6$, $\diamond j_{CO} = 9$, $\times j_{CO} = 12$, $+ j_{CO} = 15$, and $\triangle j_{CO} = 18$). These contributions have been weighted by the experimentally determined population of each j_{CO} state measured.

CH_2 is formed, within the limit of the experiment, almost exclusively in the ground vibrational state (>96%). This finding is contrary to the predictions of the zero-point motion corrected impulsive model. A slight anticorrelation between CO and CH_2 rotational energy was also found, whereas impulsive model calculations assume that these two motions are uncorrelated. Finally, the total CH_2 rotational distribution can be calculated by summing the correlated internal energy distribution of CH_2 for each CO state probed, with weighting factors that reflect the CO rotational population [190]. This total distribution is shown in Figure 20, along with the contribution that each rotational state of CO makes to the total distribution. The total CH_2 rotational distribution was found to peak at about 1 kcal mol^{-1}; the impulsive model predicts that it should peak close to 0 kcal mol^{-1}. Clearly, a simple impulsive model fails to describe this dissociation process.

As mentioned in the introduction, angular anisotropy in the dissociative PES can result in rotational distributions not predicted by the impulsive model. Where would these anisotropies arise? Ab initio calculations indicate that excited-state triplet ketene has two geometries: one is a planar C_s geometry with the C—C—O structure bent and the other has a nonplanar geometry, with the CH$_2$ moiety out of the C—C—O plane. The conversion between the two states is accomplished by two motions; a $\pi/2$ torsion about the C—C bond and an out of "C—C—H$_2$ plane" bend. These ab initio studies have estimated that the planar geometry is lower in energy, and photofragment excitation experiments have estimated that the difference in energy between the two states is about 0.7 kcal mol^{-1} [182]. At 351 nm, these two states should both be energetically accessible. It is indeed likely that we excite the molecule at high enough energy that internal notation of the CH$_2$ about the C—C bond is active. Because the conversion between the two states is accomplished by two motions: a $\pi/2$ torsion about the C—C bond and an out of "C—C—H$_2$ plane" bend, one could not excite the internal torsion without additionally exciting the out of plane bend. This out-of-plane bend correlates directly to the a-axis rotation of CH$_2$. We therefore hypothesize that as the ketene goes through these torsional and bending motions, vibrational motion in these modes will be excited, leading directly to high-energy rotational motion of CH$_2$ about its a axis. As j_{CO} increases, more vibrational motion would have been channeled into the C—C—O bending mode, leaving less energy for the torsional and bending motions mentioned above, and therefore leading to less CH$_2$ rotational excitation. This explains the anticorrelation between the CO and CH$_2$ rotational energies. What these results suggest is that more dimensions than just the C—C—O bending angle and the C—C bond distance need to be considered in order to model this dissociation.

It would be very interesting to repeat the measurements used in this experiment at a number of photolysis energies in the vicinity of 351 nm. We would predict that lowering the photolysis energy would dramatically change the X^3B_1CH$_2$ rotational distribution. Such experiments are certainly possible and are planned for in the near future.

IV. FUTURE OUTLOOK

This chapter has reviewed recent advances in the measurement of corelation of product states in photodissociation. We have seen that vector correlations are intimately involved with all of these measurements. The ion TOF pulsed-extraction and H-atom Rydberg methods have been used by several groups for quite some time, but the field is still providing new insights into the nature of photodissociation. Also, new variations of the TOF pulsed-extraction methods show that there is still a lot of new, exciting possibilities in this field. The other two methods reviewed in this chapter, FM Doppler spectroscopy and metastable TOF, are "just getting off the ground", and will certainly be extensively used in future experiments. Improvements in the FM spectroscopy include the use of pulsed molecular beams instead of running the experiment in a cell. Chang and Sears [70] have already observed HCCl spectra using FM with a pinhole jet. The CO metastable experiments were performed with an unpolarized photolysis laser. The use of a polarized laser should provide more detailed information on vector correlations and their energy release dependence. For example, the $\beta_0^2(20)$ ($\mu \cdot v$ anisotropy) parameter of a photoreaction containing CO could be resolved as a function of its velocity, similar to the work of Mordaunt et al. [173] on NH_3 dissociation.

One extension of these methods has been to use them in order to observe the dissociation of state-selected molecules, as has been done for the photofragmentation of CD_3I $|JKM\rangle = |111\rangle$ by 266-nm light. The selection was realized by passing a molecular beam through a hexapole field. The kinetic energy distribution of the nascent CD_3 in its ground vibrational state was measured via (2 + 1) REMPI for isolated rotational lines with either the core-extraction method [191] or ion imaging [192]. The TOF data shows an extreme preference for scattering in one direction. The results from the experiment can be used to calculate the detailed differential photofragmentation cross-section and also to determine elements of the transition matrix [193–195]. Knowledge of the magnitude and phase of these matrix elements is considered complete knowledge of the photodissociation process.

These methods have also been extended from the studies of dissociation to those of bimolecular reactions and reactive scattering. Recent work on the CN + ethylene reaction is an illustration of transient FM applied to gas-phase kinetics [196]. Decay curves can be followed over several orders of magnitude and the sensitivity is sufficient to detect $\langle 2 \times 10^8$ molecules per quantum state per cubic centimeter. Ion imaging of the NO molecule can be used to measure state-resolved differential cross-sections in the scattering of NO and Ar, as demonstrated by Houston and co-workers [197, 198]. Bimolecular reactions between an atom and a molecule can be initiated by producing the atomic reactant through photodissociation of a molecule, for example, HI, in a molecular beam. When linearly polarized light is used, the fragment from the photodissociation will have a well-characterized speed and angular distribution. In the next step, the fragment reacts with the other molecule in the molecular beam; a second laser then state-selectively ionizes the products that are characterized through ion imaging. Buntine et al. [199] used this technique to examine the reactions of H atoms with HI molecules while Kitsopoulos et al. [200] used D_2 molecules. The core-extraction technique has been used to study the reaction of methane with chlorine atoms,

$$Cl + CH_4 \rightarrow HCl + CH_3 \tag{42}$$

using REMPI to probe the HCl product [201, 202]. Studies of H-atom reactions have also been performed using the Rydberg tagging method [166, 167].

Chemistry has an amazing ability to stump and stymie the most persistent student of molecular motion. The dancing molecules clearly have a more vivid imagination than do we humble scientists. They continue to reveal interesting and unexpected results as new and more sophisticated measurements are devised. We hope that this chapter will help orient the interested reader in the panoply of fascinating work now being carried out.

ACKNOWLEDGMENTS

The work performed in our laboratory would not be possible without the generous grants from the National Science Foundation (CHE-9318885 and CHE-9411302). We also especially wish to thank Professor Joe I. Cline, Dr. Gregory E. Hall, Dr. Simon W. North, and Dr. Radek Uberna for providing preprints, figures, explanations about their experiments, and comments. Dr. David Chandler, Professor R. N. Dixon, Professor Michael N. R. Ashfold, Professor Hanna Reisler, Professor Paul L. Houston, and Dr. T. Suzuki are thanked for providing preprints and figures from their work.

REFERENCES

1. R. Schinke, *Photodissociation Dynamics*, P. L. K. A. Dalgarno, F. H. Read, R. N. Zare, Eds., Cambridge Monographs on Atomic, Molecular, and Chemical Physics 1, University Press, Cambridge, UK, 1993, Vol. 1.
2. R. D. Levine and J. L. Kinsey, in *Atom–Molecular Collision Theory—A Guide for the Experimentalist*, R. B. Bernstein, Ed., Plenum, New York, 1979.
3. P. Pechukas and J. C. Light, *J. Chem. Phys.*, **42**, 3281 (1965).

4. C. E. Klots, *J. Phys. Chem.*, **75**, 1526 (1971).
5. M. Quack and J. Troe, *Ber. Bunsenges. Phys. Chem.*, **78**, 240 (1974).
6. M. Quack and J. Troe, *Ber. Bunsenges. Phys. Chem.*, **79**, 469 (1975).
7. J. Troe, *J. Chem. Phys.*, **75**, 226 (1981).
8. J. Troe, *J. Chem. Phys.*, **79**, 6017 (1983).
9. C. Wittig, I. Nadler, H. Reisler, M. Noble, J. Catanzarite, and G. Radhakrishnan, *J. Chem. Phys.*, **83**, 5581 (1985).
10. G. E. Busch and K. R. Wilson, *J. Chem. Phys.*, **56**, 3629 (1972).
11. A. F. Tuck, *J. Chem. Soc. Faraday Trans. 2*, **73**, 689 (1977).
12. R. Schinke, *Comments Atomic Mol. Phys.*, **23**, 15 (1989).
13. R. Schinke and V. Staemmler, *Chem. Phys. Lett.*, **145**, 486 (1988).
14. R. Schinke, *J. Phys. Chem.*, **92**, 4015 (1988).
15. S. Leone, *Annu. Rev. Phys. Chem.*, **35**, 109 (1984).
16. I. Kovacs, *Rotational structure in the spectra of diatomic molecules*, American Elsevier, New York, 1969.
17. C. H. Greene and R. N. Zare, *J. Chem. Phys.*, **78**, 6741 (1983).
18. R. N. Dixon, *J. Chem. Phys.*, **85**, 1866 (1986).
19. R. Schinke, V. Engle, P. Andresen, D. Hausler, and G. G. Balint-Kurti, *Phys. Rev. Lett.*, **55**, 1180 (1985).
20. P. Andresen, V. Beushausen, D. Hausler, H. W. Lulf, and E. W. Rothe, *J. Chem. Phys.*, **83**, 1429 (1985).
21. V. Engel, V. Staemmler, R. L. Vander Wal, F. F. Crim, R. J. Sension, B. Hudson, P. Andresen, S. Hennig, K. Weide, and R. Schinke, *J. Phys. Chem.*, **96**, 3201 (1992).
22. H. J. Krautwald, L. Schnieder, K. H. Welge, and M. N. R. Ashfold, *Faraday Discuss. Chem. Soc.*, **82**, 99 (1986).
23. G. Theodorakopoulos, I. D. Petsalakis, and R. J. Buenker, *Chem. Phys.*, **96**, 217 (1985).
24. F. Flouquet and J. A. Horsley, *J. Chem. Phys.*, **60**, 3767 (1974).
25. B. Heumann, K. Kuehl, K. Weide, R. Dueren, B. Hess, U. Meier, S. D. Peyerimhoff, and R. Schinke, *Chem. Phys. Lett.*, **166**, 385 (1990).
26. D. H. Mordaunt, M. N. R. Ashfold, and R. N. Dixon, *J. Chem. Phys.*, **100**, 7360 (1994).
27. K. R. Wilson, in *Excited State Chemistry*, J. N. Pitts, Eds., Gordon and Breach, New York, 1970.
28. D. A. Blank, S. W. North, D. Stranges, A. G. Suits, and Y. T. Lee, *J. Chem. Phys.*, **106**, 539 (1997).
29. A. M. Wodtke and Y. T. Lee, in *Advances in Gas-Phase Photochemistry and Kinetics: Molecular Photodissociation Dynamics*, M. N. R. Ashfold, J. E. Baggot, Eds., Royal Society of Chemistry, London 1987) Series 539.6 .
30. R. N. Zare, *Mol. Photochem.*, **4**, 1 (1972).

31. N. Sivakumar, G. E. Hall, P. L. Houston, J. W. Hepburn, and I. Burak, *J. Chem. Phys.*, **88**, 3692 (1988).
32. A. U. Grunewald, K.-H. Gericke, and F. J. Comes, *J. Chem. Phys.*, **87**, 5709 (1987).
33. A. U. Grunewald, K.-H. Gericke, and F. J. Comes, *J. Chem. Phys.*, **89**, 345 (1988).
34. S. J. Klippenstein and J. I. Cline, *J. Chem. Phys.*, **103**, 5451 (1995).
35. S. W. North and G. E. Hall, *J. Chem. Phys.*, **104**, 1864 (1996).
36. C. H. Greene and R. N. Zare, *Ann. Rev. Phys. Chem.*, **33**, 119 (1982).
37. D. C. Jacobs and R. N. Zare, *J. Chem. Phys.*, **85**, 5457 (1986).
38. G. E. Hall, J. W. Hepburn, P. L. Houston, R. O. Loo, H. P. Haerri, G. K. Chawla, D. W. Chandler, I. Burak, and N. Sivakumar, *Ber. Bunsenges. Ges. Phys. Chem.*, **92**. 281 (1988).
39. G. E. Hall and P. L. Houston, *Annu. Rev. Phys. Chem.*, **40**, 375 (1989).
40. M. Mons and I. Dimicoli, *J. Chem. Phys.*, **90**, 4037 (1989).
41. A. J. Orr-Ewing and R. N. Zare, *Annu. Rev. Phys. Chem.*, **45**, 315 (1994).
42. R. Uberna, R. D. Hinchliffe, and J. I. Cline, *J. Chem. Phys.*, **103**, 7934 (1995).
43. U. Fano and J. H. Macek, *Rev. Mod. Phys.*, **45**, 553 (1973).
44. R. N. Dixon and H. Rieley, *Chem. Phys.*, **137**, 307 (1989).
45. K.-H. Gericke, H. G. Glaser, C. Maul, and F. J. Comes, *J. Chem. Phys.*, **92**, 411 (1990).
46. M. Brouard, M. T. Martinez, J. O'Mahony, and J. P. Simons, *Mol. Phys.*, **69**, 65 (1990).
47. T. J. Butenhoff, K. L. Carleton, and C. B. Moore, *J. Chem. Phys.*, **92**, 377 (1990).
48. C. E. M. Strauss, S. H. Kable, G. K. Chawla, P. L. Houston, and I. R. Burak, *J. Chem. Phys.*, **94**, 1837 (1991).
49. S. W. Noricki and R. Vadusev, *J. Chem. Phys.*, **95**, 7269 (1991).
50. G. Nan, I. Burak, and P. L. Houston, *Chem. Phys. Lett.*, **209**, 383 (1993).
51. Z. Xu, C. Wittig, and B. Koplitz, *J. Chem. Phys.*, **90**, 2692 (1989).
52. T. Haas, K.-H. Gericke, C. Maul, and F. J. Comes, *Chem. Phys. Lett.*, **202**, 108 (1993).
53. X. Xiaoxiang, L. Schnieder, H. Wallmeier, R. Boettner, K. H. Welge, and M. N. R. Ashfold, *J. Chem. Phys.*, **92**, 1608 (1990).
54. J. Biesner, L. Schnieder, G. Ahlers, X. Xiaoxiang, K. H. Welge, M. N. R. Ashfold, and R. N. Dixon, *J. Chem. Phys.*, **91**, 2901 (1989).
55. H. A. Michelsen, C. T. Rettner, D. J. Auerbach, and R. N. Zare, *J. Chem. Phys.*, **98**, 8294 (1993).
56. M. Mons and I. Dimicoli, *Chem. Phys. Lett.*, **131**, 298 (1986).
57. D. W. Chandler and P. L. Houston, *J. Chem. Phys.*, **87**, 1445 (1987).
58. G. C. Bjorklund, *Opt. Lett.*, **5**, 15 (1980).

59. J. L. Hall, L. Hollberg, T. Baer, and H. G. Robinson, *Appl. Phys. Lett.*, **39**, 680 (1981).
60. G. C. Bjorklund and M. D. Levenson, *Phys. Rev. A*, **24**, 166 (1981).
61. E. A. Whittaker, H. R. Wendt, H. E. Hunziker, and G. C. Bjorklund, *Appl. Phys. B*, **35**, 105 (1984).
62. E. A. Whittaker, B. J. Sullivan, H. R. Wendt, H. E. Hunziker, and G. C. Bjorklund, *J. Chem. Phys.*, **80**, 961 (1984).
63. E. A. Whittaker, B. J. Sullivan, G. C. Bjorklund, H. R. Wendt, and H. E. Hunziker, *J. Opt. Soc. Am. B*, **1**, 494 (1984).
64. T. J. Johnson, F. G. Wienhold, J. P. Burrows, G. W. Harris, and H. Burkhard, *J. Phys. Chem.*, **95**, 6499 (1991).
65. J. C. Bloch, R. W. Field, G. E. Hall, and T. J. Sears, *J. Chem. Phys.*, **101**, 1717 (1994).
66. B.-C. Chang Ming Wu, Gregory E. Hall, and Trevor Sears, *J. Chem. Phys.*, **101**, 9236 (1994).
67. S. W. North and G. E. Hall, *J. Chem. Phys.*, **106**, 60 (1997).
68. S. W. North and G. E. Hall, *Chem. Phys. Lett.*, **263**, 148 (1997).
69. S. W. North, X. S. Zheng, R. Fei, and G. E. Hall, *J. Chem. Phys.*, **104**, 2129 (1996).
70. B.-C. Chang and T. J. Sears, *J. Chem. Phys.*, **102**, 6374 (1995).
71. S. W. North and G. E. Hall, (1997), in press.
72. N. Nishi, H. Shinohara, and I. Hanazaki, *J. Chem. Phys.*, **77**, 246 (1982).
73. A. Fahr and A. H. Laufer, *J. Phys. Chem.*, **96**, 4217 (1992).
74. C. A. Bird and D. J. Donaldson, *Chem. Phys. Lett.*, **249**, 40 (1996).
75. J. B. Halpern and W. M. Jackson, *J. Phys. Chem.*, **86**, 973 (1982).
76. D. Eres, M. Gurnick, and J. D. McDonald, *J. Chem. Phys.*, **81**, 5552 (1984).
77. R. Lu, J. B. Halpern, and W. M. Jackson, *J. Phys. Chem.*, **88**, 3419 (1984).
78. Y. Huang, S. A. Barts, and J. B. Halpern, *J. Phys. Chem.*, **96**, 425 (1992).
79. M. Wu and G. E. Hall, *J. Photochem. Photobiol. A: Chem*, **80**, 45 (1994).
80. C. D. Jonah, R. N. Zare, and C. Ottinger, *J. Chem. Phys.*, **56**, 263 (1972).
81. E. Abramson, R. W. Field, D. Imre, K. K. Innes, and J. L. Kinsey, *J. Chem. Phys.*, **83**, 453 (1985).
82. R. L. Sundberg, E. Abramson, J. L. Kinsey, and R. W. Field, *J. Chem. Phys.*, **83**, 466 (1985).
83. H. L. Dai, C. L. Korpa, J. L. Kinsey, and R. W. Field, *J. Chem. Phys.*, **82**, 1688 (1985).
84. W. C. Wiley and I. H. McClaren, *Rev. Sci. Instrum.*, **26**, 1150 (1955).
85. R. O. Loo, H.-P. Haerri, G. E. Hall, and P. L. Houston, *J. Chem. Phys.*, **90**, 4222 (1989).

86. J. L. Tomer, M. C. Wall, B. P. Reid, and J. I. Cline, *J. Chem. Phys.*, **102**, 6100 (1995).
87. R. O. Loo, G. E. Hall, H.-P. Haerri, and P. L. Houston, *J. Phys. Chem.*, **92**, 5 (1988).
88. J. F. Black and I. Powis, *Chem. Phys.*, **125**, 375 (1988).
89. A. J. R. Heck and D. W. Chandler, *Annu. Rev. Phys. Chem.*, **46**, 335 (1995).
90. R. N. Strickland and D. W. Chandler, *Appl. Opt.*, **30**, 1811 (1991).
91. A. Sanov, C. R. Bieler, and H. Reisler, *J. Phys. Chem.*, **99**, 13637 (1995).
92. K. Mikhaylichenko, C. Riehn, I. Valachovic, A. Sanov, and C. Wittig, *J. Chem. Phys.*, **105**, 6807 (1996).
93. R. A. Hertz and J. A. Syage, *J. Chem. Phys.*, **100**, 9265 (1994).
94. J. A. Syage, *J. Phys. Chem.*, **99**, 16530 (1995).
95. J. A. Syage, *J. Chem. Phys.*, **105**, 1007 (1996).
96. J. A. Syage, *Chem. Phys. Lett.*, **245**, 605 (1995).
97. K. Tonokura and T. Suzuki, *Chem. Phys. Lett.*, **224**, 1 (1994).
98. G. Gruenefeld and P. Andresen, *Chem. Phys. Lett.*, **208**, 369 (1993).
99. T. Kinugawa and T. Arikawa, *J. Chem. Phys.*, **96**, 4801 (1992).
100. L. H. Lai, D. C. Che, and K. P. Liu, *J. Phys. Chem.*, **100**, 6376 (1996).
101. L.-H. Lai, J.-H. Wang, D. C. Che, and K. Liu, *J. Chem. Phys.*, **105**, 3332 (1996).
102. H. Ni, J. M. Serafin, and J. J. Valentini, *Chem. Phys. Lett.*, **244**, 207 (1995).
103. H. Ni, J. M. Serafin, and J. J. Valentini, *J. Chem. Phys.*, **104**, 2259 (1996).
104. H. I. Schiff, *Ann. Geophys.*, **28**, 67 (1972).
105. C. E. Fairchild, E. J. Stone, and G. M. Lawrence, *J. Chem. Phys.*, **69**, 3632 (1978).
106. R. K. Sparks, L. R. Carlson, K. Shobatake, M. L. Kowalczyk, and Y. T. Lee, *J. Chem. Phys.*, **72**, 1401 (1980).
107. J. J. Valentini, D. P. Gerrity, D. L. Phillips, J.-C. Nieh, and K. D. Tabor, *J. Chem. Phys.*, **86**, 6745 (1987).
108. A. G. Suits, R. L. Miller, L. S. Bontuyan, and P. L. Houston, *J. Chem. Soc. Faraday Trans.*, **89**, 1443 (1993).
109. R. L. Miller, A. G. Suits, P. L. Houston, R. Toumi, J. A. Mack, and A. M. Wodtke, *Science*, **265**, 1831 (1994).
110. J. Eluszkiewicz and M. Allen, *J. Geophys. Res.*, **98**, 1069 (1993).
111. J. E. Frederick and R. J. Cicerone, *J. Geophys. Res.*, **90**, 10733 (1985).
112. T. G. Slanger, L. E. Jusinski, G. Black, and G. E. Gadd, *Science*, **241**, 945 (1988).
113. C. A. Rogaski, J. M. Price, J. A. Mack, and A. M. Wodtke, *Geophys. Res. Lett.*, **20**, 2885 (1993).
114. J. M. Price, J. A. Mack, C. A. Rogaski, and A. M. Wodtke, *Chem. Phys.*, **175**, 83 (1993).

115. C. A. Rogaski, J. A. Mack, J. M. Price, and A. M. Wodtke, *Faraday Discuss.*, **100**, 229 (1996).
116. J. A. Mack, K. Mikulecky, and A. M. Wodtke, *J. Chem. Phys.*, **105**, 4105 (1996).
117. K. Takahashi, M. Kishigami, Y. Matsumi, M. Kawasaki, and A. J. Orr-Ewing, *J. Chem. Phys.*, **105**, 5290 (1996).
118. Y. Sato, Y. Matsumi, M. Kawasaki, K. Tsukiyama, and R. Bersohn, *J. Phys. Chem.*, **99**, 16307 (1995).
119. S. Nanbu, M. Aoyagi, and T. Suzuki, in press.
120. M. Katayanagi, Y. X. Mo, and T. Suzuki, *Chem. Phys. Lett.*, **247**, 571 (1995).
121. Y. Mo, H. Katayanagi, M. C. Heaven, and T. Suzuki, *Phys. Rev. Lett.*, **77**, 830 (1996).
122. R. Simonaitis, R. I. Greenberg, and J. Heicklen, *Int. J. Chem. Kinet.*, **4**, 497 (1972).
123. Y. Zhu and R. J. Gordon, *J. Chem. Phys.*, **92f**, 2897 (1990).
124. P. Felder, B.-M. Haas, and J. R. Huber, *Chem. Phys. Lett.*, **186**, 177 (1991).
125. L. L. Springsteen, S. Satyapal, Y. Matsumi, L. M. Dobeck, and P. L. Houston, *J. Phys. Chem.*, **97**, 7239 (1993).
126. T. F. Hanisco and A. C. Kummel, *J. Phys. Chem.*, **97**, 7242 (1993).
127. N. Shafer, K. Tonokura, Y. Matsumi, S. Tasaki, and M. Kawasaki, *J. Chem. Phys.*, **95**, 6218 (1991).
128. T. Suzuki, H. Katayanagi, Y. Mo, and K. Tonokura, *Chem. Phys. Lett.*, **256**, 90 (1996).
129. J. Cao, Y. Wang, and C. X. W. Qian, *J. Chem. Phys.*, **103**, 9653 (1995).
130. J. Cao, H.-P. Loock, and C. X. W. Qian, *J. Chem. Phys.*, **101**, 3395 (1994).
131. T. Suzuki, V. P. Hradil, S. A. Hewitt, P. L. Houston, and B. J. Whitaker, *Chem. Phys. Lett.*, **187**, 257 (1991).
132. V. P. Hradil, T. Suzuki, S. A. Hewitt, P. L. Houston, and B. J. Whitaker, *J. Chem. Phys.*, **99**, 4455 (1993).
133. S. Yang and R. Bersohn, *J. Chem. Phys.*, **61**, 4400 (1974).
134. H. Katagiri and S. Kato, *J. Chem. Phys.*, **99**, 8805 (1993).
135. M. M. Graff and A. F. Wagner, *J. Chem. Phys.*, **92**, 2423 (1990).
136. T. A. Spiglanin and D. W. Chandler, *Chem. Phys. Lett.*, **141**, 428 (1987).
137. S. S. Brown, H. L. Berghout, and F. F. Crim, *J. Chem. Phys.*, **105**, 8103 (1996).
138. S. S. Brown, Ph. D. Thesis, "Spectroscopy and Bond-selected Fragmentation Dynamics in the Photodissociation of Highly Excited Isocyanic Acid" University of Wisconsin-Madison, Madison, 1996.
139. M. Zyrianov, A. Sanov, T. Droz-Georget, and H. Reisler, *J. Chem. Phys.*, **105**, 8111 (1996).
140. M. Kawasaki, Y. Sato, K. Suto, Y. Matsumi, and S. H. S. Wilson, *Chem. Phys. Lett.*, **251**, 67 (1996).
141. J. V. V. Kasper and G. C. Pimentel, *Appl. Phys. Lett.*, **5**, 231 (1964).

142. S. J. Riley and K. R. Wilson, *Discuss. Faraday Soc.* **53**, 132 (1972).
143. M. Dzvonik, S. Yand, and R. Bersohn, *J. Chem. Phys.*, **61**, 4408 (1974).
144. R. K. Sparks, K. Shobatake, L. R. Carlson, and Y. T. Lee, *J. Chem. Phys.*, **75**, 3838 (1981).
145. M. D. Barry and P. A. Gorry, *Mol. Phys.*, **52**, 461 (1984).
146. G. N. A. V. Veen, T. Baller, A. E. d. Vries, and N. J. A. V. Veen, *Chem. Phys.*, **87**, 405 (1984).
147. J. F. Black and I. Powis, *J. Chem. Phys.*, **89**, 3986 (1988).
148. H. W. Hermann and S. R. Leone, *J. Chem. Phys.*, **76**, 4766 (1982).
149. R. S. Mullikan, *Phys. Rev.*, **50**, 1017 (1936).
150. D. W. Chandler, J. J. W. Thoman, M. H. M. Janssen, and D. H. Parker, *Chem. Phys. Lett.*, **156**, 151 (1989).
151. D. W. Chandler, M. H. M. Janssen, S. Stolte, R. N. Strickland, J. J. W. Thomas, and D. H. Parker, *J. Phys. Chem.*, **94**, 4839 (1990).
152. H. Eyring, J. Walter, and G. E. Kimball, *Quantum Chemistry*, Wiley, New York, 1944.
153. M. H. M. Janssen, D. H. Parker, G. O. Sitz, S. Stolte, and D. W. Chandler, *J. Phys. Chem.*, **95**, 8007 (1991).
154. W. P. Hess, D. W. Chandler, and J. J. W. Thoman, *Chem. Phys.*, **163**, 277 (1992).
155. T. Suzuki, K. Tonokura, L. S. Bontuyan, and N. Hashimoto, *J. Phys. Chem.*, **98**, 13447 (1994).
156. H. J. Hwang and M. A. El-Sayed, *J. Chem. Phys.*, **94**, 4877 (1991).
157. H. J. Hwang and M. El-Sayed, *Chem. Phys. Lett.*, **170**, 161 (1990).
158. S. W. North and J. J. Cline, personal communication.
159. R. Uberna and J. I. Cline, *J. Chem. Phys.*, **102**, 4705 (1995).
160. R. Uberna, R. D. Hinchliffe, and J. I. Cline, *J. Chem. Phys.*, **105**, 9847 (1996).
161. R. Uberna, G. N. Fickes, and J. I. Cline, in press.
162. S. Deshmukh and W. P. Hess, *J. Chem. Phys.*, **100**, 6429 (1994).
163. M. D. Person, P. W. Kash, and L. J. Butler, *J. Chem. Phys.*, **97**, 355 (1992).
164. S. W. North, D. A. Blank, and Y. T. Lee, *Chem. Phys. Lett.*, **224**, 38 (1994).
165. M. Ahmed, D. Blunt, D. Chen, and A. G. Suits, *J. Chem. Phys.*, **106**, 7617 (1997).
166. L. Schnieder, W. Meier, K. H. Welge, M. N. R. Ashfold, and C. M. Western, *J. Chem. Phys.*, **92**, 7027 (1990).
167. F. Merkt, H. Xu, and R. N. Zare, *J. Chem. Phys.*, **104**, 950 (1996).
168. H. Xu, N. E. Schafer-ray, F. Merkt, D. J. Hughes, M. Springer, R. P. Tuckett, and R. N. Zare, *J. Chem. Phys.*, **103**, 5157 (1995).
169. L. Schnieder, K. Seekamprahn, J. Borowski, E. Wrede, K. H. Welge, F. J. Aoiz, L. Banares, M. J. Dmello, V. J. Herrero, V. S. Rabanos, and R. E. Wyatt, *Science*, **269**, 207 (1995).

REFERENCES 349

170. M. N. R. Ashfold, D. H. Mordaunt, and S. H. S. Wilson, in *Advances in Photochemistry*, D. C. Neckers, D. H. Volman, and G. von Buenau, Eds., Wiley, New York, 1996, Vol. 21, 217.
171. J. S. Zhang, M. Dulligan, and C. Wittig, *J. Phys. Chem.*, **99**, 7446 (1995).
172. M. Dulligan, J. Zhang, M. Tuchler, and C. Wittig, 212th ACS National Meeting, Orlando, FL, 1996.
173. J. Zhang, C. W. Riehn, M. Dulligan, and C. Wittig, *J. Chem. Phys.*, **104**, 7027 (1996).
174. W. Yi and R. Bersohn, *Chem. Phys. Lett.*, **206**, 365 (1993).
175. D. H. Mordaunt, M. N. R. Ashfold, and R. N. Dixon, *J. Chem. Phys.*, **104**, 6460 (1996).
176. J. M. Price, A. Ludviksson, M. Nooney, M. Xu, R. M. Martin, and A. M. Wodtke, *J. Chem. Phys.*, **96**, 1854 (1992).
177. M. Drabbels, C. G. Morgan, and A. M. Wodtke, *J. Chem. Phys.*, **103**, 7700 (1995).
178. M. Drabbels, C. G. Morgan, D. S. McGuire, and A. M. Wodtke, *J. Chem. Phys.*, **102**, 611 (1995).
179. C. G. Morgan, M. Drabbels, and A. M. Wodtke, *J. Chem. Phys.*, **104**, 7460 (1996).
180. C. G. Morgan, M. Drabbels, and A. M. Wodtke, *J. Chem. Phys.*, **105**, 4550 (1996).
181. X. Zhao, Ph.D. Thesis, "Photodissociation of Cyclic Compounds in a Molecular Beam" University of California, Berkeley, 1988.
182. W. D. Allen and H. F. Schaefer, *J. Chem. Phys.*, **84**, 2212 (1984).
183. W. D. Allen and H. F. Schaefer, *J. Chem. Phys.*, **89**, 329 (1988).
184. S. K. Kim, E. R. Lovejoy, and C. B. Moore, *J. Chem. Phys.*, **102**, 3202 (1995).
185. C. C. Hayden, D. M. Neumark, K. Shobatake, R. K. Sparks, and Y. T. Lee, *J. Chem. Phys.*, **76**, 3607 185 (1982).
186. D. J. Nesbitt, H. Petek, M. F. Foltz, S. V. Filseth, D. J. Bamford, and C. B. Moore, *J. Chem. Phys.*, **83**, 223 (1985).
187. S. K. Kim, Y. S. Choi, C. D. Pibel, Q.-K. Zheng, and C. B. Moore, *J. Chem. Phys.*, **94**, 1954 (1991).
188. I.-C. Chen, W. H. Green, and C. B. Moore, *J. Chem. Phys.*, **89**, 314 (1988).
189. I. Garcia-Moreno, E. R. Lovejoy, and C. B. Moore, *J. Chem. Phys.*, **100**, 8890 (1994).
190. I. Garcia-Moreno, E. R. Lovejoy, and C. B. Moore, *J. Chem. Phys.*, **100**, 8902 (1994).
191. S. J. Klippenstein and R. A. Marcus, *J. Chem. Phys.*, **91**, 2280 (1989).
192. I.-C. Chen and C. B. Moore, *J. Phys. Chem.*, **94**, 269 (1990).
193. L. C. Pipes, D. Y. Kim, N. Brandstater, C. D. Fuglesang, and D. Baugh, *Chem. Phys. Lett.*, **247**, 564 (1995).

194. J. W. G. Masternbroek, C. A. Taatjes, K. Nauta, M. H. M. Janssen, and S. Stolte, *J. Phys. Chem.*, **99**, 4360 (1995).
195. G. G. Balint-Kurti and M. Shapiro, *Chem. Phys.*, **61**, 137 (1981).
196. T. Seideman, *J. Chem. Phys.*, **102**, 6487 (1995).
197. D. J. Leahy, K. L. Reid, and R. N. Zare, *J. Chem. Phys.*, **95**, 1757 (1991).
198. S. W. North, R. Fei, T. J. Sears, and G. E. Hall, *Int. J. Chem. Kinet.*, **29**, 127 (1997).
199. A. G. Suits, L. S. Bontuyan, P. L. Houston, and B. J. Whitaker, *J. Chem. Phys.*, **96**, 8618 (1992).
200. L. S. Bontuyan, A. G. Suits, P. L. Houston, and B. J. Whitaker, *J. Phys. Chem.*, **97**, 6342 (1993).
201. M. A. Buntine, D. P. Baldwin, R. N. Zare, and D. W. Chandler, *J. Chem. Phys.*, **94**, 4672 (1991).
202. T. N. Kitsopoulous, M. A. Buntine, D. P. Baldwin, and R. N. Zare, *Science*, **260**, 1605 (1993).
203. W. R. Simpson, A. J. Orr-Ewing, T. Rakitzis, S. A. Kandel, and R. N. Zare, *J. Chem. Phys.*, **103**, 7299 (1995).
204. W. R. Simpson, T. P. Rakitzis, S. A. Kandel, A. J. Orr-Ewing, and R. N. Zare, *J. Chem. Phys.*, **103**, 7313 (1995).

INDEX

Acetyl chloride, secondary dissociation, 325
Acrylonitrile, photolysis, 299
Alkaline earth metals:
   monovalent derivatives, 3–4
   see also Polyatomic alkaline earth containing molecules
Alkaline earth salts, flame colors, 4
Aminium cation, 123–125
Amino acids, CIDNP applications, 149–152
α-Aminoalkyl radicals, 123–124, 151–152
Anetholes:
   mixed-[2 + 2] photocycloadditions, 142
   Paterno–Büchi reactions, 136–138
Anthraquinone, time-resolved CIDNP, 118–120
Arene diazonium salts, para-substituted, 120–121
Aromatic hydrocarbons, radiationless transitions, excess energy dependence, 186–193
Aromatic molecules, see Gaseous aromatic molecules
Azabenzenes, nonradiative decay rates, excitation energy dependence, 198–200

BaOH, formation, 17–18
Benzene, channel three decay, 202–206
Benzonitrile, 121–122
Benzoquinone, cycloaddition, 137–139
Biacetyl phosphorescence, yield as function of excess vibrational energy, 190–192
Bicyclo[3.1.0]hex-2-ene, radical cation, 116

3,3′-Biisoquinoline-N,N'-dioxide, 233–234, 236–237
Bioaffinity assays, 219
Bipolar moments, renormalized, 291–292
Biradicals:
   chemically induced dynamic nuclear polarization, 106–110
   Paterno–Büchi, 142
Born–Oppenheimer electronic states, adiabatic, 181
BrNO, photodissociation, 315
Broida oven, 6–11
   Doppler and collisional broadening, 9
   experimental block diagram, 8–9
   metal flow reactor, 6–7
   millimeter wave pure rotational spectra, 10–11
   spectroscopic experiments, 9
p-tert-Butyl-calix[4]arene tetraacetamide ligand, 252

$CaBH_4$, 55–56
   Feynman–Dyson amplitude, 20–22
CaCCH, 51, 53
$CaCH_3$, 50–51
CaC_5H_5, 53
CaC_4H_4N, 54–55
Calcium and monoformamidate, 33, 35
Calcium monoformate, 33
Calixarenes:
   lanthanide complexes, 252–259
   containing chromophores, 253–255, 258

Calixarenes, lanthanide complexes (*Continued*)
  photophysical data, 256–257
  schematic representation, 253–255
Calixcrowns, 258–259
CaN$_3$, 36, 38–39
CaNC, 39–43
CaNCO, 35–37
CaNH$_2$, 46–48
CaOH, 3–4, 24–25, 27, 29
  formation, 15–16
  highly excited states, 32
  laser excitation spectrum, 31
Ca$^+$(OH$_2$), metal-ligand binding energies, 3
Carbon-heteroatom bonds, photodissociations, 129–130
Carbon monoxide, metastable experiments, 327–340
  ketene photodissociation:
    at 308 nm, 330–338
    at 351 nm, 338–340
  schematic diagram, 328–329
Carbonyl compounds, α-cleavage, 127–129
4-Carboxybenzophenone, 150–151
CaSH, 43–45
CD$_3$I, photodissociation, 318–320
Chemically induced dynamic nuclear polarization, 63–154
  advantages, 66
  applications, 114–153
    amino acids, peptides, and proteins, 149–152
    cycloadditions, cycloreversions, and isomerizations, 135–143
    electron transfer, 114–123
    fragmentations, 127–132
    hydrogen abstractions, 123–127
    inorganic and metal organic substrates, 143–144
    nucleic acids, 152–153
    photosynthesis, 154
    polymerization initiators, 144–149
    radical additions, 132–134
    radical fragmentations, 134–135
  biradicals, 106–110
    escape pathways, 107–109
  estimations of intensities and phases, 95–100
  explanation by radical pair mechanism, 91–93
  field dependence of intensities, 109–110
  flash technique, 113–114
  flip-flop transition, 92
  hyperfine coupling constants, 98–99
  information accessible, 101–102
  instrumentation and techniques, 100–101
  magnetic parameter effects, 96
  maximum intensity, 97–98
  micellar systems, 110–111
  multiplet effect, 98–99
  net effects, 103
  new methods and techniques, 112–114
  other polarization mechanisms, 111–112
  radical pair dynamics, 105
  relationship between polarizations and line intensities, 93–95
  relative intensities, 108
  spin dynamics, 102–105
CH$_3$I, photodissociation, 318–321
2-Chloro-2-nitrosopropane, photodissociation, 322–324
α-Cleavage, carbonyl compounds, 127–129
β-Cleavage, 129–130
ClNO, photodissociation, 315
Clouds, interstellar molecular, 4–5
Cryptates, lanthanide complexes, 227–238
  containing chromophores, 231–233
  crystal structures, 232, 234–235, 237
  lifetime, 232–234
  photophysical data, 230
  schematic representation, 228–229
Cyanogen, photodissociation, 299–302
Cycloadditions, CIDNP applications, 135–143
Cyclododecanone, photolysis, 106–107
Cycloreversions, CIDNP applications, 135–143

Degenerate electron-transfer reactions, 118
Density matrix:
  evolution, exact vector model, 85
  treatment, radical pair spin dynamics, 80–84
Deprotonation pathway, 125–126
Dibenzyl ketone, CIDNP experiments, 127–128
2,2-Dimethoxy-2-phenylacetophenone, 144–146
Dimethylindine dimer, cycloreversion, 135–136

N,N-Dimethyl-1-naphthylamine, 121–122
DNA hybridization assays, 221–224
Doppler effect, 293
Doppler profiles, experimental, 300–302

Electron transfer, CIDNP applications, 114–123
Electron Zeeman interaction, 69
Energy balance equation, photoreaction, 293
Energy-gap law, 166
Enzymatic immunoassays, 220–221
Eu^{3+} complex:
   applications, 219–224
   with encapsulating ligands, 225–59
      calixarene complexes, 252–259
      classes, 226
      cryptates, 227–238
      macrocyclic complexes, 238–249
      podates, 249–251
   light conversion, 215–218
Extraction time-of-flight method, 303–305

Fluoroimmunoassays, 221, 223–224
Fragmentations, CIDNP applications, 127–132
Franck–Condon factor, 172, 182–184
Frequency modulation Doppler spectroscopy, 295–302
   acrylonitrile photolysis, 299
   cyanogen photodissociation, 299–302
   schematic diagram, 296

Gaseous aromatic molecules, 165–208
   benzene, channel three decay, 202–206
   $S_n(n \geq 2) \to S_1$, 168–170
   see also Radiationless transitions

Hamiltonian:
   electronic, eigenfunctions, 180–181
   spin–orbit coupling, 70–71
Harmonic oscillator, Franck–Condon factor, 183
Hartley bands, 309, 311
Herzberg–Teller effect, 31
Hexamethyldewarbenzene, 115–116
Hexamethylprismane, radical cation, 115–116
H_2O_2, photolysis, 289

Hydrogen abstractions, CIDNP applications, 123–127
2-Hydroxy-2-propyl radicals, 132–134
Hyperfine coupling constants, 98–99
Hyperfine interaction, 69–70, 77–78
   nonsecular terms, 79

Immunoassays:
   heterogeneous, 220
   lanthanide complexes, 219
Inorganic substrates, CIDNP applications, 143–144
Interdiffusion coefficient, 87
Internal conversion, $S_n(n \geq 2) \to S_1$, 168–170
   $S_1 \to S_0$, 167
Iodobenzene, photodissociation, 322
Ion imaging, 305–306, 318–321
   secondary dissociation, 325–326
Isocyanic acid, photodissociation, 317–318
Isomerizations, CIDNP applications, 135–143
Isoquinoline:
   absorption and fluorescence excitation spectra, 194–195
   nonradiative decay rates, excess vibrational energy dependence, 196–197

Kaptein's rule, 99, 103
Ketene:
   dissociation, at 351 nm, 338–340
   photodissociation, at 308 nm, 330–338
Kinetic studies, CIDNP applications, 118–123

Lanthanide chelates, assays, 223–224
Lanthanide complexes, 213–273
   decay rate constant, 218
   immunoassays, 219
   macrocyclic polyethers, 225
   photophysical properties, 259–270
      ligand-absorption, 260–262
      ligand–metal energy-transfer efficiency, 262–263
      metal luminescence efficiency, 263–267
      metal luminescence intensity, 267–270
      singlet → triplet intersystem crossing, 216–217
   state of the art, 270–273
   see also Eu^{3+} complex; Tb^{3+} complex, 215

Laser ablation, molecular beam spectrometer, 12–13
Ligand–metal charge-transfer transitions, cryptates, 227, 231
Ligand–metal energy transfer, 216
  efficiency, lanthanide complexes, 261–263
Line intensities, relation with polarizations, 93–95
Line strength factor, recoil velocity dependent, 333

Macrocyclic ligands, lanthanide complexes, 238–249
  absorption:
    efficiency, 245
    spectra, 246
  containing chromophores, 238, 244–245
  lifetime, 245–247
  N-oxides, 244
  photophysical data, 242–243
  schematic representation, 239–241
  stability, 246
Mechanistic studies, CIDNP applications, 118–123
Metal–ligand interactions, 2
Metal luminescence:
  efficiency, 217
    lanthanide complexes, 263–267
    intensity, 267–270
Metal organic substrates, CIDNP applications, 143–144
(N-Methylanilino)acetone, photocleavage, 130–131
7-Methylenenorbornadiene, radical cation, 117
7-Methylenequadricyclane, radical cation, 117
Methylethylketone, α-cleavage, 128
Methyl iodide, photodissociation, potential energy curves, 281–282
MgCCH, 51, 53
MgNC, 40–41, 43
MgOH:
  bending potential energy function, 24, 26
  ground-state, 23–24
Micellar systems, CIDNP, 110–111
Molecular recognition, 225
Monoacetylides, 51, 53
Monoalkoxides, 32–33
Monoalkylamides, 48–49

Monoamides, 46–48
Monoazides, 36, 38–39
Monoborohydrides, 55–56
Monocarboxylates, 33–34
Monocyanides, 39–43
Monocyclopentadienides, 52–53
Monoformamidates, 33, 35
Monohydrosulfides, 43–45
Monoisocyanates, 35–37
Monoisocyanides, 39–43
Monomethylcyclopentadienides, 53
Monomethyls, 50–51
Monopyrrolates, 54–55
Monothiolates, 44–45

Naphthalene:
  decay rate, excitation energy dependence, 186–187
  fluorescence, excitation energy dependence, 187–188
  yield as function of excess vibrational energy, 190–192
Naphthalene vapor, dispersed fluorescence, 168–170
Naphthoquinone, 137, 140
β-Naphthylamine, 192–193
Nitrogen heterocyclic compounds, radiationless transitions, excess energy dependence, 194–202
  azabenzenes, 198–202
  isoquinoline, 194–197
NO_2, photodissociation, 315–317
N_2O, photodissociation, 314–315
Nuclear spin relaxation rate, temperature dependence, 120
Nucleic acids, CIDNP applications, 152–153

OCS, photodissociation, 312–313
Onium salts, photoreaction, 148–149
Oxetane, 137, 140
Ozone, photodissociation, 309–312

Paterno–Büchi reactions, anetholes, 136–138
Peptides, CIDNP applications, 149–152
Perturbations, magnetic interactions, 71
Perturbation theory, time-dependent, Golden rule formula, 171–172
Phase space theory, 283, 301–302, 335–337
Phenacylphenylsulfone, photolysis, 111

Photodissociation:
    carbon-heteroatom bonds, 129-130
    correlated product state distributions, 286-287
    cyanogen, 299-302
    isocyanic acid, 317-318
    modeling, 281
    ozone, 309-312
    potential energy surface, 283
    reaction, tetra-atomic, 286-287
Photolysis:
    acrylonitrile, 299
    cyclododecanone, 106-107
    H_2O_2, 289
    phenacylphenylsulfone, 111
Photoproduct motion, 279-342
    frequency modulation Doppler spectroscopy, 295-302
        acrylonitrile, 299
        cyanogen photodissociation, 299-302
        schematic diagram, 296
    future, 340-342
    H-atom Rydberg tagging experiments, 326-327
    product energy distributions, uncorrelated measurements, 284-287
    scalar correlation measurements, 293-294
    vector correlations, 288-292
    v · j correlation, 288-290
    Wiley-McClaren time-of-flight mass-spectrometer, 302-309
Photosynthesis, CIDNP applications, 154
Podates, lanthanide complexes, 249-251
Polarization:
    CIDNP, 111-112
    relation with line intensities, 93-95
Polyatomic alkaline earth containing molecules, 1-56
    broida oven, 6-11
    chemistry and photochemistry, 14-18
    electronic structure, 18-22
    flames as source, 5
    molecular beams, as source, 10, 12-14
    monoacetylides, 51, 53
    monoalkoxides, 32-33
    monoalkylamides, 48-49
    monoamides, 46-48
    monoazides, 36, 38-39
    monoborohydrides, 55-56
    monocarboxylates, 33-34
    monocyanides, 39-43
    monocyclopentadienides, 52-53
    monoformamidates, 33, 35
    monohydrosulfides, 43-45
    monohydroxides, 23-32
        bond lengths, 24, 26
        MgOH, 23-26
        microwave spectrum, 25, 27
        vibrational frequencies, 29
    monoisocyanates, 35-37
    monoisocyanides, 39-43
    monomethylcyclopentadienides, 53
    monomethyls, 50-51
    monopyrrolates, 54-55
    monothiolates, 44-45
    reaction dynamics with oxygen-containing molecules, 16-17
Polymerization, initiators, 144-149
Potential energy, distortion from vibronic interaction, 182-183
Potential energy surfaces, adiabatic Born-Oppenheimer, 181-182
Prefulvene, ground state, 203
Product energy distributions, uncorrelated measurements, 284-287
Proteins, CIDNP applications, 149-152
Pulsed-extraction studies:
    fragmentation of larger molecules, 321-325
    triatomic photolysis, 309-317
        other triatomics, 312-317
        ozone photodissociation, 309-312
PUMP and PROBE experiment, 284
Pyrimidine dimers, photosensitized splitting, 152-153

Quinones, photocycloadditions, 140-141

Radiationless transitions, 166
    barrier widths, 184-185
    benzene, 202-206
    decay rate:
        dependence on excess vibrational energies, 173-175
        dependence on vibrational energy, 178-179
        single-vibronic-level, 178, 180
    diabatic crossing between electronic states, 175-176
    disparity in excess energy dependence, 175-176

Radiationless transitions (*Continued*)
excess energy dependence, 171-180
aromatic hydrocarbons, 186-193
exceptional, in benzene, 205-206
molecules with nearby $n\pi^*$ and $\pi\pi^*$ states, 180-186
nitrogen heterocyclic compounds, 194-202
Franck-Condon factor, 172
limiting cases, 171
rate constant, 171-173
saturation of decay rate, 175
state-averaged rate expression, 178
Radical additions, CIDNP applications, 132-134
Radical fragmentations, CIDNP applications, 134-135
Radical ions, structures, CIDNP applications, 114-117
Radical pair:
definition, 71-72
density matrix treatment of spin dynamics, 80-84
dynamics, 105
combined with spin dynamics, 84-90
intersystem crossing, 75
mechanism, 71-90
CIDNP explanation, 91-93
qualitative explanation, 72-73
spin states and interactions, 74-80
vector model of spin states, 78-79
Renner-Teller effect, 31
Resonant-enhanced multiphoton ionization, 284, 303-306
Rubrene, 120-121

Sandwich assays, 219-220
Self-exchange reactions, 118
Spin chemistry, 65-66. *See also* Chemically induced dynamic nuclear polarization
Spin conservation, 67-68
Spin dynamics, 102-105
combined with radical pair dynamics, 84-90
Spin evolution, 68-69
Spin Hamiltonian, 69-71
Spin polarization, 82-83
Spin selectivity, 68
$SrBH_4$, 56
SrCCH, 51, 53
SrC_5H_5, 53
SrC_4H_4N, 54
SrN_3, 36, 39
SrNC, 39-41
SrOD, O-O bands, 28-29
SrOH, 4, 24-25, 29-30
SrSH, 43
Strontium monoalkylamides, 48-49
Strontium monocarboxylate, 34
Strontium monoformamidate, 33, 35
Surface electron ejection by laser excited metastables, 196

Tb^{3+} complex:
applications, 219-224
with encapsulating ligands, 225-259
calixarene complexes, 252-259
classes, 226
cryptates, 227-238
macrocyclic complexes, 238-249
podates, 249-251
light conversion, 215-218
Tetraazacyclododecane, 248

Wiley-McClaren time-of-flight mass-spectrometer, 302-309
extraction methods, 303-305
ion imaging, 305-306
other methods, 306-309

Zeeman interaction:
difference of operators, 78
external magnetic field, 76-77
sum of the operators, 77

CUMULATIVE INDEX
VOLUMES 1-23

	VOL.	PAGE
Addition of Atoms to Olefins, in Gas Phase (Cvetanovic)	1	115
Advances in the Measurement of Correlation in Photoproduct Motion (Morgan, Drabbels, and Wodtke)	23	279
AFM and STM in Photochemistry Including Photon Tunneling (Kaupp)	19	119
Alcohols, Ethers, and Amines, Photolysis of Saturated (von Sonntag and Schuchmann)	10	59
Alkanes and Alkyl Radicals, Unimolecular Decomposition and Isotope Effects of (Rabinovitch and Setser)	3	1
Alkyl Nitrites, Decomposition of and the Reactions of Alkoxyl Radicals (Heicklen)	14	177
Alternative Halocarbons, Atmospheric Photochemistry of (Francisco and Maricq)	20	79
Anthracenes, Excited State Reactivity and Molecular Topology Relationships in Chromophorically Substituted (Becker)	15	139
Anti-Stokes Fluorescence, Cooling of a Dye Solution by (Zander and Drexhage)	20	59
Aromatic Hydrocarbon Solutions, Photochemistry of (Bower)	1	23
Asymmetric Photoreactions of Conjugated Enones and Esters (Pete)	21	135
Atmospheric Reactions Involving Hydrocarbons, FTIR Studies of (Niki and Maker)	15	69
Benzene, Excitation and Deexcitation of (Cundall, Robinson, and Pereira)	10	147
Biocatalysis and Biomimetic Systems, Artificial Photosynthetic Transformations Through (Willner and Willner)	20	217
Biochromophoric Systems, Excited State Behavior of Some (De Schryver, Boens, and Put)	10	359

357

	VOL.	PAGE
Cancer Treatment, Photochemistry in (Dougherty)	17	275
Carbonyl Compounds, The Photocycloaddition of, to Unsaturated Systems: The Syntheses of Oxetanes (Arnold)	6	301
Cobalt (III) and Chromium (III) Complexes, the Photochemistry of, in Solution (Valentine, Jr.)	6	123
Complexes, Photoinitiated Reactions in Weakly Bonded (Shin, Chen, Nickolaisen, Sharpe, Beaudet, and Wittig)	16	249
Cyclic Ketones, Photochemistry of (Srinivasan)	1	83
Cyclobutanones, Solution Phase Photochemistry of (Morton and Turro)	9	197
Cyclometallated Complexes, Photochemistry and Luminescence of (Maestri, Balzani, Deuschel-Cornioley, and von Zelewsky)	17	1
α-Dicarbonyl Compounds, The Photochemistry of (Monroe)	8	77
Diffusion-Controlled Reactions, Spin-Statistical Factors in (Saltiel and Atwater)	14	1
Electron and Energy Transfer, Mimicking of Photosynthetic (Gust and Moore)	16	1
Electron Energy Transfer between Organic Molecules in Solution (Wilkinson)	3	241
Electronically Excited Halogen Atoms (Hussain and Donovan)	8	1
Electron Spin Resonance Spectroscopy, Application of to Photochemistry (Wan)	9	1
Electron Transfer, Photoinduced in Organic Systems, Control of Back Electron Transfer of (Fox)	13	237
Electron Transfer Luminescence in Solution (Zweig)	6	425
Elementary Photoprocesses in Designed Chromophore Sequence on α-Helical Polypeptides (Sisido)	22	197
Ethylenic Bonds, Present Status of the Photoisomerization About (Arai and Tokumaru)	20	1
Excimers, What's New in (Yakhot, Cohen, and Ludmer)	11	489
Excited Electronic States, Electronic and Resonance Raman Spectroscopy Determination of Molecular Distortions in (Zink and Shin)	16	119
Free Radical and Molecule Reactions in Gas Phase, Problems of Structure and Reactivity (Benson)	2	1
FTIR Emission Studies, Time Resolved, of Photochemical Reactions (Hancock and Heard)	18	1
Gas Phase, Addition of Atoms of Olefins in (Cvetanovic)	1	115
Gas Phase Reactions, Photochemical, in Hydrogen-Oxygen System (Volman)	1	43
Gas Phase Reactions, Involving Hydroxyl and Oxygen Atoms, Mechanisms and Rate Constants of (Avramenko and Kolesnika)	2	25
Halogenated Compounds, Photochemical Processes in (Major and Simons)	2	137
Heterogeneous Catalysts, the Question of Artificial Photosynthesis of Ammonia on (Davies, Boucher, and Edwards)	19	235
Hydrogen-Oxygen Systems, Photochemical Gas Phase Reactions in (Volman)	1	43

	VOL.	PAGE
Hydroxyl and Oxygen Atoms, Mechanisms and Rate Constants of Elementary Gas Phase Reactions Involving (Avramenko and Kolesnikova)	2	25
Hydroxyl Radical with Organic Compounds in the Gas Phase, Kinetics and Mechanisms of the Reactions of (Atkinson, Darnall, Winer, Lloyd, and Pitts)	11	375
Hypohalites, Developments in Photochemistry of (Akhtar)	2	263
Imaging Systems, Organic Photochemical (Delzenne)	11	1
Intramolecular Proton Transfer in Electronically Excited Molecules (Klöpffer)	10	311
Ionic States, in Solid Saturated Hydrocarbons, Chemistry of (Kevan and Libby)	2	183
Isotopic Effects in Mercury Photosensitization (Gunning and Strausz)	1	209
Ketone Photochemistry, a Unified View of (Formosinho and Arnaut)	16	67
Lanthanide Complexes of Encapsulating Ligands at Luminescent Devices (Sabbatini, Guardigli, and Manet)	23	213
Mechanism of Energy Transfer, in Mercury Photosensitization (Gunning and Strausz)	1	209
Mechanistic Organic Photochemistry, A New Approach to (Zimmerman)	1	183
Mercury Photosensitization, Isotopic Effects and the Mechanism of Energy Transfer in (Gunning and Strausz)	1	209
Metallocenes, Photochemistry in the (Bozak)	8	227
Methylene, Preparation, Properties, and Reactivities of (De More and Benson)	2	219
Molecular Distortions in Excited Electronic States, Electronic and Resonance Raman Spectroscopy Determination of (Zink and Shin)	16	119
Neutral Oxides and Sulfides of Carbon, Vapor Phase Photochemistry of the (Fileeth)	10	1
Nitric Oxide, Role in Photochemistry (Heicklen and Cohen)	5	157
Noyes, W. A., Jr., a Tribute (Heicklen)	13	vii
Nucleic Acid Derivatives, Advances in the Photochemistry of (Burr)	6	193
Olefins, Photolysis of Simple, Chemistry of Electronic Excited States or Hot Ground States? (Collin)	14	135
Onium Salts, Photochemistry and Photophysics of (DeVoe, Olofson, and Sahyun)	17	313
Organic Molecules, Photochemical Rearrangements of (Chapman)	1	323
Organic Molecules in Adsorbed or Other Perturbing Polar Environments, Photochemical and Spectroscopic Properties of (Nicholls and Leermakers)	8	315
Organic Molecules in their Triplet States, Properties and Reactions of (Wagner and Hammond)	5	21
Organic Nitrites, Developments in Photochemistry of (Akhtar)	2	263
Organic Photochemical Refractive-Index Image Recording Systems (Tomlinson and Chandross)	12	201
Organized Media on Photochemical Reactions, A Model for the Influence of (Ramamurthy, Weiss, and Hammond)	18	67

	VOL.	PAGE
Organo-Transition Metal Compounds, Primary Photoprocesses of (Bock and von Gustorf)	10	221
Perhalocarbons, Gas Phase Oxidation of (Heicklen)	7	57
Phenyl Azide, Photochemistry of (Schuster and Platz)	17	69
Phosphorescence and Delayed Fluorescence from Solutions (Parker)	2	305
Phosphorescence-Microwave Multiple Resonance Spectroscopy (El-Sayed)	9	311
Photoassociation in Aromatic Systems (Stevens)	8	161
Photochemical Mechanisms, Highly Complex (Johnston and Cramarossa)	4	1
Photochemical Oxidation of Aldehydes by Molecular Oxygen, Kinetics and Mechanism of (Niclause, Lemaire, and Letort)	4	25
Photochemical Reactivity, Reflections on (Hammond)	7	373
Photochemical Rearrangements of Conjugated Cyclic Ketones: The Present State of Investigations (Schaffner)	4	81
Photochemical Transformations of Polyenic Compounds (Mousseron)	4	195
Photochemically Induced Dynamic Nuclear Polarization (Goez)	23	63
Photochemistry in Cyclodextrin Cavities (Bortolus and Monti)	21	1
Photochemistry of Conjugated Dienes and Trienes (Srinivasan)	4	113
Photochemistry of Rhodopsins, The (Ottolenghi)	12	97
Photochemistry of Simple Aldehydes and Ketones in the Gas Phase (Lee and Lewis)	12	1
Photochemistry of the Troposphere (Levy)	9	369
Photochemistry of Vitamin D and Its Isomers and of Simple Trienes (Jacobs and Havinga)	11	305
Photochemistry, Vocabulary of (Pitts, Wilkinson, Hammond)	1	1
Photochromism (Dessauer and Paris)	1	275
Photodissociation Dynamics of Hydride Molecules: H Atom Photofragment Translational Spectroscopy (Ashfold, Mordaunt, and Wilson)	21	217
Photo-Fries Rearrangement and Related Photochemical (1,j) Shifts of (j = 3,5,7) of Carbonyl and Sulfonyl Groups (Bellus)	8	109
Photography, Silver Halide, Chemical Sensitization, Spectral Sensitization, Latent Image Formation (James)	13	329
Photoionization and Photodissociation of Aromatic Molecules, by Ultraviolet Radiation (Terenin and Vilessov)	2	385
Photoluminescence Methods in Polymer Science (Beavan, Hargreaves, and Phillips)	11	207
Photolysis of the Diazirines (Frey)	4	225
Photooxidation Reactions, Gaseous (Hoare and Pearson)	3	83
Photooxygenation Reactions, Type II, in Solution (Gollnick)	6	1
Photophysics of Gaseous Aromatic Molecules: Excess Vibrational Energy Dependence of Radiationless Processes (Lim)	23	165
Photopolymerization, Dye Sensitized (Eaton)	13	427
Photoreactive Organic Thin Films in the Light of Bound Electromagnetic Waves (Sekkat and Knoll)	22	117
Photosensitized Reactions, Complications in (Engel and Monroe)	8	245
Photosynthetic Electron and Energy Transfer, Mimicking of (Gust and Moore)	16	1

	VOL.	PAGE
Phytochrome, Photophysics and Photochemistry of (Schaffner, Branslavsky, and Holzwarth)	15	229
Polymers, Photochemistry and Molecular Motion in Solid Amorphous (Guillet)	14	91
Primary Processes and Energy Transfer: Consistent Terms and Definitions (Porter, Balzani and Moggi)	9	147
Quantized Matter, Photochemistry and Photoelectrochemistry of: Properties of Semiconductor Nanoparticles in Solution and Thin-Film Electrodes (Weller and Eychmüller)	20	165
Quantum Theory of Polyatomic Photodissociation (Kreslin and Lester)	13	95
Radiationless Transitions, Isomerization as a Route for (Phillips, Lemaire, Burton, and Noyes, Jr.)	5	329
Radiationless Transitions in Photochemistry (Jortner and Rice)	7	149
Semiconductor Nanoclusters, Photophysical and Photochemical Processes of (Wang)	19	179
Silver Halides, Photochemistry and Photophysics of (Marchetti and Eachus)	17	145
Single Crystals, Photochemical Mechanism in: FTIR Studies of Diacyl Peroxides (Hollingsworth and McBride)	15	279
Singlet Molecular Oxygen (Wayne)	7	311
Singlet Molecular Oxygen, Bimolecular Reactivity of (Gorman)	17	217
Singlet Molecular Oxygen, Physical Quenchers of (Bellus)	11	105
Singlet and Triple States: Benzene and Simple Aromatic Compounds (Noyes and Unger)	4	49
Small Molecules, Photodissociation of (Jackson and Okabe)	13	1
Solid Saturated Hydrocarbons, Chemistry of Ionic States in (Kevan and Libby)	2	183
Solvation, Ultrafast Photochemical Intramolecular Charge Transfer and Excited State (Barbara and Jarzeba)	15	1
Spectroscopy and Photochemistry of Polyatomic Alkaline Earth Containing Molecules (Bernath)	23	1
Spin Conservation (Matsen and Klein)	7	1
Stilbenes, Bimolecular Photochemical Reactions of (Lewis)	13	165
Stilbenes and Stilbene-Like Molecules, *Cis–Trans* Photoisomerization of (Görner and Kuhn)	19	1
Sulfur Atoms, Reactions of (Gunning and Strausz)	4	143
Sulfur and Nitrogen Heteroatomic Organic Compounds, Photochemical Reactions of (Mustafa)	2	63
Surfactant Solutions, Photochemistry in (von Bünau and Wolff)	14	273
The Photochemistry of Indoles (Weedon)	22	229
Theory and Applications of Chemically Induced Magnetic Polarization in Photochemistry (Wan)	12	283
Transition Metal Complexes, Primary Processes in (Forster)	16	215
Triatomic Free Radicals, Spectra and Structures of (Herzberg)	5	1

	VOL.	PAGE
Ultraviolet Photochemistry, Vacuum (McNesby and Okabe)	3	157
Ultraviolet Photodissociation Studies of Organosulfur Molecules and Radicals: Energetics, Structure Identification, and Internal State Distribution (Cheuk-Yiu Ng)	22	1
Ultraviolet Radiation, Photoionization and Photodissociation of Aromatic Molecules by (Terenin and Vilessov)	2	385
Up-Scaling Photochemical Reactions (Braun, Jakob, Oliveros, Oller do Nascimento)	18	235
Weakly Bonded Complexes, Photoinitiated Reactions in (Shin, Chen, Nic kolaisen, Sharpe, Beaudet, and Wittig)	16	249
Xanthine Dyes, Photochemistry of the (Neckers and Valdes-Aguilera)	18	315